Y0-BVQ-966

TRANSLATION SERIES IN MATHEMATICS AND ENGINEERING

TRANSLATION SERIES IN MATHEMATICS AND ENGINEERING

M.I. Yadrenko
Spectral Theory of Random Fields
1983, viii + 259 pp.
ISBN 0-911575-00-6 Optimization Software, Inc.
ISBN 0-387-90823-4 Springer-Verlag New York Berlin Heidelberg Tokyo
ISBN 3-540-90823-4 Springer-Verlag Berlin Heidelberg New York Tokyo

G.I. Marchuk
Mathematical Models In Immunology
1983, xxv + 353 pp.
ISBN 0-911575-01-4 Optimization Software, Inc.
ISBN 0-387-90901-X Springer-Verlag New York Berlin Heidelberg Tokyo
ISBN 3-540-90901-X Springer-Verlag Berlin Heidelberg New York Tokyo

A.A. Borovkov, Ed.
Advances In Probability Theory: Limit Theorems and Related Problems
1984, xiv + 378 pp.
ISBN 0-911575-03-0 Optimization Software, Inc.
ISBN 0-387-90945-1 Springer-Verlag New York Berlin Heidelberg Tokyo
ISBN 3-540-90945-1 Springer-Verlag Berlin Heidelberg New York Tokyo

V.A. Dubovitskij
The Ulam Problem of Optimal Motion of Line Segments
1985, xiv + 114 pp.
ISBN 0-911575-04-9 Optimization Software, Inc.
ISBN 0-387-90946-X Springer-Verlag New York Berlin Heidelberg Tokyo
ISBN 3-540-90946-X Springer-Verlag Berlin Heidelberg New York Tokyo

N.V Krylov, R.S. Liptser, and A.A. Novikov, Eds.
Statistics and Control of Stochastic Processes
1985, xiv + 507 pp.
ISBN 0-911575-18-9 Optimization Software, Inc.
ISBN 0-387-96101-1 Springer-Verlag New York Berlin Heidelberg Tokyo
ISBN 3-540-96101-1 Springer-Verlag Berlin Heidelberg New York Tokyo

Yu. G. Evtushenko
Numerical Optimization Techniques
1985, approx. 450 pp.
ISBN 0-911575-07-3 Optimization Software, Inc.
ISBN 0-387-90949-4 Springer-Verlag New York Berlin Heidelberg Tokyo
ISBN 3-540-90949-4 Springer-Verlag Berlin Heidelberg New York Tokyo

Continued on page 505

STEKLOV SEMINAR, 1984

Edited by
N. V. Krylov, R. Sh. Lipster,
and A. A. Novikov

STATISTICS AND CONTROL OF STOCHASTIC PROCESSES

OPTIMIZATION SOFTWARE, INC.
Publications Division, New York

Editors

N.V. KRYLOV
Department of Mechanics
and Mathematics
Moscow State University

R.Sh. LIPTSER
Institute of Control Sciences
USSR Academy of Sciences

A.A. Novikov
Steklov Institute of Mathematics
USSR Academy of Sciences

Series Editor

A.V. BALAKRISHNAN
School of Engineering
University of California
Los Angeles, CA 90024
USA

Library of Congress Cataloging in Publication Data

Steklov Seminar (1984: Steklov Institute of
Mathematics)
Statistics and control of stochastic processes.

(Translations series in mathematics and engineering)
"A.N. Shiryaev anniversary volume."
Translation of *Statistika i upravlenie sluchainykh protsessov.*
1. Mathematical statistics--Congresses. 2. Stochastic processes--Congresses. I. Krylov, N.V. (Nikolai Vladimirovich) II. Liptser, R.Sh. (Robert Shevilevich) III. Novikov, A.A. (Aleksandr Aleksandrovich) IV. Shiriaev, Al'bert Nikolaevich. V. Title. VI. Series.
QA276.A1S7513 1984 519.5 85-8851
ISBN 0-911575-18-9

Worldwide Distribution Rights by Springer-Verlag New York, Inc.,
175 Fifth Avenue, New York, New York 10010, USA and
Springer-Verlag Berlin Heidelberg New York Tokyo,
Heidelberg Platz 3, Berlin-Wilmersdorf-33, West Germany.

ISBN 0-911575-18-9 Optimization Software, Inc.
ISBN 0-387-96101-1 Springer-Verlag New York Berlin Heidelberg Tokyo
ISBN 3-540-96101-1 Springer-Verlag Berlin Heidelberg New York Tokyo

Proceedings of the Seminar
held at the Steklov Institute
of Mathematics, USSR Academy
of Sciences, Moscow, 1984

This volume is dedicated to
Albert Nikolaevich Shiryaev
on his 50th birthday

Albert Nikolaevich Shiryaev

Steklov Institute of Mathematics
USSR Academy of Sciences
Moscow, 1984

Professor ALBERT NIKOLAEVICH SHIRYAEV, born October 12, 1934 in Moscow, has been with the Steklov Institute of Mathematics since 1957. He defended his Candidate dissertation entitled "Optimal Methods of a Quickest Detection Problem" in 1961 under the supervision of A.N. Kolmogorov, and his doctoral dissertation entitled "Contribution to Sequential Analysis, Optimal Control and Filtering" in 1967. He was appointed Professor at the Moscow State University in 1969. He currently serves on the Editorial Boards of several mathematical journals. He is a member of the Council of the Bernoulli Society and Academic Secretary of the Moscow Mathematical Society.

PREFACE

This volume is based on the papers presented at the 1984 Seminar on Statistics and Control of Stochastic Processes, held under the guidance of Professor Nikolaj Vladimirovich Krylov and Professor Albert Nikolaevich Shiryaev (who founded the Seminar in 1966), at the Steklov Institute of Mathematics in Moscow.

The main topics featured at the Seminar are Statistical and Sequential Analysis, Filtering, Stochastic Control of Random Processes, and related Stochastic Calculus (Martingales, Stochastic Differential Systems).

This Seminar keeps close ties with the Seminars of Professor R.J. Chitashvili in Tbilisi, Professor I.I. Gikhman in Donetsk, Professor B.I. Grigelionis in Vilnius, and Professor A.V. Skorokhod in Kiev, as well as with several Seminars outside the Soviet Union.

This volume is dedicated to Professor Albert Nikolaevich Shiryaev on his 50th birthday. All the participants of the Seminar have been enriched by his ideas and have freely partaken of his vast store of knowledge. They can attest to his infectious enthusiasm and energy, in matters of science as well as in the fine art of skiing.

We all wish Albert Nikolaevich Shiryaev even higher peaks to climb--and continued success throughout his life.

N.V. Krylov,
R.Sh. Liptser,
A.A. Novikov

CONTENTS

S.V. ANULOVA

THE BELLMAN EQUATION IN AN OPTIMAL CONTROL PROBLEM FOR A DIFFUSION PROCESS IN IN A REGION WITH A REFLECTING BOUNDARY

We consider an optimal control problem for a non-branching diffusion process in a region. On reaching the boundary the process may either be reflected back into the interior or continue the motion along the boundary. The diffusion in the interior of the region and the tangential diffusion on the boundary are assumed to be uniformly nondegenerate. The following problem is posed: maximize the expectation of an integral along the paths of the process. Assuming the Hölder continuity of the coefficients, we prove that the reward function belongs to the Hölder space $C^{2+\alpha}$ and satisfies the Bellman equation. The formulation of the problem and the result have been prompted by N.V. Krylov [1], and the proofs essentially exploit the ideas, methods and results of [1] and of M.V. Safonov [2].

Constants appearing in Propositions and Proofs may be denoted by the same letter N, even though they may differ in value in different formulas. The notation $N = N(p_1, \ldots, p_n)$ implies that the constant N depends on parameters $p_1, \ldots, p_n$ and only on them.

Fix an integer $d \geq 2$, $\kappa \in (0,1]$ and $\alpha \in (0,1)$.

Let D be an open convex bounded set in R^d such that for some function $\psi \in C^{2+\alpha}(R^d)$ with $||\psi||_{C^{2+\alpha}(R^d)} \leq \kappa^{-1}$

$$D = \{x \in R^d : \psi(x) > 0\} , \qquad |\psi_x| \geq \kappa \quad \text{on} \quad \partial D .$$

The unit inner normal to the boundary ∂D at a point x is denoted $n(x)$.

We use tensor notation:

$$a^{ij}\xi^i\xi^j = \sum_{i,j=1}^{d} a^{ij}\xi^i\xi^j , \qquad b^i\xi^i = \sum_{i=1}^{d} b^i\xi^i .$$

For $x \in \bar{D}$ we denote by $L(x)$ the set of triples $\{a,b,c\}$, where $a = (a^{ij})$ is a symmetric $d \times d$ matrix with real elements, $b = (b^i) \in R^d$, c a number, satisfying the following conditions:

- •1. norms of a, b and c do not exceed κ^{-1};
- •2. if $x \in D$, then for every $\xi \in R^d$

$$\kappa|\xi|^2 \leq a^{ij}\xi^i\xi^j \leq \kappa^{-1}|\xi|^2 ;$$

- •3. if $x \in \partial D$, then for every $\xi \in R^d$

$$\kappa|\xi_\tau|^2 \leq a^{ij}\xi^i\xi^j \leq \kappa^{-1}|\xi_\tau|^2 ,$$

where $\xi_\tau = \xi - (\xi, n(x))n(x)$;

- •4. if $x \in \partial D$, then $(b, n(x)) \geq 0$;
- •5. $c \geq \kappa$.

For $s = 1,2, \ldots$ suppose the functions $c^s: \bar{D} \to [0,\infty)$, $f^s: \bar{D} \to R_1$ and let the linear operators L_0^s, L^s on $C^2(\bar{D})$ be defined:

$$L_0^s u(x) = (a^s)^{ij}(x)\, u_{x_i x_j}(x) + (b^s)^i(x)\, u_{x_i}(x) \quad ,$$

$$L^s u = (L_0^s - c^s)u \quad ,$$

with $\{a(x), b(x), c(x)\} \in L(x)$ and $|f^s(x)| \le \kappa^{-1}$ for every $x \in \bar{D}$.

Suppose that for all s, $C^\alpha(D)$ and $C^\alpha(\partial D)$ norms of a^s, b^s, c^s and f^s do not exceed κ^{-1} (we emphasize that these functions are continuous on D and ∂D separately, but on $\bar{D}$ they are, in general, discontinuous).

We call a pair $((x_t), (F_t))$, where (x_t) is a continuous process with values in $\bar{D}$, (F_t) a filtration defined on some probability space, an L_0-process, if there exists a measurable (F_t)-adapted process (s_r) with positive integer values such that for every $g \in C^2(\bar{D})$ the process

$$g(x_t) - \int_0^t L_0^{s_r} g(x_r)\, dr$$

is an (F_t)-martingale.

One may regard an L_0-process (x_t) as a controlled solution of a martingale problem in the sense of [3], with the process (s_r) playing the role of a control policy.

Define a reward function

$$v(x) = \sup M \int_0^\infty \exp\left(- \int_0^t c^{s_r}(x_r)dr\right) f^{s_t}(x_t)\, dt \quad ,$$

where supremum is taken over all L_0-processes (x_t), starting from a point x, $x \in \bar{D}$.

THEOREM 1. There exists an $\bar{\alpha} = \bar{\alpha}(\kappa,d) \in (0,1)$ such that, provided $\alpha \in (0,\bar{\alpha})$, $v \in C^{2+\alpha}(\bar{D})$ and satisfies the Bellman equation

$$\sup_{s=1,2,\dots} (L^s v(x) + f^s(x)) = 0 \quad , \qquad x \in \bar{D} \quad .$$

This theorem can easily be deduced from the following Theorem 2 which states the existence of a $C^{2+\alpha}$ solution of the Bellman equation. Indeed, with the help of ε-optimal policies it is easy to show that the solution of the Bellman equation is just the reward function (cf. [4]).

THEOREM 2. There exists an $\bar{\alpha} = \bar{\alpha}(\kappa,d) \in (0,1)$ such that, provided $\alpha \in (0,\bar{\alpha})$, the problem

$$\sup_{s=1,2,\dots} (L^s u(x) + f^s(x)) = 0 \quad , \qquad x \in \bar{D} \quad ,$$

has a solution in $C^{2+\alpha}(\bar{D})$ and $\|u\|_{C^{2+\alpha}(\bar{D})} \le N(\kappa,d,\alpha)$.

The proof of Theorem 2 exploits essentially the following lemma, which is an extension of a well-known result by Krylov-Safonov (see [5]) to the case of reflected processes (see [6]). The lemma is stated in a specific way suiting our purposes.

In this lemma and in the proof of Theorem 2 we shall need new coordinates in some inner neighborhood of the boundary ∂D, "straightening" the boundary. Denote

$$B^+(r) = \left\{ x \in R^d \,:\, x^1 \in [0,r),\ \sum_{i=2}^{d} (x^i)^2 < r^2 \right\} ,$$

$$\delta B^+(r) = \{x \in B^+(r) \,:\, x^1 = 0\} \quad .$$

Fix in some inner neighborhood of the boundary ∂D a finite atlas c_i, setting a $C^{2+\alpha}$-diffeormorphism (with respect to coordinates in R^d) between parts of the neighborhood and cylinders $B^+(R_i)$, $R_i \in (0,1]$, mapping the boundary onto the foundations of the cylinders. We may take for granted that in each chart c mapping a part of the neighborhood on $B^+(R)$, $R \in (0,1]$:

for $s = 1,2,\ldots$, $C^\alpha((B^+\setminus\delta B^+)(R))$ and $C^\alpha(\delta B^+(R))$ norms of local coefficients $a^{s,c}$ and $b^{s,c}$ of the operator L^s do not exceed a certain $N = N(\kappa,d)$;

the operator $\frac{\partial}{\partial n}$ has the form of $\frac{\partial}{\partial x^1}$.

<u>LEMMA 1</u>. Let an operator L acting on $C^2(\bar{D})$,

$$Lu(x) = a^{ij}(x)\, u_{x_i x_j}(x) + b^i(x)\, u_{x_i}(x) - c(x)u(x) \quad ,$$

such that $\{a(x),\, b(x),\, c(x)\} \in L(x)$, $x \in \bar{D}$, and a function $u \in C^2(\bar{D})$, such that $u \geq 0$ and $Lu(x) \leq 0$ in $\bar{D}$, be given. Suppose in a chart c, mapping a part of some inner neighborhood of the boundary ∂D on $B^+(R)$, $R \in (0,1]$, for some $\xi > 0$ $\text{mes}\,\{x\colon x \in \delta B^+(R/2),\ u(x) \geq 1\} \geq \xi\, \text{mes}\, \delta B^+(R/2)$ (mes denotes $(d-1)$-dimensional Lebesgue measure). Then

$$\inf\,\{u;\ B^+(R/2)\} \geq \beta(\kappa,d,\xi) > 0 \quad .$$

<u>Proof of Theorem 2</u>. The technique developed in [1] allows us to reduce the problem only to the proof of an <u>a priori</u> $C^{2+\alpha}(\bar{D})$ estimate of the solution u provided that it belongs to $C^4(\bar{D})$ and $\psi \in C^\infty(R^d)$.

1°. First of all, observe that since u satisfies the Bellman equation, there exist measurable functions a, b, c and f

such that for every $x \in \bar{D}$, $\{a(x), b(x), c(x)\} \in L(x)$,

$$\sup\{|f|;\ \bar{D}\} \le \kappa^{-1}$$

and

$$\begin{aligned} Lu(x) + f(x) &= a^{ij}(x)\,u_{x_i x_j}(x) + b^i(x)\,u_{x_i}(x) - c(x)u(x) \\ &\quad + f(x) \\ &= 0 \end{aligned}$$

(see e.g., [1]).

2°. Consider in detail the Bellman equation on the boundary ∂D. Set for $s = 1,2,\dots$

$$b_n^s(x) = (b^s(x),\ n(x)) ,$$

$$\tilde{f}^s(x) = f^s(x) + b_n^s(x)\,\frac{\partial u}{\partial n}(x) , \qquad x \in \partial D ,$$

and define a linear operator L_τ^s acting on $C^2(\partial D)$, $L_\tau^s = L^s - b_n^s \frac{\partial}{\partial n}$. On ∂D the function u satisfies the equation

$$\sup_{s=1,2,\dots} (L_\tau^s u + \tilde{f}^s) = 0 .$$

In 3° - 7° we fix a chart c, mapping a part of some inner neighborhood of the boundary ∂D on a cylinder $B^+ = B^+(R)$, $R \in (0,1]$. In 3° - 6° we assume local coefficients of L^s in B^+, $a^{s;c}$ and $b^{s;c}$ and functions c^s and f^s independent of x separately on each of the sets δB^+ and $B^+\backslash\delta B^+$, $s = 1,2,\dots$

3°. We shall prove that for some $\alpha_1 = \alpha_1(\kappa,d) \in (0,1)$

$$\|u_{x_i}\|_{C^{\alpha_1}(B^+(R/2))} \le N(\kappa,d) , \qquad i = 2,\dots,d .$$

Following the argument of [1] or [2], one can show that

$$Lu_{x_i}(x) = 0 \qquad \text{in } B^+, \quad i = 2,\dots,d \ .$$

Using Lemma 1 and interpolation inequalities for Hölder spaces, we obtain

$$\|u_{x_i}\|_{C^{\alpha_1}(B^+(R/2))} \le N(\kappa, d, \|u\|_{C(B^+)}) \ , \qquad i = 2,\dots,d$$

(see [5]).

The maximum principle yields $\|u\|_{C(\bar{D})} \le \kappa^{-2}$; thus we get the needed estimate.

4°. 1° and the estimate of tangent derivatives in 3° provide that for some $\alpha_2 = \alpha_2(\kappa, d) \in (0,1)$

$$\left\|\frac{\partial u}{\partial n}\right\|_{C^{\alpha_2}(\delta B^+(R/4))} \le N(\kappa, d) \ .$$

This result can be found in [7]; a close statement yields Theorem 4.2 [8].

5°. Now we shall show that for some $\alpha_3 = \alpha_3(\kappa, d) \in (0,1)$,

$$\|u\|_{C^{2+\alpha_3}(\delta B^+(R/8))} \le N(\kappa, d) \ .$$

By virtue of 4°

$$\left\|b_n^s \frac{\partial u}{\partial n}\right\|_{C^{\alpha_2}(\delta B^+(R/4))} \le N(\kappa, d) \ , \qquad s = 1,2,\dots$$

Therefore, in notations of 2°, on δB^+ the function u satisfies the equation

$$\sup_{s=1,2,\dots} (L_\tau^s u + \tilde{f}^s) = 0$$

with

$$\|\tilde{f}\|_{C^{\alpha_2}(\delta B^+(R/4))} \leq N(\kappa,d) \quad , \qquad s = 1,2,\ldots \; .$$

From this, using results of [2], we infer the desired statement.

6°. 5° and the theorem of [2] provide that for some $\alpha_4 = \alpha_4(\kappa,d) \in (0,1)$

$$\|u\|_{C^{2+\alpha_4}(B^+(R/16))} \leq N(\kappa,d) \quad .$$

7°. By means of the "frozen coefficients" method the estimate of 6° can be extended to the Hölder continuous functions $a^{s;c}$, $b^{s;c}$, c^s, f^s: if $\alpha \in (0,\alpha_4)$, then

$$\|u\|_{C^{2+\alpha}(B^+(R/16))} \leq N(\kappa,d,\alpha)$$

(see [2]).

8°. 6° implies that if $\alpha \in (0,\alpha_4)$ then

$$\|u\|_{C^{2+\alpha}(\partial D)} \leq N(\kappa,d,\alpha) \quad .$$

Applying the theorem of [2], we obtain the existence of an appropriate $\bar{\alpha} = \bar{\alpha}(\kappa,d) \in (0,1)$ and the required <u>a priori</u> estimate of $\|u\|_{C^{2+\alpha}(\bar{D})}$.

The author wishes to express her deep gratitude to M.V. Safonov for his friendly help in her work on this paper.

REFERENCES

[1] Krylov, N.V. "Boundedly Nonhomogeneous Elliptic and Parabolic Equations," *Math. USSR Izvestiya*, vol.20, no.3 (1983): 459-492.

[2] Safonov, M.V. "Classical Solutions of the Elliptic Bellman Equation." *Doklady Akademii Nauk SSSR*, vol.278, no.4 (1984): 810-813.

[3] Stroock, D.W., and Varadhan, S.R.S. *Multidimensional Diffusion Processes*. Berlin Heidelberg New York: Springer-Verlag Inc., 1979.

[4] Krylov, N.V. *Controlled Diffusion Processes*. New York Berlin Heidelberg: Springer-Verlag Inc., 1980.

[5] Krylov, N.V., and Safonov, M.F. "An Estimate of the Probability that a Diffusion Process Hits a Set of Positive Measure." *Soviet Math. Doklady*, vol.20, no.2 (1979): 253-255.

[6] Anulova, S.V. "The Lower Bound of the Probability that a Degenerate Diffusion Process Hits a Positive Measure Set." *Abstracts of Commun., 4th USSR-Japan Symposium on Probability Theory and Mathem. Statistics*, pp. 108-109, vol.I. Tbilisi: Metzniereba, 1982.

[7] Krylov, N.V. *Nelinejnye ellipticheskie and parabolicheskie uravneniya vtorogo poryadka* (Nonlinear Second-order Elliptic and Parabolic Equations). Moskva: Nauka, 1984.

[8] Krylov, N.V. "Boundedly Nonhomogeneous Elliptic and Parabolic Equations in a Domain." *Math. USSR Izvestiya*, vol.22, no.1 (1984): 67-97.

R.J. CHITASHVILI and N.V. ELBAKIDZE

OPTIMAL STOPPING BY TWO PLAYERS

INTRODUCTION

A minimax version of the optimal stopping problem is described by:

•1. a probability space (Ω, F, P) equipped with a nondecreasing family of σ-algebras $F = \{F_n\}$, $n \geq 0$, characterizing information available to observers at the moment $n \geq 0$;

•2. two classes Σ and Π, Σ (resp. Π) being the sets of admissible strategies for the first (resp. second) player $\alpha = \{a_n, n \geq 0\}$ (resp. $\beta = \{b_n, n \geq 0\}$), where for each $n \geq 0$ $a_n = a_n(\omega)$ and $b_n = b_n(\omega)$ are F_n-measurable random variables taking value in $R^{(1)}$;

•3. real-valued random variables $P_n^{a,b} = P_n^{a,b}(\omega)$, $0 \leq P_n^{a,b} \leq 1$, and $r_n^{a,b} = r_n^{a,b}(\omega)$ for each $n \geq 0$ adapted to F and a random variable η with $E|\eta| < \infty$.

The above $P_n^{a,b}$ can be interpreted as the probability of stopping at $n \geq 0$ when the decisions of players are a and b, respectively, $r_n^{a,b}$ denotes the reward of the first player and η is the reward if observation is continued indefinitely.

From now on we assume that the following condition is fulfilled:

$$E \sup_{n\geq 0} \sup_{\alpha\in\Sigma} \sup_{\beta\in\Pi} \sum_{k=n}^{\infty} \prod_{i=n}^{n-1} \left(1 - P_i^{a_i,b_i}\right) \cdot \left|r_k^{a_k,b_k}\right| < \infty . \qquad (1)$$

For each pair of admissible strategies α and β let us consider the following expected reward of the first player:

$$S_\eta^{\alpha,\beta}(n) = E\left\{\eta_n^{\alpha,\beta} + \prod_{k=n}^{\infty} \left(1 - P_k^{a_k,b_k}\right)\eta \;\middle|\; F_n\right\} , \qquad (2)$$

where

$$\eta_n^{\alpha,\beta} = \sum_{k=n}^{\infty} \prod_{i=n}^{n-1} \left(1 - P_i^{a_i,b_i}\right) r_k^{a_k,b_k} .$$

Define the lower and upper values of the game

$$\underline{S}_\eta(n) = \sup_\alpha \inf_\beta S_\eta^{\alpha,\beta}(n)$$

and

$$\overline{S}_\eta(n) = \inf_\beta \sup_\alpha S_\eta^{\alpha,\beta}(n) .$$

The following notation will be used:

$$\sup_{\alpha\in I} X_\alpha = \operatorname{ess\,sup}_{\alpha\in I} X_\alpha , \qquad \inf_{\alpha\in I} X_\alpha = \operatorname{ess\,inf}_{\alpha\in I} X_\alpha$$

and

$$\prod_{i=n}^{n-1} \left(1 - P_i^{a_i,b_i}\right) = 1 .$$

The main problems are to prove the existence of the value $S_\eta(n) = \underline{S}_\eta(n) = \overline{S}_\eta(n)$ and to construct ε-optimal strategies, i.e., such α_ε and β_ε that

$$\sup_{\alpha} S_{\eta}^{\alpha,\beta_{\varepsilon}}(n) - \varepsilon \leq S_{\eta}(n) \leq \inf_{\beta} S_{\eta}^{\alpha_{\varepsilon},\beta}(n) + \varepsilon \quad \text{a.s.}$$

for all $n \geq 0$.

We say that the sequence of random variables $\{v_n, n \geq 0\}$ belongs to the class $\mathbb{R}$ if there exists such a random variable ξ with $E|\xi| < \infty$ that for all strategies $\alpha = \{a_n, n \geq 0\} \in \Sigma$ and $\beta = \{b_n, n \geq 0\} \in \Pi$

$$\prod_{i=n}^{k-1} \left(1 - P_i^{a_i,b_i}\right) \cdot |v_k| \leq E\{\xi \mid \mathcal{F}_k\} \quad \text{a.s.}$$

for all $n \geq 0$ and $k \geq n$.

The solution of the above problems is closely associated with the Bellman equation for the nonlinear operator defined by the formula

$$U v_n = \sup_{a} \inf_{b} [(1 - P_n^{a,b})E\{v_{n+1} \mid \mathcal{F}_n\} + r_n^{a,b}] . \tag{3}$$

This formulation yields several problems which are generalizations of the standard stopping rule problem in different ways.

The standard stopping (by one player) problem (see [1], [2]) corresponds to the case $P_n^{a,b} = P_n^a$, $\left[\inf_a P_n^a, \sup_a P_n^a\right] = [0,1]$, $r_n^{a,b} = P_n^a g_n$, where every nonrandomized strategy α for which $P_n^{a_n}$ takes only the values 0 or 1 is related to the Markov time $\tau = \min \{n: P_n^{a_n} = 1\}$, the random variable g_n acquires the meaning of a reward for the stopping at n and the expected reward can be represented as

$$S_{\eta}^{\alpha,\beta}(n) = S_{\eta}^{\alpha}(n) = S_{\eta}^{\tau}(n) = E\{g_{\tau} I_{[\tau<\infty]} + \eta I_{[\tau=\infty]} \mid \mathcal{F}_n\} .$$

The choice of the final reward η of a special type $\underline{\eta} = \overline{\lim_{n\to\infty}} g_n$ enables one to associate the finite time stopping problem with the problem allowing all stopping times:

$$S(n) = \sup_{\tau<\infty} E\{g_\tau \mid F_n\} = \sup_\tau S^\tau_{\underline{\eta}}(n) = S_{\underline{\eta}}(n) \quad ;$$

and instead of the condition, difficult to check, that the cost $S(n)$ is the smallest solution of the Bellman-Wald equation

$$S(n) = U\, S(n) = \max\, [g_n,\ E\{S(n+1 / F_n\}] \quad , \tag{4}$$

we characterize the cost function as a solution of (4) with the boundary condition (1)

$$\lim_n S(n) = \underline{\eta} \qquad \text{a.s.} \ .$$

The inclusion of general expressions for the final reward in the study made it possible to give a complete description of all solutions of equation (4) as a set of costs characterized by the boundary conditions (see [1], [3])

$$\lim_{n\to\infty} S_\eta(n) = \max\, (\underline{\eta}, \eta) \qquad \text{a.s.} \ .$$

Hence the random variable $\underline{\eta}$ is the lower bound of possible boundary conditions for the solution of equation (4).

Furthermore, taking into account the restrictions upon the choice of the probability of stopping when

$$\left[\operatorname*{Inf}_a P^a_n,\ \operatorname*{Sup}_a P^a_n\right] \subset [0,1]$$

and the consideration of the general expression for the current and final rewards r_n^a, η leads to Bellman operators having the form

$$U v_n = \phi_n(E\{v_{n+1} / F_n\}) ,$$

where $\phi_n(y) = \phi_n(\omega, y)$ is an F-adapted and convex monotone function with respect to y, satisfying the Lipschitz condition $|\phi_n(y) - \phi_n(y')| \le |y - y'|$. In fact (see [4]) any such operator can be represented as a Bellman operator

$$U v_n = \operatorname*{Sup}_a [(1 - P_n^a) E\{v_{n+1} / F_n\} + r_n^a] . \tag{5}$$

The set of all costs $S_\eta(n) = \operatorname*{Sup}_\alpha S_\eta^\alpha(n)$, with respect to all η, exhausts the class of all solutions of the equation $v_n = U v_n$ and to characterize the cost by means of the boundary condition one has to consider the partition of Ω into the sets

$$\Omega_d = \left\{\omega : \sum_{n=0}^{\infty} \operatorname*{Inf}_a P_n^a = \infty\right\} ,$$
$$\Omega_c = \left\{\omega : \sum_{n=0}^{\infty} \operatorname*{Sup}_a P_n^a < \infty\right\} , \tag{6}$$
$$\Omega_0 = \left\{\omega : \sum_{n=0}^{\infty} \operatorname*{Inf}_a P_n^a < \infty, \quad \sum_{n=0}^{\infty} \operatorname*{Sup}_a P_n^a = \infty\right\}$$

corresponding to the elementary events for which: any strategy leads to a stopping in a finite number of steps (Ω_d); a stopping or an infinite extension of the observation is possible (Ω_0); and a stopping under any strategy is impossible (Ω_c).

The solutions are uniquely characterized by the given boundary conditions only on $\Omega_0 + \Omega_c$ and

$$\lim_{n\to\infty} S_\eta(n) = \max(\underline{\eta}, \eta) \qquad [\text{a.s. } \Omega_0],$$
$$\lim_{n\to\infty} S_\eta(n) = \eta \qquad [\text{a.s. } \Omega_c], \tag{7}$$

where $\underline{\eta}$ is a random variable which plays here the same role as $\underline{\eta} = \overline{\lim_{n\to\infty}}\, g_n$ in the standard stopping problem (see [4]).

Moreover, the expressions of the boundary values of the cost enable an effective construction of ε-optimal strategies. It is well known (see [5]) that for the strategy α to be optimal the "conservation" property, i.e., achieving the maximum in the Bellman equation

$$\Delta_n^{a_n} = S_\eta(n) - \left(1 - P_n^{a_n}\right) E\left\{S_\eta(n+1) / F_n\right\} + r_n^{a_n} = 0 ,$$

is not sufficient.

As the strategy defect formula (see [5]) implies

$$S_\eta(n) - S_\eta^\alpha(n) = \mathbf{E}\left\{\sum_{k=n}^{\infty} \prod_{i=n}^{k-1} 1 - P_i^{a_i} \Delta_k^{a_k} + \prod_{k=n}^{\infty}\left(1 - P_k^{a_k}\right)\left(\lim_{n\to\infty} S_\eta(n) - \eta\right) / F_n\right\}, \tag{8}$$

the "equalizing" property is also needed, i.e., when the second summand on the right-hand side of (8) becomes zero.

The selection of the conserving (or ε-conserving) strategy having this additional property presents the main difficulty in optimal (or ε-optimal) strategy construction.

In [4] for the construction of ε-optimal control the following fact is used: as it follows from (7) the "equalizing" condition for the strategy α is equivalent to the condition

$$\sum_{n=0}^{\infty} P_n^{a_n} < \infty \qquad [\text{a.s. } \Omega_0 \cap \{\underline{\eta} > n\}]$$

and the lower boundary evidently has the form $\underline{\eta} = \overline{\lim\limits_{n\to\infty}}\, \eta_n^{\alpha}$.

Various special problems with restriction on the probability of stopping were considered in [6], [7].

Finally, one more approach to the generalization of the problem was considered in [8] by a study of a special case of the stopping problem, which was later on developed in [9], [10], [11], [12], [13] when the standard version of the optimal stopping by two players led to the consideration of the natural generalization of the operator (4)

$$U\, v_n = \min \{G_n,\ \max\, (g_n,\ E\{v_{n+1} / \mathcal{F}_n\})\}$$

being a special case of the operator of the form

$$U\, v_n = \phi_n(E\{v_{n+1} / \mathcal{F}_n\}) \ , \tag{9}$$

where $\phi_n(y)$ is an F-adapted, for each $y \in R^{(1)}$, monotone function satisfying the Lipschitz condition $|\phi_n(y) - \phi_n(y')| \le |y - y'|$.

In the special case associated with the stopping of the Markov chain $X = \{x_n, \mathcal{F}_n, P_x\}$, $n \ge 0$, with state space $(E, \mathcal{B})$, when $\phi_n(y) = \phi(x_n, y)$, $v_n = v(x_n)$ operator (9) takes the form

$$U\, v(x) = (x,\ Tv(x)) \ , \qquad x \in E \ , \tag{10}$$

where $Tv(x) = \int v(y)\, P(dy/x)$ is an integral operator connected with the transition probability $P(\cdot,x)$ of the Markov chain, $\phi(x,y)$, $v(x)$ are B-measurable functions, and represents the general form of a monotone, contractive Hammerstein operator.

Conversely, every operator of the form (9) can be represented as a Bellman operator (3).

Introduce the operator U acting on $v = \{v_n, n \geq 0\}$ by the formula $U v_n = (U v)_n = \phi_n(E\{v_{n+1} / F_n\})$. The equation related with such an operator has the form

$$v_n = U v_n \quad . \tag{11}$$

It turns out that $\phi_n(y)$ can be represented as

$$\phi_n(y) = \operatorname*{Sup}_a \operatorname*{Inf}_b [(1 - P_n^{a,b})y + r_n^{a,b}] \quad , \tag{12}$$

where $P_n^{a,b}$ and $r_n^{a,b}$ are real-valued F_n-measurable random variables and for all $n \geq 0$, $a,b \in R^{(1)}$, $P_n^{a,b} \in [0,1]$ a.s. . It is obvious that

$$\phi_n(y) = \operatorname*{Sup}_a \operatorname*{Inf}_b \left[\frac{\phi_n(b) - \phi_n(a)}{b - a} (y - a) + \phi_n(a) \right] \quad \text{a.s.},$$

and supposing

$$P_n^{a,b} = 1 - \frac{\phi_n(b) - \phi_n(a)}{b - a} \quad , \qquad r_n^{a,b} = \frac{b\phi_n(a) - a\phi_n(b)}{b - a} \quad , \tag{13}$$

we obtain (12). Thus we establish that equation (11) can be represented as the following Bellman equation:

$$v_n = \operatorname*{Sup}_{a \in R^{(1)}} \operatorname*{Inf}_{b \in R^{(1)}} [(1 - P_n^{a,b}) E\{v_{n+1} / F_n\} + r_n^{a,b}] \quad . \tag{14}$$

Further, since (y,y) is a saddle-point for

$$\Phi_n^{a,b}(y) = \frac{\phi_n(b) - \phi_n(a)}{b - a}\, y + \frac{b\phi_n(a) - a\phi_n(b)}{b - a}$$

(i.e., $\operatorname*{Sup}_a \Phi_n^{a,y}(y) = \operatorname*{Inf}_b \Phi_n^{y,b}(y)$), for all $n \geq 0$ there exist $\hat{a}_n$ and $\hat{b}_n$ such that

$$\operatorname*{Sup}_a \left[\left(1 - P_n^{a,\hat{b}_n}\right) E\{v_{n+1} / \mathcal{F}_n\} + r_n^{a,\hat{b}_n}\right]$$

$$\leq v_n$$

$$\leq \operatorname*{Inf}_b \left[\left(1 - P_n^{\hat{a}_n,b}\right) E\{v_{n+1} / \mathcal{F}_n\} + r_n^{\hat{a}_n,b}\right] .$$

The statements below are generalizations of the above facts for the general stopping problem, associated with operator (3). Specifically, a lemma on the representation of the general solution of equation (14) in the form of the cost (Lemma 3) is given. The cost existence is proved, its construction is given as well as its characterization by boundary conditions (Theorem 6) in terms of the partition of Ω, ε-optimal strategies are constructed (Theorem 7).

In [15] the cost existence was proved in the special case when $\Omega = \Omega_d^{1,2}$ (the definition of $\Omega_d^{1,2}$ is given below).

Note also that condition (1) used by us is analogous to the condition

$$E \operatorname*{Sup}_n |g(x_n)| < \infty$$

in the standard stopping problem.

1. NOTATION

The following disjoint partition of the set Ω is obvious:

$$\Omega_d = \left\{ \omega : \sum_{n=0}^{\infty} \operatorname*{Inf}_a \operatorname*{Inf}_b P_n^{a,b} = \infty \right\},$$

$$\Omega^1 = \left\{ \omega : \sum_{n=0}^{\infty} \operatorname*{Inf}_a \operatorname*{Sup}_b P_n^{a,b} < \infty, \ \sum_{n=0}^{\infty} \operatorname*{Sup}_a \operatorname*{Inf}_b P_n^{a,b} = \infty \right\},$$

$$\Omega^2 = \left\{ \omega : \sum_{n=0}^{\infty} \operatorname*{Sup}_a \operatorname*{Inf}_b P_n^{a,b} < \infty, \ \sum_{n=0}^{\infty} \operatorname*{Inf}_a \operatorname*{Sup}_b P_n^{a,b} = \infty \right\},$$

$$\Omega_d^{1,2} = \left\{ \omega : \sum_{n=0}^{\infty} \operatorname*{Sup}_a \operatorname*{Inf}_b P_n^{a,b} = \infty, \ \sum_{n=0}^{\infty} \operatorname*{Inf}_a \operatorname*{Sup}_b P_n^{a,b} = \infty, \ \sum_{n=0}^{\infty} \operatorname*{Inf}_a \operatorname*{Inf}_b P_n^{a,b} < \infty \right\},$$

$$\Omega_c^{1,2} = \left\{ \omega : \sum_{n=0}^{\infty} \operatorname*{Sup}_a \operatorname*{Inf}_b P_n^{a,b} < \infty, \ \sum_{n=0}^{\infty} \operatorname*{Inf}_a \operatorname*{Sup}_b P_n^{a,b} < \infty, \ \sum_{n=0}^{\infty} \operatorname*{Sup}_a \operatorname*{Sup}_b P_n^{a,b} = \infty \right.$$

and

$$\Omega_c = \left\{ \omega : \sum_{n=0}^{\infty} \operatorname*{Sup}_a \operatorname*{Sup}_b P_n^{a,b} < \infty \right\}.$$

Define

$$A_d = \left\{ \alpha = \{a_n, \ n \geq 0\} : \sum_{n=0}^{\infty} \inf_b P_n^{a_n, b} = \infty \right\},$$

$$A_c = \left\{ \alpha : \sum_{n=0}^{\infty} \sup_b P_n^{a_n, b} < \infty \right\},$$

$$A_0 = \left\{ \alpha : \sum_{n=0}^{\infty} \inf_b P_n^{a_n, b} < \infty \, , \, \sum_{n=0}^{\infty} \sup_b P_n^{a_n, b} = \infty \right\} .$$

Analogously define B_d, B_c and B_0.

The following notation will be used:

$$\underline{\eta} = \begin{cases} \inf\limits_{\beta \in B_d + B_0} \sup\limits_{\alpha \in D_\beta} \overline{\lim\limits_{N\to\infty}} \eta_N^{\alpha,\beta}, & \text{if } \omega \in \Omega \setminus \Omega_c , \\ -\infty , & \text{if } \omega \in \Omega_c , \end{cases}$$

and

$$\overline{\eta} = \begin{cases} \sup\limits_{\alpha \in A_d + A_0} \inf\limits_{\beta \in D_\alpha} \underline{\lim\limits_{N\to\infty}} \eta_N^{\alpha,\beta}, & \text{if } \omega \in \Omega \setminus \Omega_c , \\ +\infty , & \text{if } \omega \in \Omega_c , \end{cases}$$

where $D_\beta = \left\{ \alpha : \sum\limits_{n=0}^{\infty} P_n^{a_n, b_n} = \infty \right\}$, $D_\alpha = \left\{ \beta : \sum\limits_{n=0}^{\infty} P_n^{a_n, b_n} = \infty \right\}$.

2. AUXILIARY RESULTS

The proofs of main results will be based on the following lemma which is an immediate consequence of Lemma 1 (see [4]).

LEMMA 1. Let $\alpha = \{a_n, n \geq 0\}$ and $\beta = \{b_n, n \geq 0\}$ be some strategies of the first and second players, respectively, and the sequence $\{v_n, n \geq 0\} \in \mathbb{R}$. Then:

- 1. if for all $n \geq 0$ the following inequality is fulfilled:

$$v_n \geq \left(1 - P_n^{a_n, b_n}\right) E\{v_{n+1} / F_n\} + r_n^{a_n, b_n} \qquad \text{a.s.} ,$$

then $\lim_{N\to\infty} v_N$ exists $\left[\text{a.s.} \sum_{n=0}^{\infty} P_n^{a_n,b_n} < \infty\right]$, the inequality

$$v_n \geq E\left\{\eta_n^{\alpha,\beta} + \lim_{N\to\infty} v_N \prod_{k=n}^{\infty}\left(1 - P_k^{a_k,b_k}\right) / F_n\right\} \quad \text{a.s.}$$

is true and

$$\underline{\lim}_{N\to\infty} v_N \geq \underline{\lim}_{N\to\infty} \eta_N^{\alpha,\beta} \qquad \left[\text{a.s.} \sum_{n=0}^{\infty} P_n^{a_n,b_n} = \infty\right],$$

where

$$\eta_N^{\alpha,\beta} = \sum_{k=N}^{\infty} \prod_{i=N}^{k-1} \left(1 - P_i^{a_i,b_i}\right) r_k^{a_k,b_k} \quad ;$$

- 2. if for all $n \geq 0$

$$v_n \leq \left(1 - P_n^{a_n,b_n}\right) E\{v_{n+1} / F_n\} + r_n^{a_n,b_n} \qquad \text{a.s.,}$$

then $\lim_{N\to\infty} v_N$ exists $\left[\text{a.s.} \sum_{n=0}^{\infty} P_n^{a_n,b_n} < \infty\right]$,

$$v_n \leq E\left\{\eta_n^{\alpha,\beta} + \lim_{N\to\infty} v_N \prod_{k=n}^{\infty}\left(1 - P_k^{a_k,b_k}\right) / F_n\right\} \quad \text{a.s.}$$

and

$$\overline{\lim}_{N\to\infty} v_N \leq \overline{\lim}_{N\to\infty} \eta_N^{\alpha,\beta} \qquad \left[\text{a.s.} \sum_{n=0}^{\infty} P_n^{a_n,b_n} = \infty\right].$$

<u>COROLLARY</u>. Let $\hat{\alpha} = \{\hat{a}_n, n \geq 0\}$ and $\hat{\beta} = \{\hat{b}_n, n \geq 0\}$ be such strategies that for all $n \geq 0$ a.s.

$$\operatorname*{Sup}_{a} \left[\left(1 - P_n^{a,\hat{b}_n}\right) E\{v_{n+1} / F_n\} + r_n^{a,\hat{b}_n} \right]$$

$$\leq v_n$$

$$\leq \operatorname*{Inf}_{b} \left[\left(1 - P_n^{\hat{a}_n,b}\right) E\{v_{n+1} / F_n\} + r_n^{\hat{a}_n,b} \right] .$$

Then the limit $\lim_{N\to\infty} v_N$ exists $\left[\text{a.s.} \left\{\sum_{n=0}^{\infty} \operatorname*{Inf}_{b} P_n^{\hat{a}_n,b} < \infty\right\} + \left\{\sum_{n=0}^{\infty} \operatorname*{Inf}_{a} P_n^{a,\hat{b}_n} < \infty\right\}\right]$,

$$\operatorname*{Sup}_{\alpha} E \left\{\eta_n^{\alpha,\hat{\beta}} + \lim_{N\to\infty} v_N \prod_{k=n}^{\infty} \left(1 - P_k^{a_k,\hat{b}_k}\right) / F_n\right\}$$

$$\leq v_n$$

$$\leq \operatorname*{Inf}_{\beta} E \left\{\eta_n^{\hat{\alpha},\beta} + \lim_{N\to\infty} v_N \prod_{k=n}^{\infty} \left(1 - P_k^{\hat{a}_k,b_k}\right) / F_n\right\} \quad \text{a.s.},$$

$$\overline{\lim_{N\to\infty}} v_N \leq \operatorname*{Inf}_{\beta\in D_{\hat{\alpha}}} \overline{\lim_{N\to\infty}} \eta_N^{\hat{\alpha},\beta} \qquad [\text{a.s. } \{\hat{\alpha} \in A_0 + A_d\}]$$

and

$$\underline{\lim_{N\to\infty}} v_N \geq \operatorname*{Sup}_{\alpha\in D_{\hat{\beta}}} \underline{\lim_{N\to\infty}} \eta_N^{\alpha,\hat{\beta}} \qquad [\text{a.s. } \{\hat{\beta} \in B_0 + B_d\}] .$$

<u>LEMMA 2</u> (see Lemma 3 in [4]). For all admissible strategies $\hat{\alpha}$ and $\hat{\beta}$ the following equalities are true:

$$\operatorname*{Sup}_{\alpha\in D_{\hat{\beta}}} \underline{\lim_{N\to\infty}} \eta_N^{\alpha,\hat{\beta}} = \operatorname*{Sup}_{\alpha\in D_{\hat{\beta}}} \overline{\lim_{N\to\infty}} \eta_N^{\alpha,\hat{\beta}} \qquad [\text{a.s. } \{\hat{\beta} \in B_0\}]$$

and

$$\operatorname*{Inf}_{\beta\in D_{\hat{\alpha}}} \underline{\lim}_{N\to\infty} \eta_N^{\hat{\alpha},\beta} = \operatorname*{Inf}_{\beta\in D_{\hat{\alpha}}} \overline{\lim}_{N\to\infty} \eta_N^{\hat{\alpha},\beta} \qquad [\text{a.s. } \{\hat{\alpha}\in A_0\}] .$$

Let us next denote

$$\underline{\eta}^{\beta} = \operatorname*{Sup}_{\alpha\in D_{\beta}} \overline{\lim}_{N\to\infty} \eta_N^{\alpha,\beta} \quad \text{and} \quad \overline{\eta}^{\alpha} = \operatorname*{Inf}_{\beta\in D_{\alpha}} \underline{\lim}_{N\to\infty} \eta_N^{\alpha,\beta} .$$

The following two theorems are immediate consequences of Theorems 1 and 2 (see [4]).

THEOREM 1. Let $\{\beta = \{b_n, n\geq 0\}$ be a fixed strategy of the second player. Then:

- 1. if v_n is a solution of the equation

$$v_n = \operatorname*{Sup}_{a} \left[\left(1 - P_n^{a,b_n}\right) E\{v_{n+1} / \mathcal{F}_n\} + r_n^{a,b_n}\right] \tag{15}$$

and $\{v_n, n\geq 0\} \in \mathbb{R}$, then the following representation holds:

$$v_n = \operatorname*{Sup}_{\alpha\in\Sigma} E\left\{\eta_n^{\alpha,\beta} + \lim_{N\to\infty} v_N \prod_{k=n}^{\infty} \left(1 - P_k^{a_k,b_k}\right) / \mathcal{F}_n\right\} \quad \text{a.s.};$$

- 2. the value

$$S_\eta^{\beta}(n) = \operatorname*{Sup}_{\alpha\in\Sigma} S_\eta^{\alpha,\beta}(n)$$

is a solution of equation (15) and

$$\lim_{N\to\infty} S_\eta^{\beta}(N) = \max(\underline{\eta}^{\beta}, \eta) \qquad [\text{a.s. } \{\beta\in B_0\}] .$$

THEOREM 2. Let $\beta = \{b_n, n\geq 0\}$ be a fixed strategy of the second player and let there exist a strategy of the first player

$\alpha^* = \{a^*_n, n \geq 0\}$ such that $\alpha^* \in D_\beta$. Then:

• 1. the value

$$\underline{S}^\beta_\eta(n) = \sup_{\alpha \in D_\beta} S^{\alpha,\beta}_\eta(n)$$

is a solution of equation (15) and

$$\lim_{N\to\infty} \underline{S}^\beta_\eta(N) = \underline{\eta}^\beta \qquad [\text{a.s. } \{\beta \in B_0\}] \quad ;$$

• 2. if U_β is an operator acting on v_n defined by the formula

$$U_\beta v_n = \sup_{a \in R^{(1)}} \left[\left(1 - P_n^{a,b_n}\right) E\{v_{n+1} / F_n\} + r_n^{a,b_n}\right] \quad , \tag{16}$$

then

$$\underline{S}^\beta_\eta(n) = \lim_{N\to\infty} U^N_\beta \, S^{\alpha^*,\beta}_\eta(n) \quad , \tag{17}$$

where $U^N_\beta = \underbrace{U_\beta \dots U_\beta}_{N}$.

In the next lemma all solutions of the Bellman equation (14) are represented as the value in some minimax version of the optimal stopping problem and the limit behavior of these solutions is investigated.

LEMMA 3 (see [15]). Let v_n be a solution of equation (14) and $\{v_n, n \geq 0\} \in \mathbb{R}$. Then:

• 1. the following representation holds:

$$\begin{aligned} v_n &= \operatorname*{Sup}_{\alpha\in\Sigma} \operatorname*{Inf}_{\beta\in\Pi} E\left\{\eta_n^{\alpha,\beta} + \lim_{N\to\infty} v_N \prod_{k=n}^{\infty} \left(1 - P_k^{a_k,b_k}\right) / F_n\right\} \\ &= \operatorname*{Inf}_{\beta\in\Pi} \operatorname*{Sup}_{\alpha\in\Sigma} E\left\{\eta_n^{\alpha,\beta} + \lim_{N\to\infty} v_N \prod_{k=n}^{\infty} \left(1 - P_k^{a_k,b_k}\right) / F_n\right\} \; ; \end{aligned} \tag{18}$$

•2. [a.s. $\Omega^1 + \Omega^2 + \Omega_c + \Omega_c^{1,2} + \Omega_d^{1,2}$ $(\underline{\eta} \le \overline{\eta})$] there exists a limit $\lim_{N\to\infty} v_N$,

$$\underline{\lim}_{N\to\infty} v_N \ge \underline{\eta} \qquad [\text{a.s. } \{\hat{\beta} \in B_0\}] , \tag{19}$$

$$\overline{\lim}_{N\to\infty} v_N \le \overline{\eta} \qquad [\text{a.s. } \{\hat{\alpha} \in A_0\}] , \tag{20}$$

$$\overline{\lim}_{N\to\infty} v_N \le \underline{\eta} \qquad [\text{a.s. } \{\hat{\alpha} \in A_d\}] , \tag{21}$$

$$\underline{\lim}_{N\to\infty} v_N \ge \overline{\eta} \qquad [\text{a.s. } \{\hat{\beta} \in B_d\}] , \tag{22}$$

and

$$\underline{\eta} \le \lim_{N\to\infty} v_N \le \overline{\eta} \quad [\text{a.s. } \Omega_d^{1,2}\{\underline{\eta} \le \overline{\eta}\}] , \tag{23}$$

where $\hat{\alpha} = \{\hat{a}_n, n \ge 0\}$ and $\hat{\beta} = \{\hat{b}_n, n \ge 0\}$ are such strategies of the first and second players, respectively, that for all $n \ge 0$

$$\begin{aligned}
&\operatorname*{Sup}_{a \in R^{(1)}} \left[\left(1 - P_n^{a,\hat{b}_n}\right) E\{v_{n+1} / \mathcal{F}_n\} + r_n^{a,\hat{b}_n}\right] \\
&\le v_n \\
&\le \operatorname*{Inf}_{b \in R^{(1)}} \left[\left(1 - P_n^{\hat{a}_n,b}\right) E\{v_{n+1} / \mathcal{F}_n\} + r_n^{\hat{a}_n,b}\right] \quad \text{a.s.}
\end{aligned} \tag{24}$$

<u>COROLLARY</u>. The following relations hold:

$$\lim_{N\to\infty} v_N = \overline{\eta} \qquad [\text{a.s. } \{\hat{\alpha} \in A_0, \hat{\beta} \in B_d\}] ,$$

$$\lim_{N\to\infty} v_N = \underline{\eta} \qquad [\text{a.s. } \{\hat{\alpha} \in A_d, \ \hat{\beta} \in B_0\}] ,$$

$$\overline{\eta} \le \underline{\lim}_{N\to\infty} v_N \le \overline{\lim}_{N\to\infty} v_N \le \underline{\eta} \qquad [\text{a.s. } \{\hat{\alpha} \in A_d, \ \hat{\beta} \in B_d\}]$$

and

$$\underline{\eta} \le \underline{\lim}_{N\to\infty} v_N \le \overline{\lim}_{N\to\infty} v_N \le \overline{\eta} \qquad [\text{a.s. } \{\hat{\alpha} \in A_0, \ \hat{\beta} \in B_0\}] .$$

LEMMA 4. Let the solutions of equation (14), v_n^1 and v_n^2, satisfy the following conditions:

- i. $\{v_n^i, \ n \ge 0\} \in \mathbb{R}, \quad i = 1,2;$
- ii. $\lim_{N\to\infty} v_N^1 = \lim_{N\to\infty} v_N^2 \quad [\text{a.s. } \Omega_c + \Omega_c^{1,2} + \Omega^1 + \Omega^2 + \Omega_d^{1,2}\{\underline{\eta}\le\overline{\eta}\}].$

Then a.s. $v_n^1 = v_n^2, \quad n \ge 0.$

Proof. Let $\alpha^i = \{a_n^i, \ n \ge 0\}$ and $\beta^i = \{b_n^i, \ n \ge 0\}$, $i = 1,2,$ be strategies of players such that for each $n \ge 0$, a_n^i and b_n^i are such that

$$\begin{aligned} &\operatorname*{Sup}_a \left[\left(1 - P_n^{a,b_n^i}\right) E\left\{v_{n+1}^i / \mathcal{F}_n\right\} + r_n^{a,b_n^i}\right] \\ &= v_n^i \qquad (25) \\ &= \operatorname*{Inf}_b \left[\left(1 - P_n^{a_n^i,b}\right) E\left\{v_{n+1}^i / \mathcal{F}_n\right\} + r_n^{a_n^i,b}\right] \quad \text{a.s.} \end{aligned}$$

Let $H_{dd}^i = \{\alpha^i \in A_d, \ \beta^i \in B_d\}$, $H_{d0}^i = \{\alpha^i \in A_d, \ \beta^i \in B_0\}$, $H_{0d}^i = \{\alpha^i \in A_0, \ \beta^i \in B_d\}$ and $H_{00}^i = \{\alpha^i \in A_0, \ \beta^i \in B_0\}$, $i = 1,2.$

Note that from the definition of sets A_c, B_c and $\Omega_d^{1,2}$ for each strategy $\alpha \in \Sigma$ or $\beta \in \Pi$ we have

$$\Omega_d^{1,2}\{\alpha \in A_c\} = \Omega_d^{1,2}\{\beta \in B_c\} = \emptyset . \tag{26}$$

It is easy to show that

$$P\{(\underline{\eta} > \overline{\eta})\ H_{00}^{i}\} = 0 , \qquad i = 1,2 . \tag{27}$$

Indeed, using (26) we have (i = 1,2)

$$\underline{\eta} \le \varliminf_{N\to\infty} v_N^i \le \varlimsup_{N\to\infty} v_N^i \le \overline{\eta} \qquad [\text{a.s. } H_{00}^i]$$

and consequently (27) is true.

Further, in agreement with (25), a.s.

$$v_n^1 = \operatorname*{Sup}_{a} \left[\left(1 - P_n^{a,b_n^1} \right) E \left\{ v_{n+1}^1 / \mathcal{F}_n \right\} + r_n^{a,b_n^1} \right] .$$

Therefore, by Theorem 1 the following representation holds:

$$v_n^1 = \operatorname*{Sup}_{\alpha\in\Sigma} E\left\{ \eta_n^{\alpha,\beta^1} + \lim_{N\to\infty} v_N^1 \prod_{k=n}^{\infty} \left(1 - P_k^{a_k,b_k^1} \right) / \mathcal{F}_n \right\} \qquad \text{a.s. .}$$

From this, taking into account

$$P\left\{ \left[\prod_{k=n}^{\infty} \left(1 - P_k^{a_k,b_k^1} \right) > 0 \right] \cap \{\beta^1 \in B_d\} \right\} = 0$$

(see Section 4 in [4]) and the conditions of the Lemma, we obtain that a.s.

$$v_n^1 = \sup_{\alpha\in\Sigma} E\left\{\eta_n^{\alpha,\beta^1} + \prod_{k=n}^{\infty}\left(1 - P_k^{a_k,b_k^1}\right)\right.$$

$$\times \left[I_{\Omega_c+\Omega_c^{1,2}+\Omega^1+\Omega^2+\Omega_d^{1,2}\{\underline{\eta}\le\overline{\eta}\}} \lim_{N\to\infty} v_N^2 \right. \tag{28}$$

$$\left.\left. + I_{\Omega_d^{1,2}\{\underline{\eta}\le\overline{\eta}\}} I_{\{\beta^1\in B_0\}} \lim_{N\to\infty} v_N^1 \right] \Big/ F_N \right\} .$$

Let us show now that a.s.

$$I_{\Omega_d^{1,2}\{\underline{\eta}>\overline{\eta}\}\{\beta^1\in B_0\}} \overline{\lim_{N\to\infty}} v_N^1 \ge I_{\Omega_d^{1,2}\{\underline{\eta}>\overline{\eta}\}\{\beta^1\in B_0\}} \overline{\lim_{N\to\infty}} v_N^2 . \tag{29}$$

Using (26) and (27) we have

$$\overline{\lim_{N\to\infty}} v_N^1 \, I_{\Omega_d^{1,2}\{\underline{\eta}>\overline{\eta}\}\{\beta^1\in B_0\}}$$

$$= \overline{\lim_{N\to\infty}} v_N^1 \cdot I_{\Omega_d^{1,2}\{\underline{\eta}>\overline{\eta}\}} I_{H_{d0}^1} \tag{30}$$

$$= \overline{\lim_{N\to\infty}} v_N^1 \cdot I_{\Omega_d^{1,2}\{\underline{\eta}>\overline{\eta}\}\{\beta^1\in B_0\}} \left[I_{H_{d0}^1H_{d0}^2} + I_{H_{d0}^1H_{0d}^2} + I_{H_{0d}^1H_{dd}^2} \right] .$$

However, due to (22),

$$\lim_{N\to\infty} v_N^1 = \lim_{N\to\infty} v_N^2 = \underline{\eta} \qquad [\text{a.s. } H_{d0}^1H_{d0}^2]$$

and combining (21) and (22), we obtain

$$\lim_{N\to\infty} v_N^1 = \underline{\eta} > \overline{\eta} = \lim_{N\to\infty} v_N^2 \qquad [\text{a.s. } \{\underline{\eta}>\overline{\eta}\}H_{d0}^1H_{0d}^2] .$$

Further, due to (22) and (23)

$$\lim_{N\to\infty} v_N^1 = \underline{\eta} \ge \overline{\lim_{N\to\infty}} v_N^2 \qquad [\text{a.s. } H_{d0}^1H_{dd}^2] .$$

The last three inequalities and (30) yield (29). Consequently, from (28) we have

$$v_n^1 \geq \operatorname*{Sup}_{\alpha\in\Sigma} E\left\{\eta_n^{\alpha,\beta^1} + \lim_{N\to\infty} v_N^2 \prod_{k=n}^{\infty}\left(1 - P_k^{a_k,b_k^1}\right) / \mathcal{F}_n\right\}$$

$$\geq \operatorname*{Inf}_{\beta\in\Pi} \operatorname*{Sup}_{\alpha\in\Sigma} E\left\{\eta_n^{\alpha,\beta} + \lim_{N\to\infty} v_N^2 \prod_{k=n}^{\infty}\left(1 - P_k^{a_k,b_k^1}\right) / \mathcal{F}_n\right\} \quad \text{a.s.}\ .$$

In the same way, using the equality

$$v_n^1 = \operatorname*{Inf}_{b}\left[\left(1 - P_n^{a_n^1,b}\right) E\{v_{n+1}^1 / \mathcal{F}_n\} + r_n^{a_n^1,b}\right] \quad \text{a.s.},$$

the inverse inequality can be proved and the desired conclusion follows.

3. CONSTRUCTION OF THE VALUE

Let U be an operator acting on $\{v_n, n\geq 0\} \in \mathbb{R}$ by the formula

$$U v_n = \operatorname*{Sup}_{a} \operatorname*{Inf}_{b} [(1 - P_n^{a,b}) E\{v_{n+1} / \mathcal{F}_n\} + r_n^{a,b}] .$$

Denote

$$U^N = \underbrace{U \dots U}_{N} .$$

<u>THEOREM 3</u>. Let $\alpha^* = \{a_n^*, n\geq 0\}$ be a strategy of the first player satisfying the following conditions:

$$\alpha^* \in A_d \qquad [\text{a.s. } \Omega_d^{1,2} + \Omega_d + \Omega^1] ,$$

$$\alpha^* \in A_0 \qquad [\text{a.s. } \Omega^2 + \Omega_c^{1,2}] ,$$

$$\alpha^* \in A_c \qquad [\text{a.s. } \Omega_c] .$$

Then $\underset{\sim}{S}_\eta(n) = \lim_{N\to\infty} U^N S_\eta^{\alpha^*}(n)$, where

$$S_\eta^{\alpha^*}(n) = \operatorname*{Inf}_{\beta\in\Pi} S_\eta^{\alpha^*,\beta}(n) ,$$

is a solution of equation (14). Moreover,

$$\lim_{N\to\infty} \underset{\sim}{S}_\eta(N) \le \eta \qquad [\text{a.s. } \Omega_c + \Omega^2 + \Omega_c^{1,2}] , \tag{31}$$

$$\lim_{N\to\infty} \underset{\sim}{S}_\eta(N) = \underline{\eta} \qquad [\text{a.s. } \Omega^1 + \Omega_d^{1,2}\{\underline{\eta}\le\overline{\eta}\}] \tag{32}$$

and

$$\overline{\lim_{N\to\infty}} \underset{\sim}{S}_\eta(N) \le \underline{\eta} \qquad [\text{a.s. } \Omega^1 + \Omega_d^{1,2}] . \tag{33}$$

<u>Proof</u>. It is not difficult to show that the sequence $\{U^N S_\eta^{\alpha^*}(n), N\ge 0\}$ is monotone nondecreasing. Indeed, from Theorem 1 (if we consider the minimum problem) we have that the cost $S_\eta^{\alpha^*}(n)$ satisfies the equality

$$S_\eta^{\alpha^*}(n) = \operatorname*{Inf}_{b} \left[\left(1 - P_n^{a^*_n,b}\right) E\{S_\eta^{\alpha^*}(n+1) / \mathcal{F}_n\} + r_n^{a^*_n,b}\right] \quad \text{a.s.}$$

and, consequently, a.s.

$$S_\eta^{\alpha^*}(n) \le U S_\eta^{\alpha^*}(n) .$$

Therefore, since the operator U is monotone, we have

$$U^{N-1} S_\eta^{\alpha^*}(n) \le U^N S_\eta^{\alpha^*}(n) , \qquad N \ge 1 , \quad \text{a.s.}.$$

Hence, the limit $\underset{\sim}{S}_\eta(n) = \lim_{N\to\infty} U^N S_\eta^{\alpha^*}(n)$ exists and satisfies equation (14). From (1) we establish that $\{\underset{\sim}{S}_\eta(n),\ n\geq 0\} \in \mathbb{R}$.

Now let us fix the strategy of the second player $\beta = \{b_n,\ n\geq 0\} \in \Pi$ and consider the sequence $\{U_\beta^N\ S_\eta^{\alpha^*,\beta}(n),\ N\geq 0\}$, where the operator U_β is defined from (16). By Theorem 2

$$\mathcal{U}_\eta(n) = \lim_{N\to\infty} U_\beta^N\ S_\eta^{\alpha^*,\beta}(n)$$

is a solution of the equation $v_n = U_\beta\ v_n$ and

$$\lim_{N\to\infty} \mathcal{U}_\eta(N) = \underline{\eta}^\beta \qquad [\text{a.s. } \{\beta \in B_0\}] \ . \tag{34}$$

Moreover, since $S_\eta^{\alpha^*}(n) \leq S_\eta^{\alpha^*,\beta}(n)$ it follows that a.s.

$$U\ S_\eta^{\alpha^*}(n) \leq U\ S_\eta^{\alpha^*,\beta}(n) \leq U_\beta\ S_\eta^{\alpha^*,\beta}(n) \ .$$

Consequently, for each $N \geq 1$, a.s.

$$U^N\ S_\eta^{\alpha^*}(n) \leq U_\beta^N\ S_\eta^{\alpha^*,\beta}(n) \ .$$

It is then quite obvious that a.s. $\underset{\sim}{S}_\eta(n) \leq \mathcal{U}_\eta(n)$.

Thus, taking into account (28) for all $\beta \in \Pi$, we get

$$\lim_{N\to\infty} \underset{\sim}{S}_\eta(N) \leq \underline{\eta}^\beta \qquad [\text{a.s. } \{\beta \in B_0\}] \ . \tag{35}$$

Now we shall prove

$$\overline{\lim_{N\to\infty}}\ \underset{\sim}{S}_\eta(N) \leq \underline{\eta}^\beta \qquad [\text{a.s. } \{\beta \in B_d\}] \ . \tag{36}$$

Since $\underset{\sim}{S}_\eta(n)$ is a solution of equation (14), we have

$$\underset{\sim}{S}_\eta(n) \leq \operatorname*{Sup}_a \left[\left(1 - P_n^{a,b_n}\right) E\{\underset{\sim}{S}_\eta(n+1)\ /\ \mathcal{F}_n\} + r_n^{a,b_n}\right] \quad \text{a.s.} \ . \tag{37}$$

Let us fix $\varepsilon > 0$ and let $\{\varepsilon_n, n\geq 0\}$ be a sequence of positive numbers such that $\sum_{n=0}^{\infty} \varepsilon_n < \varepsilon$. Due to (37) there exists $a_n^{\varepsilon_n}$ such that for each $n \geq 0$ a.s.

$$\underset{\sim}{S}_\eta(n) \leq \left(1 - P_n^{a_n^{\varepsilon_n}, b_n}\right) E\{\underset{\sim}{S}_\eta(n+1) / F_n\} + r_n^{a_n^{\varepsilon_n}, b_n} + \varepsilon_n .$$

Recall the proof of Lemma 1 (see [4]). Then

$$\overline{\lim_{N\to\infty}}\, \underset{\sim}{S}_\eta(N) \leq \overline{\lim_{N\to\infty}}\, \eta_N^{\alpha^\varepsilon, \beta} \qquad [\text{a.s. } \{\beta \in B_d\}] ,$$

where $\alpha^\varepsilon = \{a_n^{\varepsilon_n}, a_{n+1}^{\varepsilon_{n+1}}, \ldots\}$. The last inequality implies (36). Consequently putting (35) and (36) together, we obtain (33). On the other hand, from the Corollary of Lemma 3 we have

$$\lim_{N\to\infty} \underset{\sim}{S}_\eta(N) \geq \underline{\eta} \qquad [\text{a.s. } \Omega^1 + \Omega_d^{1,2}\{\underline{\eta} \leq \overline{\eta}\}] .$$

This verifies property (32).

It remains to prove that (31) holds.

Let us consider a strategy of the second player $\tilde{\beta} = \{\tilde{b}_n, n\geq 0\}$ for which

$$\tilde{\beta} \in B_c \qquad [\text{a.s. } \Omega_c + \Omega_c^{1,2} + \Omega^2] .$$

From the Corollary of Theorem 1.5 (see [14]) we get

$$\lim_{N\to\infty} U_{\tilde{\beta}}^N S_\eta^{\alpha^*, \tilde{\beta}}(n) = S_{\eta_1}(n) \qquad \text{a.s.},$$

where

$$\eta_1 = \begin{cases} \eta, & \text{if } \omega \in \left\{ \sum_{n=0}^{\infty} P_n^{a_n^*, \tilde{b}_n} < \infty \right\}, \\ \underline{\lim}_{N\to\infty} S_\eta^{\alpha^*, \tilde{\beta}}(N), & \text{if } \omega \in \left\{ \sum_{n=0}^{\infty} P_n^{a_n^*, \tilde{b}_n} = \infty \right\}. \end{cases}$$

First, applying Theorem 1, we have

$$\lim_{N\to\infty} S_{\eta_1}(N) = \begin{cases} \eta_1 & [\text{a.s. } \{\tilde{\beta} \in B_c\}], \\ \max(\underline{\eta}^{\tilde{\beta}}, \eta_1) & [\text{a.s. } \{\tilde{\beta} \in B_0\}], \end{cases}$$

so that from the definition of the strategy $\tilde{\beta}$ we obtain

$$I_{[\Omega_c + \Omega_c^{1,2} + \Omega^2]} \lim_{N\to\infty} S_{\eta_1}(N) = I_{[\Omega_c + \Omega_c^{1,2} + \Omega^2]} \cdot \eta_1$$

$$= I_{[\Omega_c + \Omega_c^{1,2} + \Omega^2]} \cdot \eta \quad \text{a.s.} .$$

Consequently

$$\lim_{N\to\infty} S_{\eta_1}(N) = \eta \qquad [\text{a.s. } \Omega_c + \Omega_c^{1,2} + \Omega^2] . \tag{38}$$

On the other hand, one easily sees that a.s.

$$\underset{\sim}{S}_\eta(n) = \lim_{N\to\infty} U^N S_\eta^{\alpha^*}(n) \leq \lim_{N\to\infty} U^N S_\eta^{\alpha^*, \tilde{\beta}}(n)$$

$$\leq \lim_{N\to\infty} U_{\tilde{\beta}}^N S_\eta^{\alpha^*, \tilde{\beta}}(n) = S_{\eta_1}(n) ;$$

and, taking into account (38), we obtain finally

$$\lim_{N\to\infty} \underset{\sim}{S}_\eta(N) \leq \lim_{N\to\infty} S_{\eta_1}(N) = \eta \quad [\text{a.s. } \Omega_c + \Omega_c^{1,2} + \Omega^2] . \ ///$$

Now consider the strategy of the second player $\hat{\beta} = \{\hat{b}_n, n \geq 0\}$, where for each $n \geq 0$, $\hat{b}_n$ is such that a.s.

$$\underset{\sim}{S}_\eta(n) = \operatorname*{Sup}_{a \in R^{(1)}} \left| \left(1 - P_n^{a,\hat{b}_n}\right) E\{\underset{\sim}{S}_\eta(n+1) / F_n\} + r_n^{a,\hat{b}_n} \right|$$

$$= \operatorname*{Sup}_{a \in R^{(1)}} \operatorname*{Inf}_{b \in R^{(1)}} \left| \left(1 - P_n^{a,b}\right) E\{\underset{\sim}{S}_\eta(n+1) / F_n\} + r_n^{a,b} \right| . \quad (39)$$

The proofs of main results are based on the following three lemmas.

LEMMA 5. Let $\hat{\beta} = \{\hat{b}_n, n \geq 0\}$ be the strategy of the second player defined by (39). Then

$$\underline{\eta}^{\beta} \leq \eta \qquad [\text{a.s. } \Omega_c^{1,2}\{\hat{\beta} \in B_0\} + \Omega^2\{\hat{\beta} \in B_0\}] , \quad (40)$$

$$\overline{\eta} \leq \eta \qquad [\text{a.s. } \Omega^2\{\hat{\beta} \in B_d\}] , \quad (41)$$

$$\underline{\eta} = \overline{\eta} \qquad [\text{a.s. } \Omega_d^{1,2}\{\underline{\eta} \leq \overline{\eta}\}\{\hat{\beta} \in B_d\}] \quad (42)$$

and

$$\underline{\eta} = \underline{\eta}^{\hat{\beta}} \qquad [\text{a.s. } \{\hat{\beta} \in B_0\}\Omega_d^{1,2} + \Omega^1] . \quad (43)$$

Proof. From Lemma 1 we have

$$\lim_{N\to\infty} \underset{\sim}{S}_\eta(N) \geq \underline{\eta}^{\hat{\beta}} \qquad [\text{a.s. } \{\hat{\beta} \in B_0\}]$$

and due to Theorem 3

$$\lim_{N\to\infty} \underset{\sim}{S}_\eta(N) \leq \eta \qquad [\text{a.s. } \Omega_c^{1,2} + \Omega^2] . \quad (44)$$

Putting these results together we obtain

$$\underline{\eta}^{\hat{\beta}} \quad \leq \quad \eta \qquad\qquad [\text{a.s.} \quad \Omega_c^{1,2}\{\hat{\beta}\in B_0\} + \Omega^2\{\hat{\beta}\in B_0\}] \ .$$

Further, using the same Lemma, we have

$$\lim_{N\to\infty} S_{\underset{\sim}{\eta}}(N) \quad \geq \quad \overline{\eta} \qquad\qquad [\text{a.s.} \quad \{\hat{\beta}\in B_d\}] \ .$$

Recalling (44), it is then quite obvious that (42) holds. From Lemma 3 it turns out that

$$\underline{\lim}_{N\to\infty} \ S_{\underset{\sim}{\eta}}(N) \quad \geq \quad \overline{\eta} \qquad\qquad [\text{a.s.} \quad \{\hat{\beta}\in B_d\}]$$

and by Theorem 3

$$\lim_{N\to\infty} S_{\underset{\sim}{\eta}}(N) \quad = \quad \underline{\eta} \qquad\qquad [\text{a.s.} \quad \Omega_d^{1,2}\{\underline{\eta}\leq\overline{\eta}\}] \ .$$

Hence, combining the last formulas, we get (42).

It remains to check (43). From the definition of the strategy $\hat{\beta} = \{\hat{b}_n, \ n\geq 0\}$ we have a.s.

$$S_{\underset{\sim}{\eta}}(n) \quad = \quad \operatorname*{Sup}_{a} \left[\left(1 - P_n^{a,\hat{b}_n}\right) E\{S_{\underset{\sim}{\eta}}(n+1) \ / \ \mathcal{F}_n\} + r_n^{a,\hat{b}_n}\right] \ .$$

Therefore from Lemma 1 we obtain

$$\lim_{N\to\infty} S_{\underset{\sim}{\eta}}(N) \quad \geq \quad \underline{\eta}^{\hat{\beta}} \qquad\qquad [\text{a.s.} \quad \{\hat{\beta}\in B_0\}] \ .$$

Furthermore, by Theorem 3

$$\overline{\lim_{N\to\infty}} \ S_{\underset{\sim}{\eta}}(N) \quad \leq \quad \underline{\eta} \qquad\qquad [\text{a.s.} \quad \Omega^1 + \Omega_d^{1,2}]$$

and, consequently, as $\Omega^1 + \Omega_d^{1,2}\{\hat{\beta}\in B_0\} \in \{\hat{\beta}\in B_0\}$, we get

$$\underline{\eta} \geq \underline{\eta}^{\hat{\beta}} \qquad [\text{a.s. } \Omega^1 + \Omega_d^{1,2}\{\hat{\beta}\in B_0\}] \quad .$$

Finally, the property of (43) follows easily from the fact that an inverse inequality is always true.

LEMMA 6. Let v_n be a solution of equation (14) and $\{v_n, n\geq 0\} \in \mathbb{R}$. Then

$$\underline{\lim}_{N\to\infty} v_N \leq \eta \qquad [\text{a.s. } \Omega^2\{\hat{\beta}\in B_d\}] \tag{45}$$

and

$$\overline{\lim}_{N\to\infty} v_N \leq \max(\underline{\eta}, \eta) \qquad [\text{a.s. } \Omega_d^{1,2}\{\hat{\beta}\in B_d\}] \quad , \tag{46}$$

where the strategy $\hat{\beta} \doteq \{\hat{b}_n, n\geq 0\}$ is defined from (39).

Proof. By Lemma 3 we have

$$\underline{\lim}_{N\to\infty} v_n \leq \overline{\eta} \qquad [\text{a.s. } \Omega^2] \quad .$$

Moreover, due to Lemma 5, $\overline{\eta} \leq \eta$ $[\text{a.s. } \Omega^2\{\hat{\beta}\in B_d\}]$. This proves (45).

Further

$$\overline{\lim}_{N\to\infty} v_N \cdot I_{\Omega_d^{1,2}\{\hat{\beta}\in B_d\}} = \overline{\lim}_{N\to\infty} v_N \cdot I_{\Omega_d^{1,2}\{\hat{\beta}\in B_d\}} [I_{[\underline{\eta}\leq\overline{\eta}]} + I_{[\underline{\eta}>\overline{\eta}]}] \; . \tag{47}$$

From the Corollary of Lemma 3

$$\lim_{N\to\infty} v_N \cdot I_{[\underline{\eta}\leq\overline{\eta}]} \leq \overline{\eta}\, I_{[\underline{\eta}\leq\overline{\eta}]} \qquad \text{a.s. } .$$

Consequently, using (42), we may conclude that a.s.

$$\lim_{N\to\infty} v_N \cdot I_{\Omega_d^{1,2}\{\hat{\beta}\in B_d\}\{\underline{\eta}\leq\overline{\eta}\}} \leq \underline{\eta}\, I_{\Omega_d^{1,2}\{\hat{\beta}\in B_d\}\{\underline{\eta}\leq\overline{\eta}\}} \quad . \tag{48}$$

Now let $L = \{\tilde{a}_n, n \geq 0\} \in \Sigma$ be a strategy of the first player, such that for each $n \geq 0$

$$\inf_b \sup_a \left[\left(1 - p_n^{a,b}\right) E\{v_{n+1} / F_n\} + r_n^{a,b}\right]$$

$$= \inf_b \left[\left(1 - p_n^{\tilde{a}_n,b}\right) E\{v_{n+1} / F_n\} + r_n^{\tilde{a}_n,b}\right] \quad \text{a.s.} \; .$$

Then, as $A_c \neq \emptyset$ on the set $\Omega_d^{1,2}$, it is clear that

$$\overline{\lim_{N\to\infty}} v_N \cdot I_{\Omega_d^{1,2}\{\underline{\eta}>\overline{\eta}\}} = \overline{\lim_{N\to\infty}} v_N \cdot I_{\Omega_d^{1,2}\{\underline{\eta}>\overline{\eta}\}} \left[I_{\{\tilde{\alpha}\in A_0\}} + I_{\{\tilde{\alpha}\in A_d\}}\right]$$

$$\text{a.s.} \; . \tag{49}$$

Moreover, due to Lemma 3

$$\overline{\lim_{N\to\infty}} v_N \cdot I_{\{\tilde{\alpha}\in A_0\}} \leq \overline{\eta}\, I_{\{\tilde{\alpha}\in A_0\}} \qquad \text{a.s.}$$

and thus we obtain

$$\overline{\lim_{N\to\infty}} v_N \cdot I_{\Omega_d^{1,2}\{\underline{\eta}>\overline{\eta}\}\{\tilde{\alpha}\in A_0\}} \leq \underline{\eta}\, I_{\Omega_d^{1,2}\{\underline{\eta}>\overline{\eta}\}\{\tilde{\alpha}\in A_0\}} \qquad \text{a.s.} \tag{50}$$

Further, using Lemma 3, we get

$$\overline{\lim_{N\to\infty}} v_N \cdot I_{\{\tilde{\alpha}\in A_d\}} \leq \underline{\eta}\, I_{\{\tilde{\alpha}\in A_d\}} \qquad \text{a.s.}$$

and therefore

$$\lim_{N\to\infty} v_N \cdot I_{\Omega_d^{1,2}\{\underline{\eta}>\overline{\eta}\}\{\tilde{\alpha}\in A_d\}} \leq \underline{\eta}\, I_{\Omega_d^{1,2}\{\underline{\eta}>\overline{\eta}\}\{\tilde{\alpha}\in A_d\}} \qquad \text{a.s.} \; .$$

Now taking into account the last inequality and (50), from (49) it immediately follows that a.s.

$$\overline{\lim_{N\to\infty}} v_N \cdot I_{\Omega_d^{1,2}\{\underline{\eta}>\overline{\eta}\}} \leq \underline{\eta}\, I_{\Omega_d^{1,2}\{\underline{\eta}>\overline{\eta}\}} \quad . \tag{51}$$

Hence, due to (47) and (48) we obtain a.s.

$$\overline{\lim_{N\to\infty}} v_N \cdot I_{\Omega_d^{1,2}\{\hat{\beta}\in B_d\}} \leq \underline{\eta}\, I_{\Omega_d^{1,2}\{\hat{\beta}\in B_d\}} \leq \max(\underline{\eta},\eta)\, I_{\Omega_d^{1,2}\{\hat{\beta}\in B_d\}},$$

which proves (46). ///

<u>LEMMA 7</u>. Let $\hat{\beta} = \{\hat{b}_n, n\geq 0\}$ be a strategy of the second player defined from (39) and

$$S_\eta^{\hat{\beta}}(n) = \sup_{\alpha\in\Sigma} S_\eta^{\alpha,\hat{\beta}}(n) \quad .$$

Then

$$\lim_{N\to\infty} S_\eta^{\hat{\beta}}(N) = \eta \qquad [\text{a.s. } \Omega_c + \Omega_c^{1,2} + \Omega^2\{\hat{\beta}\in B_0+B_c\}] \tag{52}$$

and

$$\lim_{N\to\infty} S_\eta^{\hat{\beta}}(N) = \max(\underline{\eta},\eta) \qquad [\text{a.s. } \Omega^1 + \Omega_d^{1,2}\{\hat{\beta}\in B_0\}] \ . \tag{53}$$

<u>Proof</u>. By Theorem 1 we have

$$\lim_{N\to\infty} S_\eta^{\hat{\beta}}(N) = \begin{cases} \eta & [\text{a.s. } \{\hat{\beta}\in B_c\}] \ , \\ \max(\underline{\eta}^{\hat{\beta}},\eta) & [\text{a.s. } \{\hat{\beta}\in B_0\}] \ . \end{cases} \tag{54}$$

Then since $\Omega_c \subseteq \{\beta\in B_c\}$ for each $\beta\in\Pi$ we get

$$\lim_{N\to\infty} S_\eta^{\hat{\beta}}(N) = \eta \qquad [\text{a.s. } \Omega_c] \quad . \tag{55}$$

Taking into account the fact that $B_d \neq \emptyset$ on the set $\Omega_c^{1,2}$, one easily sees that

$$\lim_{N\to\infty} S_\eta^{\hat\beta}(N)\cdot I_{\Omega_c^{1,2}} = \lim_{N\to\infty} S_\eta^{\hat\beta}(N)\cdot I_{\Omega_c^{1,2}\{\hat\beta\in B_0\}} + \lim_{N\to\infty} S_\eta^{\hat\beta}(N)\cdot I_{\Omega_c^{1,2}\{\hat\beta\in B_c\}} \quad \text{a.s.} \ . \tag{56}$$

Moreover, from (53) we have

$$\lim_{N\to\infty} S_\eta^{\hat\beta}(N)\cdot I_{\Omega_c^{1,2}\{\hat\beta\in B_0\}} = \max(\underline{\eta}^{\hat\beta},\eta)\cdot I_{\Omega_c^{1,2}\{\hat\beta\in B_0\}} \quad \text{a.s.} \ .$$

However due to Lemma 5

$$I_{\Omega_c^{1,2}\{\hat\beta\in B_0\}}\,\underline{\eta}^{\hat\beta} \le \eta\, I_{\Omega_c^{1,2}\{\hat\beta\in B_0\}} \quad \text{a.s.} \ ,$$

so that from the above equality we obtain a.s.

$$\lim_{N\to\infty} S_\eta^{\hat\beta}(N)\cdot I_{\Omega_c^{1,2}\{\hat\beta\in B_0\}} = \eta\, I_{\Omega_c^{1,2}\{\hat\beta\in B_0\}} \ .$$

The last equality, (53) and (55) imply

$$\lim_{N\to\infty} S_\eta^{\hat\beta}(N)\cdot I_{\Omega_c^{1,2}} = \eta\, I_{\Omega_c^{1,2}} \quad \text{a.s.} \ . \tag{57}$$

Further, using (53), we get

$$\lim_{N\to\infty} S_\eta^{\hat\beta}(N) = \eta \qquad [\text{a.s. } \Omega^2\{\hat\beta\in B_c\}]$$

and

$$\lim_{N\to\infty} S_\eta^{\hat\beta}(N)\cdot I_{\Omega^2\{\hat\beta\in B_0\}} = \max(\underline{\eta}^{\hat\beta},\eta)\cdot I_{\Omega^2\{\hat\beta\in B_0\}} \quad \text{a.s.} \ .$$

Hence due to (40) we obtain

$$\lim_{N\to\infty} S_\eta^{\hat\beta}(N)\cdot I_{\Omega^2\{\hat\beta\in B_c\}+\Omega^2\{\hat\beta\in B_0\}} = \eta\, I_{\Omega^2\{\hat\beta\in B_c\}+\Omega^2\{\hat\beta\in B_0\}} \quad \text{a.s.} \ .$$

Combining the last equality and (57), we obtain (52).

Next let us prove (53).

Since $\Omega^1 \subseteq \{\hat{\beta} \in B_0\}$, from (54) it is clear that

$$\lim_{N\to\infty} S_\eta^{\hat{\beta}}(N) = \max(\underline{\eta}^{\hat{\beta}}, \eta) \qquad [\text{a.s. } \Omega^1 + \Omega_d^{1,2}\{\hat{\beta} \in B_0\}] ,$$

which, using (43), implies (53).

THEOREM 4. The sequence $\{U^N S_\eta^{\hat{\beta}}(n),\ N \geq 0\}$ is monotone nonincreasing, $S_\eta^*(n) = \lim_{N\to\infty} U^N S_\eta^{\beta}(n)$ satisfies equation (14) and $\{S_\eta^*(n),\ n \geq 0\} \in \mathbb{R}$. Moreover,

$$\lim_{N\to\infty} S_\eta^*(N) \leq \eta \qquad [\text{a.s. } \Omega_c + \Omega_c^{1,2} + \Omega^2] \tag{58}$$

and

$$\overline{\lim_{N\to\infty}}\, S_\eta^*(N) \leq \max(\underline{\eta}, \eta) \qquad [\text{a.s. } \Omega^1 + \Omega_d^{1,2}] . \tag{59}$$

Proof. From Theorem 1 we have a.s.

$$S_\eta^{\hat{\beta}}(n) = \sup_a \left[\left(1 - P_n^{a,\hat{b}_n}\right) E\{S_\eta^{\hat{\beta}}(n+1) / F_n\} + r_n^{a,\hat{b}_n}\right] \quad \text{a.s.} .$$

It is then quite obvious that

$$S_\eta^{\hat{\beta}}(n) \geq U\, S_\eta^{\hat{\beta}}(n) \qquad \text{a.s.} .$$

Therefore, since the operator U is monotone for all $N \geq 1$, we obtain

$$U^{N-1} S_\eta^{\hat{\beta}}(n) \geq U^N S_\eta^{\hat{\beta}}(n) \qquad \text{a.s.} .$$

Hence, there exists a limit $S_\eta^*(n) = \lim_{N\to\infty} U^N S_\eta^{\hat{\beta}}(n)$ which satisfies equation (14). From condition (1) it follows that

$\{S^*_\eta(n),\ n\geq 0\} \in \mathbb{R}$.

Further, by Lemma 6

$$\lim_{N\to\infty} S^*_\eta(N) \leq \eta \qquad [\text{a.s. } \Omega^2\{\hat\beta\in B_d\}]$$

and

$$\overline{\lim_{N\to\infty}}\ S^*_\eta(N) \leq \max(\underline{\eta},\eta) \qquad [\text{a.s. } \Omega^{1,2}_d\{\hat\beta\in B_d\}]\ .$$

Since $S^*_\eta(n) \leq S^{\hat\beta}_\eta(n)$ a.s., the above inequalities and Lemma 7 yield (58) and (59). ///

Now let us fix a strategy of the second player $\beta^* = \{b^*_n,\ n\geq 0\}$ satisfying the following condition:

$$\beta^* \in B_d \qquad [\text{a.s. } \Omega^{1,2}_d + \Omega_d + \Omega^2]\ ,$$

$$\beta^* \in B_0 \qquad [\text{a.s. } \Omega^1 + \Omega^{1,2}_d]\ ,$$

$$\beta^* \in B_c \qquad [\text{a.s. } \Omega_c]\ ;$$

and let us construct $\tilde S_\eta(n) = \lim_{N\to\infty} {}^N S^{\beta^*}_\eta(n)$, where

$$S^{\beta^*}_\eta(n) = \operatorname*{Sup}_{\alpha\in\Sigma} S^{\alpha,\beta^*}_\eta(n)\ .$$

In exactly the same way as in Theorem 3 it can be proved that $\tilde S_\eta(n)$ satisfies equation (14).

Define the strategy of the first player $\tilde\alpha = \{\hat a_n,\ n\geq 0\}$ in such a way that

$$\tilde S_\eta(n) = \operatorname*{Inf}_{b}\left[\left(1-P^{\hat a_n,b}_n\right) E\{\tilde S_\eta(n+1)/\mathcal{F}_n\} + r^{\hat a_n,b}_n\right]$$

$$= \operatorname*{Sup}_{a}\operatorname*{Inf}_{b}\left[\left(1-P^{a,b}_n\right) E\{\tilde S_\eta(n+1)/\mathcal{F}_n\} + r^{a,b}_n\right]\ . \qquad (60)$$

Using the same techniques as in Theorem 4, in exactly the same way it can be proved that the following theorem is true.

THEOREM 5. Let $\hat{\alpha} = \{\hat{a}_n, n \geq 0\}$ be the strategy of the first player defined from (60). Then the function $S^{**}_\eta(n) = \lim_{N\to\infty} U^N S^{\hat{\alpha}}_\eta(n)$, where

$$S^{\hat{\alpha}}_\eta(n) = \operatorname{Inf}_\beta S^{\hat{\alpha},\beta}_\eta(n) \quad ,$$

is a solution of equation (14), $\{S^{**}_\eta(n), n \geq 0\} \in \mathbb{R}$,

$$\lim_{N\to\infty} S^{**}_\eta(N) \geq \eta \qquad [\text{a.s. } \Omega_c + \Omega_c^{1,2} + \Omega^1]$$

and

$$\underline{\lim}_{N\to\infty} S^{**}_\eta(N) \geq \min(\overline{\eta}, \eta) \qquad [\text{a.s. } \Omega^2 + \Omega_d^{1,2}] \quad .$$

In the following theorem an iterative scheme of construction of the value and its limit characterization are given.

THEOREM 6. There exists a value $S_\eta(n) = \underline{S}_\eta(n) = \overline{S}_\eta(n)$ and

$$\lim_{N\to\infty} S_\eta(N) = \begin{cases} \eta & [\text{a.s. } \Omega_c + \Omega_c^{1,2}] \quad , \\ \max(\underline{\eta}, \eta) & [\text{a.s. } \Omega^1] \quad , \\ \min(\overline{\eta}, \eta) & [\text{a.s. } \Omega^2] \quad , \\ [\eta]_{\underline{\eta}}^{\overline{\eta}} & [\text{a.s. } \Omega_d^{1,2}\{\underline{\eta} \leq \overline{\eta}\}] \quad , \end{cases} \tag{61}$$

where $[\eta]_{\underline{\eta}}^{\overline{\eta}} = \max\{\underline{\eta}, \min(\eta, \overline{\eta})\} = \min\{\overline{\eta}, \max(\eta, \underline{\eta})\}$ [a.s. $\{\underline{\eta} \leq \overline{\eta}\}$].

Moreover $S_\eta(n) = S^*_\eta(n) = S^{**}_\eta(n)$.

Proof. First let us show that for all $\varepsilon > 0$ a.s.

$$S^*_\eta(n) \geq \overline{S}_\eta(n) - \varepsilon \quad . \tag{62}$$

Consider a sequence of positive random variables $\{\varepsilon_k, k \geqslant 0\}$ such that $\sum_{n=0}^{\infty} \varepsilon_k < \varepsilon$. Then for all $n \geq 0$ there exists such $b_n^{\varepsilon_n}$ that

$$U\, S_\eta^{\hat\beta}(n) = \underset{a}{\operatorname{Sup}}\ \underset{b}{\operatorname{Inf}} \left[\left(1 - P_n^{a,b}\right) E\{S_\eta^{\hat\beta}(n+1) / \mathcal{F}_n\} + r_n^{a,b}\right]$$

$$\geq \underset{a}{\operatorname{Sup}} \left[\left(1 - P_n^{a,b_n^{\varepsilon_n}}\right) E\{S_\eta^{\hat\beta}(n+1) / \mathcal{F}_n\} + r_n^{a,b_n^{\varepsilon_n}}\right] - \varepsilon_n$$

$$\geq \underset{\alpha\in\Sigma}{\operatorname{Sup}}\ S_\eta^{\alpha,\beta^n}(n) - \varepsilon_n \qquad \text{a.s.},$$

where $\beta^n = \{b_n^{\varepsilon_n}, \hat b_{n+1}, \hat b_{n+2}, \ldots\}$. Further, let us recurrently construct a strategy $\beta^N = \{b_n^{\varepsilon_n}, \ldots, b_N^{\varepsilon_N}, \hat b_{n+1}, \hat b_{N+2}, \ldots\}$ such that

$$U^N\, S_\eta^{\hat\beta}(n) \geq \underset{\alpha\in\Sigma}{\operatorname{Sup}}\ S_\eta^{\alpha,\beta^N}(n) - \sum_{k=n}^{N} \varepsilon_k$$

Hence

$$U^N\, S_\eta^{\hat\beta}(n) \geq \underset{\beta\in\Pi}{\operatorname{Inf}}\ \underset{\alpha\in\Sigma}{\operatorname{Sup}}\ S_\eta^{\alpha,\beta}(n) - \sum_{k=n}^{N} \varepsilon_k = \overline{S}_\eta(n) - \sum_{k=n}^{N} \varepsilon_n \qquad \text{a.s.},$$

so going to the limit as $N \to \infty$, we obtain (62).

Now let us show that

$$S_\eta^*(n) \leq \overline{S}_\eta(n) \qquad \text{a.s.}\ . \tag{63}$$

Due to Theorem 4 $S_\eta^*(n)$ is a solution of equation (14) and $\{S_\eta^*(n), n \geq 0\} \in \mathbb{R}$. Consequently, using Lemma 3, we get the following representation:

$$S_\eta^*(n) = \underset{\beta\in\Pi}{\operatorname{Inf}}\ \underset{\alpha\in\Sigma}{\operatorname{Sup}}\ E\left\{\eta_n^{\alpha,\beta} + \lim_{N\to\infty} S_\eta^*(N) \cdot \prod_{k=n}^{\infty} \left(1 - P_k^{a_k,b_k}\right) / \mathcal{F}_n\right\} \text{ a.s.},$$

and since for all $\beta = \{b_n, n \geq 0\} \in \Pi$

$$P\left\{\left[\prod_{k=n}^{\infty}\left(1 - P_k^{a_k, b_k}\right) > 0\right] \cap \{\beta \in B_d\}\right\} = 0 , \qquad n \geq 0 ,$$

one easily sees that

$$S_\eta^*(n) = \operatorname*{Inf}_{\beta \in \Pi} \operatorname*{Sup}_{\alpha \in \Sigma} E\left\{\eta_n^{\alpha,\beta} + \prod_{k=n}^{\infty}\left(1 - P_k^{a_k, b_k}\right) \times \lim_{N \to \infty} S_\eta^*(N) \left[I_{\{\beta \in B_c\}} + I_{\{\beta \in B_0\}}\right] / F_n\right\} \quad \text{a.s.} . \tag{64}$$

But from the definition of the sets Ω_c, $\Omega_c^{1,2}$, Ω^1, Ω^2, $\Omega_d^{1,2}$ and Ω_d it is obvious that for each strategy of the second player $\beta = \{b_n, n \geq 0\} \in \Pi$, $\Omega_c \subseteq \{\beta \in B_c\}$, $\Omega^1 \subseteq \{\beta \in B_0\}$ and $\{\beta \in B_c\}(\Omega^1 + \Omega_d^{1,2} + \Omega_d) = \{\beta \in B_0\}(\Omega_c + \Omega_d) \neq \emptyset$. Hence

$$\{\beta \in B_c\} = \Omega_c + \Omega_c^{1,2}\{\beta \in B_c\} + \Omega^2\{\beta \in B_c\}$$

and

$$\{\beta \in B_0\} = \Omega^1 + \Omega_c^{1,2}\{\beta \in B_0\} + \Omega^2\{\beta \in B_0\} + \Omega_d^{1,2}\{\beta \in B_0\} . \tag{65}$$

Therefore by Theorem 4 for each $\beta \in \Pi$, $\lim_{N \to \infty} S_\eta^*(N) \leq \eta$ [a.s. $\{\beta \in B_c\}$] and $\overline{\lim}_{N \to \infty} S_\eta^*(N) \leq \max(\underline{\eta}^\beta, \eta)$ [a.s. $\{\beta \in B_0\}$]. Now using the last two inequalities and (64), we obtain that

$$S_\eta^*(n) \leq \operatorname*{Inf}_{\beta} \operatorname*{Sup}_{\alpha} E\left\{\eta_n^{\alpha,\beta} + \prod_{k=n}^{\infty}\left(1 - P_k^{a_k, b_k}\right) \times \left[I_{\{\beta \in B_c\}}\eta + I_{\{\beta \in B_0\}} \max(\underline{\eta}^\beta, \eta)\right] / F_n\right\} \quad \text{a.s.} . \tag{66}$$

On the other hand, for each $\beta = \{b_n, n \geq 0\} \in \Pi$

$$S_\eta^\beta(n) = \operatorname*{Sup}_{\alpha\in\Sigma} E\left\{\eta_n^{\alpha,\beta} + \eta \prod_{k=n}^{\infty}\left(1 - P_k^{a_k,b_k}\right) \Big/ F_n\right\} \quad \text{a.s.}$$

is a value in the generalized optimal stopping problem (see [4]); so by Theorem 1

$$S_\eta^\beta(n) = \operatorname*{Sup}_{\alpha\in\Sigma} E\left\{\eta_n^{\alpha,\beta} + \prod_{k=n}^{\infty}\left(1 - P_k^{a_k,b_k}\right)\right.$$

$$\left. \times \left[\eta I_{\{\beta\in B_c\}} + \max(\underline{\eta}^\beta, \eta)\, I_{\{\beta\in B_0\}}\right] \Big/ F_n\right\} \quad \text{a.s. .}$$

Therefore, applying (66) we establish

$$S_\eta^*(n) \le \operatorname*{Inf}_{\beta\in\Pi} \operatorname*{Sup}_{\alpha\in\Sigma} E\left\{\eta_n^{\alpha,\beta} + \eta \prod_{k=n}^{\infty}\left(1 - P_k^{a_k,b_k}\right) \Big/ F_n\right\} = \overline{S}_\eta(n) \quad \text{a.s.,}$$

and taking into account (62) we obtain $S_\eta^*(n) = \overline{S}_\eta(n)$ a.s. .

Now from Theorem 4 we get

$$\overline{\lim_{N\to\infty}}\, \overline{S}_\eta(N) \le \begin{cases} \eta & [\text{a.s. } \Omega_c + \Omega_c^{1,2} + \Omega^2] \; , \\ \max(\underline{\eta}, \eta) & [\text{a.s. } \Omega^1 + \Omega_d^{1,2}] \; . \end{cases} \tag{67}$$

Also in the same way applying Theorem 5, it follows that $S_\eta^{**}(n) = \underline{S}_\eta(n)$ a.s. and

$$\underline{\lim_{N\to\infty}}\, \underline{S}_\eta(N) \ge \begin{cases} \eta & [\text{a.s. } \Omega_c + \Omega_c^{1,2} + \Omega^1] \; , \\ \min(\overline{\eta}, \eta) & [\text{a.s. } \Omega^2 + \Omega_d^{1,2}] \; . \end{cases} \tag{68}$$

Further, since $\overline{S}_\eta(n) \ge \underline{S}_\eta(n)$ a.s., (67) and (68) imply $\lim_{N\to\infty} \underline{S}_\eta(N) = \lim_{N\to\infty} \overline{S}_\eta(N) = \eta$ $[\text{a.s. } \Omega_c + \Omega_c^{1,2}]$. Moreover, from (67) and (68) we obtain

$$\eta \le \lim_{N\to\infty} \underline{S}_\eta(N) \le \overline{\lim_{N\to\infty}} \overline{S}_\eta(N) \le \max(\underline{\eta}, \eta) \qquad [\text{a.s. } \Omega^1] .$$

But since $\underline{S}_\eta(n)$ and $\overline{S}_\eta(n)$ are solutions of equation (14), due to Lemma 3 the limits $\lim_{N\to\infty} \underline{S}_\eta(N)$, $\lim_{N\to\infty} \overline{S}_\eta(N)$ exist and $\lim_{N\to\infty} \underline{S}_\eta(N) \ge \underline{\eta}$ [a.s. Ω^1]. Consequently, from the last inequality we get

$$\lim_{N\to\infty} \underline{S}_\eta(N) = \lim_{N\to\infty} \overline{S}_\eta(N) = \max(\underline{\eta}, \eta) \qquad [\text{a.s. } \Omega^1] .$$

In the same way we prove that $\lim_{N\to\infty} \underline{S}_\eta(N) = \lim_{N\to\infty} \overline{S}_\eta(N) = \min(\overline{\eta}, \eta)$ [a.s. Ω^2].

Next from the Corollary of Lemma 3 we have $\underline{\eta} \le \lim_{N\to\infty} \underline{S}_\eta(N) \le \lim_{N\to\infty} \overline{S}_\eta(N) \le \overline{\eta}$ [a.s. $\Omega_d^{1,2}\{\underline{\eta}\le\overline{\eta}\}$]. But by (67) and (68)

$$\min(\overline{\eta}, \eta) \le \lim_{N\to\infty} \underline{S}_\eta(N) \le \lim_{N\to\infty} \overline{S}_\eta(N) \le \max(\underline{\eta}, \eta) \qquad [\text{a.s. } \Omega_d^{1,2}\{\underline{\eta}\le\overline{\eta}\}]$$

and hence

$$\lim_{N\to\infty} \underline{S}_\eta(N) = \lim_{N\to\infty} \overline{S}_\eta(N) = [\eta]_{\underline{\eta}}^{\overline{\eta}} \qquad [\text{a.s. } \Omega_d^{1,2}\{\underline{\eta}\le\overline{\eta}\}] .$$

Thus we have proved that

$$\lim_{N\to\infty} \underline{S}_\eta(N) = \lim_{N\to\infty} \overline{S}_\eta(N) \qquad [\text{a.s. } \Omega_c + \Omega_c^{1,2} + \Omega^1 + \Omega^2 + \Omega_d^{1,2}\{\underline{\eta}\le\overline{\eta}\}] ,$$

from which by Lemma 4 we obtain

$$\underline{S}_\eta(n) = \overline{S}_\eta(n) \qquad \text{a.s.},$$

i.e., there exists a value $S_\eta(n) = \underline{S}_\eta(n) = \overline{S}_\eta(n)$.

LEMMA 8 (see Lemma 2 in [4]). Let $\{v_n, n \geq 0\} \in \mathbb{R}$ and for the strategy $\alpha^\varepsilon = \{a_n^\varepsilon, n \geq 0\}$ let the following inequality hold $(0 < \delta \leq \varepsilon)$:

$$\operatorname*{Sup}_a \left[\left(1 - P_n^{a,b_n}\right) v_n + r_n^{a,b_n} + \delta P_n^{a,b_n}\right]$$

$$\leq \left(1 - P_n^{a_n^\varepsilon, b_n}\right) v_n + r_n^{a_n^\varepsilon, b_n} + \varepsilon P_n^{a_n^\varepsilon, b_n} \qquad \text{a.s.}$$

for all $\beta = \{b_n, n \geq 0\}$.

Then

$$\overline{\lim_{N\to\infty}}\, v_N \geq \underline{\eta}^\beta + \delta \tag{69}$$

$$\left[\text{a.s.} \sum_{n=0}^{\infty} \operatorname*{Inf}_a P_n^{a,b_n} < \infty, \quad \sum_{n=0}^{\infty} \operatorname*{Sup}_a P_n^{a,b_n} = \infty, \quad \sum_{n=0}^{\infty} P_n^{a_n^\varepsilon, b_n} < \infty\right].$$

THEOREM 7. For every $n \geq 0$ let a_n^ε be defined by the inequality

$$S_\eta(n) \leq \operatorname*{Sup}_a \left[\left(1 - P_n^{a,b_n}\right) E\{S_\eta(n+1) / \mathcal{F}_n\} + r_n^{a,b_n} + \delta P_n^{a,b_n}\right]$$
$$\leq \left(1 - P_n^{a_n^\varepsilon, b_n}\right) E\{S_\eta(n+1) / \mathcal{F}_n\} + r_n^{a_n^\varepsilon, b_n} + \varepsilon P_n^{a_n^\varepsilon, b_n} \qquad \text{a.s.,} \tag{70}$$

where $\beta = \{b_n, n \geq 0\}$ is an arbitrary strategy of the second player. Then the strategy $\alpha^\varepsilon = \{a_n^\varepsilon, n \geq 0\}$ is ε-optimal, i.e.,

$$S_\eta(n) \leq \operatorname*{Inf}_\beta S_\eta^{\alpha^\varepsilon,}(n) + \varepsilon \qquad \text{a.s. .} \tag{71}$$

Proof. From (70) it is clear that $\{\eta_k(n), \mathcal{F}_k\}$, $k \geq n$, where

$$\eta_k = \sum_{i=n}^{k-1} \left(1 - P_i^{a_i^\varepsilon, b_i}\right) S_\eta(k) + \sum_{i=n}^{k-1} \prod_{j=n}^{i-1} \left(1 - P_j^{a_j^\varepsilon, b_j}\right)\left[r_i^{a_i^\varepsilon, b_i} + \varepsilon P_i^{a_i^\varepsilon, b_i}\right],$$

is a submartingale. Also it is not difficult to show that

$$\sum_{i=n}^{\infty} \prod_{j=n}^{i-1} \left(1 - P_j^{a_j^\varepsilon, b_j}\right) P_i^{a_i^\varepsilon, b_i} \leq 1 \qquad \text{a.s.} .$$

Consequently, for each $\beta = \{b_n, n \geq 0\}$

$$S_\eta(n) \leq E\left\{\eta_n^{\alpha^\varepsilon, \beta} + \lim_{N\to\infty} S_\eta(N) \prod_{k=n}^{\infty}\left(1 - P_k^{a_k^\varepsilon, b_k}\right) \mid F_n\right\} + \varepsilon \qquad \text{a.s.}$$

and hence a.s.

$$S_\eta(n) \leq \operatorname{Inf}_\beta E\left\{\eta_n^{\alpha^\varepsilon, \beta} + \lim_{N\to\infty} S_\eta(N) \prod_{k=n}^{\infty}\left(1 - P_k^{a_k^\varepsilon, b_k}\right) \times \left[I_{\Omega_c+\Omega_c^{1,2}} + I_{\Omega^1} + I_{\Omega^2} + I_{\Omega_d^{1,2}}\right] \mid F_n\right\} .$$

From this, using (69), (51), (61) and (65), we obtain (71). ///

4. SOME RESULTS FOR THE MINIMAX VERSION OF THE STANDARD OPTIMAL STOPPING PROBLEM

Theorems 3, 4 and 6 unify several special cases that have been previously studied in [8], [10], [15] and [16].

In order to illustrate assertions of these theorems, let us consider the case where

$$r_n^{a,b} = a(1-b)x_n + b(1-a)y_n + abz_n , \qquad n \geq 0, \quad a,b \in [0,1] ,$$

where x_n, y_n and z_n are F_n-measurable random variables satisfying the following condition: $x_n \leq z_n \leq y_n$, $n \geq 0$ a.s. Also, instead of $P_n^{a,b}$ we take $1 - (1-a)(1-b)$, where $a,b \in [0,1]$, $n \geq 0$, i.e.,

$$S_\eta^{\alpha,\beta}(n) = E\left\{\sum_{k=n}^{\infty} \prod_{i=n}^{k-1} (1-a_i)(1-b_i)[a_k(1-b_k)x_k + b_k(1-a_k)y_k + a_k b_k z_k]\right.$$

$$\left. + \ \eta \prod_{k=n}^{\infty} (1-a_k)(1-b_k) \ \middle| \ F_n\right\} .$$

Here F_n characterizes the information of observers at $n \geq 0$. The following stopping problem is considered: an observation can be stopped at stage n by the first or second player. If it is stopped by the first, the second or both of the players, the first player gets, respectively, the reward x_n, y_n or z_n. If the observation continues infinitely, the first player gets reward η.

The mixed strategies of players are defined by $\alpha = \{a_n, n\geq 0\}$ and $\beta = \{b_n, n\geq 0\}$, where a_n (b_n) is interpreted as a probability of stopping by the first (second) player at stage $n \geq 0$.

The minimax version of the optimal stopping problem was considered for the first time in [8] for the case where

$$x_n = \tilde{x}_n I_{E_n^+} + E\left\{\sup_{k\geq n} |\tilde{x}_k| \ \middle| \ F_n\right\} I_{E_n^-} ,$$

$$y_n = \tilde{x}_n I_{E_n^-} + E\left\{\sup_{k\geq n} |\tilde{x}_n| \ \middle| \ F_n\right\} I_{E_n^+} ,$$

where $E_n^+ \cup E_n^- = \Omega$ and $E_n^+ \cap E_n^- = \emptyset$, $n \geq 0$, $\eta = 0$. In [10] and [16] the last conditions were removed. For a Markov process with continuous time the optimal stopping problem was considered in [17] - [19].

Note that in the problem considered here equation (14) takes the following form:

$$v_n = [E\{v_{n+1} \mid \mathcal{F}_n\}]_{x_n}^{y_n} , \qquad (72)$$

where $[u_x^y] = \min\{y, \max(u,x)\} = \max\{x, \min(u,y)\}$.

It turns out that in this case we need only consider pure strategies and the cost function $S_\eta(n)$ can be represented as

$$S_\eta(n) = \sup_{\alpha\in\Sigma_n} \inf_{\beta\in\Pi_n} S_\eta^{\alpha,\beta}(n) \qquad \text{a.s.} ,$$

where Σ_n is the class of strategies of the first player $\alpha = \{a_n, n\geq 0\}$ such that for each $k \geq n$, a_k takes a value from the set $\{0,1\}$ (the class Π_n is defined in the same way).

Further, as shown in [15], the cost takes the form

$$S_\eta(n) = \sup_{\tau\in\mathcal{M}_n} \inf_{\sigma\in\mathcal{M}_n} S_\eta(n,\tau,\sigma) = \inf_{\sigma\in\mathcal{M}_n} \sup_{\tau\in\mathcal{M}_n} S_\eta(n,\tau,\sigma) ,$$

where

$$S_\eta(n,\tau,\sigma) = E\{x_\tau I_{[\tau<\sigma]} + y_\sigma I_{[\sigma<\tau]} + z_\tau I_{[\tau=\delta<\infty]} + \eta I_{[\tau=\sigma=\infty]} \mid \mathcal{F}_n\}$$

and $\mathcal{M}_n$ is a class of F-adapted Markov times τ such that $P\{\tau\geq n\} = 1$. Here τ (σ) denotes the moment when the observation is stopped by the first (second) player.

Also it is easy to show (see [15]) that $\underline{\eta} = \overline{\lim}_{N\to\infty} x_N$ and $\overline{\eta} = \underline{\lim}_{N\to\infty} y_N$.

Now suppose $r_n' = x_n$, $n \geq 0$. Then due to Lemma 3 we obtain that

$$\underset{\sim}{S}(n) = \lim_{N\to\infty} r_n^N ,$$

where

$$r_n^N = [E\{r_{n+1}^{N-1} \mid \mathcal{F}_n\}]_{x_n}^{y_n}, \qquad N > 1, \quad n \geq 0,$$

is a minimal solution of equation (72). From Theorem 1 (see [11]) we have

$$\underset{\sim}{S}(n) = \sup_{\tau\in\mathcal{M}_n} \inf_{\sigma\in\mathcal{M}_n} E\Big\{x_\tau I_{[\tau<\sigma]} + y_\sigma I_{[\sigma<\tau]} + z_\tau I_{[\tau=\sigma<\infty]}$$
$$+ \overline{\lim_{N\to\infty}}\, x_N \cdot I_{[\tau=\sigma=\infty]} \mid \mathcal{F}_n\Big\} .$$

Next let us define

$$\hat{\sigma} = \inf \{k \geq n : \underset{\sim}{S}(k) = y_k\} .$$

The next theorem is a consequence of Theorems 5 and 6.

<u>THEOREM 8</u>. The game has a value: $S_\eta(n) = \underline{S}_\eta(n) = \overline{S}_\eta(n)$, $S_\eta(n) = \lim_{N\to\infty} r_n^N$, where for all $n \geq 0$

$$r_n^N = \Big[E\Big\{r_{n+1}^{N-1} \mid \mathcal{F}_n\Big\}\Big]_{x_n}^{y_n}, \qquad N > 1 ,$$

$$r_n^1 = \sup_{\tau\in\mathcal{M}_n} S_\eta(n,\tau,\hat{\sigma}) ,$$

and

$$\lim_{N\to\infty} S_\eta(N) = [\eta]_{\overline{\lim}_{N\to\infty} x_N}^{\underline{\lim}_{N\to\infty} y_N} \qquad \Big[\text{a.s.} \quad \overline{\lim_{N\to\infty}}\, x_N \leq \underline{\lim_{N\to\infty}}\, y_N\Big] .$$

REFERENCES

[1] Shiryaev, A.N. *Statistical Sequential Analysis*. Providence, R.I.: American Mathematical Society, 1973.

[2] Chow, Y.S., Robbins, H., and Siegmund, D. *Great Expectations: The Theory of Optimal Stopping*. Boston: Houghton, Mifflin, 1971.

[3] Lazrieva, N.L. "Solutions of the Wald-Bellman Equation." *Lithuanian Math. Transactions*, vol.14, no.2 (1974): 233-246.

[4] Elbakidze, N.V. "Optimal Stopping of a Random Sequence and Hammerstein's Operators." *Theory Probab. Applications*, vol.27, no.2 (1983): 338-356.

[5] Yushkevich, A.A., and Chitashvili, R.J. "Controlled Random Sequences." *Russian Math. Surveys*, vol.37, no.6 (1982): 239-274.

[6] Yasuda, M., Nakagami, N., and Kurano, M. "Multivariate Stopping Problems with a Monotone Rule." *J. Operat. Research Soc. Japan*, vol.25 (1982): 334-350.

[7] Yasuda, M. "On a Stopping Problem Involving Refusal and Forced Stopping." *J. Appl. Probab.*, vol.20, no.1 (1983): 71-81.

[8] Dynkin, E.B. "Game Variant of a Problem on Optimal Control." *Soviet Math. Doklady*, vol.10, no.2 (1969): 270-274.

[9] Kifer, Yu.I. "Optimal Stopped Games." *Theory Probab. Applications*, vol.16, no.1 (1971): 185-189.

[10] Elbakidze, N.V. "Construction of the Cost and Optimal Policies in a Game Problem of Stopping a Markov Process." *Theory Probab. Applications*, vol.21, no.1 (1976): 163-168.

[11] Elbakidze, N.V. "On the Optimal Stopping Rules in a Discrete-time Game." *Soobshcheniya Akademii Nauk Gruzinskoj SSR*, vol. 78, no.2 (1975): 289-292.

[12] Elbakidze, N.V. "Probabilistic Representation of the Solution of Hammerstein's Equation." *Soobshcheniya Akademii Nauk Gruzinskoj SSR*, vol.96, no.2 (1979): 293-296.

[13] Andre, C. "Un jeu stochastique a deux personnes en tempt continu." *C.R. Acad. Sci.*, vol.275, no.18 (1972): A839-A840.

[14] Elbakidze, N.V. "On the Construction of the Cost Policy in the Generalized Problem of Stopping a Random Process." In *Teoriya sluchainykh protsessov*, p. 15, vol.10. Kiev: Naukova Dumka, 1982.

[15] Chitashvili, R.J., and Elbakidze, N.V. "The Construction of the Cost and Optimal Policies in a Game Problem of Stopping a Markov Sequence." *Issledovaniya po teorii veroyatnostei i matematich. statistike*, pp.163-195. Tbilisi: Metzniereba, 1982.

[16] Domanskii, V.K. "A Game-Theoretic Problem of Stopping a Markov Chain." *Theory Probab. Applications*, vol.23, no.4 (1978): 830-832.

[17] Krylov, N.V. "On a Problem with Two Free Boundaries for an Elliptic Equation and Optimal Stopping of a Markov Process." *Soviet Math., Doklady*, vol.11, no.5 (1970): 1370-1372.

[18] Friedman, A. "Stochastic Games and Variational Inequalities." *Archive for Rational Mechanics and Analysis*, vol.51, no.5 (1973): 321-346.

[19] Bismut, J.M. "Contrôle de processus alternants et applications." *Z. Wahrscheinlichkeitstheorie und Verw. Gebiete*, vol.47 (1979): 241-288.

B.S. DARKHOVSKIJ and G.G. MAGARIL-ILLYAEV

A LINEAR MINIMAX ESTIMATE OF A PERIODIC FUNCTION IN WHITE NOISE

1. INTRODUCTION AND STATEMENT OF THE PROBLEM

The following problem of nonparametric estimation is considered. On the segment $[-\pi,\pi]$ the deterministic function $u(\cdot)$ from some class and the random process $\xi(\cdot)$ with zero mean are given. It is necessary to find the estimate (best in the class) of the function $u(\cdot)$ from the observed realization of the process

$$v(t) = u(t) + \xi(t) , \qquad t \in [-\pi,\pi] .$$

Different versions of this problem were considered by many authors (see e.g., [1] - [3]). The main results in our direction are compiled in the monograph [1] where for an observation scheme of diffusion-type and a smooth function $u(\cdot)$ projection-type estimates based on an approximation of the unknown function by Fourier sums with the linear functionals on the observation as coefficients have been given and studied.

In this paper, for the observed process $y(\cdot)$ with the stochastic differential

$$dy(t) = x(t)dt + \sigma d\tilde{w}(t) , \qquad t \in [-\pi,\pi] ,$$

where $\tilde{w}(t) = w(t+\pi)$, $w(t)$ is a standard Wiener process for any $A > 0$, $\sigma > 0$ and integer $n \geq 1$, an explicit solution of this problem

$$\sup_{x(\cdot)\in W} \left\| E\left\{x(\cdot) - \int_{-\pi}^{\pi} K(\cdot - \tau)dy(\tau)\right)^2\right\}^{\frac{1}{2}} \Bigg\|_{C([-\pi,\pi])} \to \inf_{K(\cdot)\in M}, \qquad (1)$$

is found. Here:

$$W = W_\infty^n([-\pi,\pi], A)$$

is the collection of the 2π-periodic functions where the $(n-1)^{th}$ derivative is absolutely continuous and $\|x^{(n)}\|_\infty \leq A$; M is the collection of continuous 2π-periodic functions such that

$$\int_{-\pi}^{\pi} K(t)\,dt = 1 , \qquad K(0) \geq 0 .$$

Note that the choice of the function from M as the smoothing kernel guarantees unbiased estimate of constants and nonnegativeness of the mathematical expectation of the estimate, if the function under estimation is nonnegative. Those properties of the estimate seem to be natural for applications.

2. THE FORMULATION OF THE MAIN RESULT

We need some definitions. Denote by $\phi_n(\cdot)$, $n \geq 1$ the Favard functions:

$$\phi_n(t) = \frac{4}{\pi} \sum_{j=0}^{\infty} \frac{\sin\left((2j+1)t - \frac{n\pi}{2}\right)}{(2j+1)^{n+1}} .$$

It is known (see [4] - [5]) that $\phi_n(\cdot)$ is a 2π-periodic function with zero mean uniquely determined by the condition:

$$\phi_n^{(n)}(t) = \text{sign } \sin t \quad ;$$

and besides, its norm in $C([-\pi,\pi])$ is the Favard constant $\mathcal{X}_n$, which is equal to

$$\mathcal{X}_n = \frac{4}{\pi} \sum_{j=0}^{\infty} \frac{(-1)^{j(n+1)}}{(2j+1)^{n+1}} .$$

It follows from the definition of $\phi_n(\cdot)$ that $\phi_n\left(t + \frac{n+1}{2}\pi\right)$ is even, monotone decreasing on $[0,\pi]$ and

$$\phi_n\left(\frac{n+1}{2}\pi\right) = \mathcal{X}_n .$$

Let

$$\psi_n(t) = \phi_n\left(t + \frac{n+1}{2}\pi\right) - \mathcal{X}_n .$$

Denote by $T = T(W^2, M)$ the value defined in (1).

<u>THEOREM</u>. Let $A > 0$, $\sigma > 0$ and $n \geq 1$ be integer. Then if $\sigma^2 \geq A^2\left(2\pi\mathcal{X}_n^2 - \|\phi_n\|_2^2\right)$,

$$J = \sigma\sqrt{\frac{A^2\mathcal{X}_n^2}{\sigma^2 + A^2\|\phi_n\|_2^2} + \frac{1}{2\pi}} ,$$

and the unique solution of the problem (1) is the function

$$\hat{K}(t) = \frac{A^2\mathcal{X}_n}{\sigma^2 + A^2\|\phi_n\|_2^2}\phi_n\left(t + \frac{n+1}{2}\pi\right) + \frac{1}{2\pi} .$$

If $\sigma^2 < A^2\left(2\pi\mathcal{X}_n^2 - \|\phi_n\|_2^2\right)$, then $J = \sigma\sqrt{\beta}$ and the unique solution of the problem (1) is the function

$$\hat{K}(t) = \begin{cases} \alpha\phi_n\left(t + \frac{n+1}{2}\pi\right) + \beta - \alpha X_n , & |t| \le t_0 \\ 0 , & t_0 \le |t| \le \pi , \end{cases}$$

where t_0 is the unique (on $(0,\pi)$) root of the equation

$$\int_0^t \psi_n^2(\tau)\, d\tau - \psi_n(t)\int_0^t \psi_n(\tau)\, d\tau = \frac{-\sigma^2}{2A^2} , \qquad (2)$$

and $\alpha > 0$, $\beta > 0$ are the unique solutions of the linear system

$$\begin{cases} \left[1 + \frac{2A^2}{\sigma^2}\int_0^{t_0} \psi_n^2(\tau)\, d\tau\right]\alpha + \frac{2A^2}{\sigma^2}\left[\int_0^{t_0} \psi_n(\tau)\, d\tau\right]\beta = 0 , \\ \left[\int_0^{t_0} \psi_n(\tau)\, d\tau\right]\alpha + t\,\beta = \frac{1}{2} . \end{cases} \qquad (3)$$

3. PROOF OF THE THEOREM

By properties of stochastic integrals and periodicity of $K(\cdot)$ we reduce the problem (1) to the form

$$\sup_{x(\cdot)\in W} \left\| \left(\left[x(\cdot) - \int_{-\pi}^{\pi} K(\cdot - \tau)\, x(\tau)\, d\tau \right]^2 + \sigma^2 \int_{-\pi}^{\pi} K^2(\tau)\, d\tau \right)^{\frac{1}{2}} \right\|_{C([-\pi,\pi])} \to \inf_{K(\cdot)\in M} , \qquad (4)$$

We start with the "internal" problem:

$$\left\| \left(\left(x(\cdot) - \int_{-\pi}^{\pi} K(\cdot - \tau)\, x(\tau)\, d\tau \right)^2 + \sigma^2 \int_{-\pi}^{\pi} K^2(\tau)\, d\tau \right)^{\frac{1}{2}} \right\|_{C([-\pi,\pi])} \to \sup ,$$

$$x(\cdot) \in W \quad . \tag{5}$$

LEMMA 1. A solution of the problem (5) exists.

Proof. It is evident that it suffices to prove the existence of the solution in the problem

$$I(x(\cdot)) = \left\| x(\cdot) - \int_{-\pi}^{\pi} K(\cdot - \tau)\, x(\tau)\, d\tau \right\|_{C([-\pi,\pi])} \to \sup ,$$

$$x(\cdot) \in W \quad . \tag{6}$$

Let S be the problem (6)-value and let $\{x_j(\cdot)\}_{j \geq 1}$ be the minimizing sequence, i.e., $x_j(\cdot) \in W$ and $I(x_j(\cdot)) \to S$ as $j \to \infty$. Since

$$\int_{-\pi}^{\pi} K(\tau)\, d\tau = 1 ,$$

then without loss of generality we assume that

$$\int_{-\pi}^{\pi} x_j(\tau)\, d\tau = 0 .$$

Put

$$(Tx)(t) = x(t) - \int_{-\pi}^{\pi} K(t - \tau)\, x(\tau)\, d\tau .$$

Since the norm is invariant with respect to the shift (for periodic functions), we assume that $|(Tx_j)(\cdot)|$, $j \geq 1$, attains the maximum at the zero. Multiplying, if necessary, $x_j(\cdot)$ by -1, we get $(Tx_j))0) = 0$. Thus $I(x_j(\cdot)) = (Tx_j)(0)$, $j \geq 1$.

The sequence $\{x_j^{(n)}(\cdot)\}_{j\geq 1}$ is contained in the ball with radius A in $L_\infty([-\pi,\pi])$ and so one can extract a weakly (*) convergent subsequence. We assume that the same sequence $\{x_j^{(n)}(\cdot)\}_{j\geq 1}$ converges weakly (*) to the function $z(\cdot) \in L_\infty([-\pi,\pi])$. Since the ball is weakly (*) closed, $\|z\|_\infty \leq A$. Since the functions $x_j(\cdot)$, $j \geq 1$, are on the average equal to zero, then (see [5], p. 80)

$$x_j(t) = \frac{1}{\pi} \int_{-\pi}^{\pi} D_n(t-\tau)\, x_j^{(n)}(\tau)\, d\tau \quad , \qquad j \geq 1 \, , \tag{7}$$

where

$$D_n(t) = \sum_{k=1}^{\infty} \frac{\cos\left(kt - \frac{\pi n}{2}\right)}{k^n} \quad .$$

Since $D_n(\cdot) \in L_1([-\pi,\pi])$, then for each $t \in [-\pi,\pi]$ the sequence $\{x_j(t)\}$ has the finite limit

$$\tilde{x}(t) = \frac{1}{\pi} \int_{-\pi}^{\pi} D_n(t-\tau)\, z(\tau)\, d\tau \quad . \tag{8}$$

Now we prove that $\tilde{x}(\cdot)$ is the solution of the problem (6). By differentiation of (8) we have that $\tilde{x}(\cdot)$ is an absolutely continuous function and $\tilde{x}^{(n)}(t) = z(t)$ a.e., that is $\tilde{x}(\cdot) \in W$. From (7), according to Hölder's inequality, we have that the functions $x_j(\cdot)$, $j \geq 1$, are uniformly bounded and so for each $\tau \in [-\pi,\pi]$, $x_j(\tau) \to \tilde{x}(\tau)$ as $j \to \infty$ and $K(\cdot) \in C([-\pi,\pi])$. Then using the Lebesgue theorem on bounded convergence, we obtain

$$S = \lim_{j\to\infty} I(x_j(\cdot)) = \lim_{j\to\infty} (Tx_j)(0)$$

$$= \lim_{j\to\infty} \left\{ x_j(0) - \int_{-\pi}^{\pi} K(-\tau)\, x_j(\tau)\, d\tau \right\}$$

$$= \tilde{x}(0) - \int_{-\pi}^{\pi} K(-\tau)\, \tilde{x}(\tau)\, d\tau = (T\tilde{x})(0) ,$$

i.e., S is finite and $\tilde{x}(\cdot)$ is the solution of the problem (6). ///

We shall establish that for the solution of the problem (4) it is sufficient to consider only the even functions from M. Indeed, it is easy to verify that for the functions $K_1(\cdot)$ and $K_2(t) = K_1(-t)$ which belongs to M, the values of the functional $F(K(\cdot))$ to be minimized in (4) coincide, and since F is convex we obtain

$$F(\tfrac{1}{2}(K_1(\cdot) + K_2(\cdot))) \leq F(K_1(\cdot)) .$$

That is why we shall consider below only even functions $K(\cdot) \in M$ in the problem (4).

Next we reduce the problem (6) to the standard problem of optimal control. In proving Lemma 1 one establishes that the problem (6) is equivalent to the following problem:

$$(Tx)(0) = x(0) - \int_{-\pi}^{\pi} K(-\tau)\, x(\tau)\, d\tau \to \sup , \qquad x(\cdot) \in W . \tag{9}$$

Consider the even function

$$\hat{x}(t) = \tfrac{1}{2}(\tilde{x}(t) + \tilde{x}(-t)) - \tilde{x}(0) ,$$

where $\tilde{x}(\cdot)$ is the solution of the problem (9). It is clear that

$$(T\tilde{x})(0) = (T\hat{x})(0) = -2\int_0^\pi K(\tau)\,\hat{x}(\tau)\,d\tau ,$$

and that is why it is also the solution of the problem (9). Due to evenness and periodicity of $\hat{x}(\cdot)$ it is necessary that

$$\hat{x}^{(2j-1)}(0) = \hat{x}^{(2j-1)}(\pi) = 0 , \quad j = 1,\ldots,[\tfrac{n}{2}]$$

(for $n=1$ these conditions vanish). Thus it follows that there exists the even solution $\hat{x}(\cdot)$ of the problem (9) (and hence of the problem (6)), which on $[0, \pi]$ is the solution of the following optimal control problem:

$$-2\int_0^\pi K(\tau)\,x(\tau)\,d\tau \to \sup , \quad x(0) = 0, \quad x^{(2j-1)}(0) = x^{(2j-1)}(\pi) = 0 , \quad j = 1,\ldots,[\tfrac{n}{2}] , \quad \|x^{(n)}\|_\infty \le A . \tag{10}$$

By the maximum principle (see e.g., [6]) there exists a constant $\lambda \le 0$ and a function $p(\cdot)$ with the $(n-1)^{th}$ derivative being absolutely continuous and such that $|\lambda| + |p(t)| \not\equiv 0$ and almost everywhere on $[0,\pi]$

$$\begin{cases} p^{(n)}(t) = 2\lambda(-1)^n K(t) , \\ x^{(n)}(t) = A \operatorname{sign} p(t) \end{cases} \tag{11}$$

and

$$p^{(n-1)}(\pi) = 0 , \quad p^{(n-2j-1)}(0) = p^{(n-2j-1)}(\pi) = 0 , \quad j = 1,\ldots,[\tfrac{n-1}{2}] \tag{12}$$

(for $n = 1,2,$ the conditions (12) vanish).

Since in the problem (10) (which is also a convex programming problem) Slater's condition is evidently satisfied, $\lambda < 0$.

LEMMA 2. The function $A\left(\phi_n\left(t + \frac{n+1}{2}\pi\right) - \chi_n\right) = A\psi_n(t)$ is the solution of the problem (10).

Proof. The statement of Lemma 2 is based on the following proposition:

if $K(\cdot) > 0$ then

$$\operatorname{sign} p(t) = \operatorname{sign} \sin\left(t + \frac{n+1}{2}\pi\right), \quad t \in [0,\pi]\ . \tag{13}$$

It is easy to check that

$$\operatorname{sign} \sin\left(t + \frac{n+1}{2}\pi\right) = (-1)^{\frac{n+1}{2}}$$

if n is odd and

$$\operatorname{sign} \sin\left(t + \frac{n+1}{2}\pi\right) = \begin{cases} (-1)^{\frac{n}{2}}, & 0 < t < \frac{\pi}{2} \\ (-1)^{\frac{n-2}{2}}, & \frac{\pi}{2} < t < \pi \end{cases}$$

if n is even.

Let us prove that the functions $p^{(n-2j-1)}(\cdot)$, $j = 0,1,\ldots,[\frac{n-1}{2}]$, have no zeros on $(0,\pi)$ and $\operatorname{sign} p^{(n-2j-1)}(t) = (-1)^{n-j}$, and the functions $p^{(n-2j)}(\cdot)$, $j = 1,\ldots,[\frac{n}{2}]$, are strictly monotone, have the zero on $(0,\pi)$ and $\operatorname{sign} p^{(n-2j)}(0) = (-1)^{n-j}$ ($p^{(0)}(\cdot) = p(\cdot)$).

Indeed, let $n \geq 3$ be odd. Then from (11) and the condition $K(\cdot) > 0$ it follows that $p^{(n-1)}(\cdot)$ is monotone increasing, and since $p^{(n-1)}(\pi) = 0$ then $p^{(n-1)}(t) < 0$ on $(0,\pi)$. It means that $p^{(n-2)}(\cdot)$ is monotone decreasing. Furthermore, $p^{(n-2)}(\cdot)$ has the zero on $(0,\pi)$ (and so $p^{(n-2)}(0) > 0$), otherwise, the function $p^{(n-3)}(\cdot)$ would be monotone, which contradicts the condition $p^{(n-3)}(0) = p^{(n-3)}(\pi) = 0$. Thus the function $p^{(n-3)}(\cdot)$ has no zeros on $(0,\pi)$, since otherwise its derivative would have two different zeros on $(0,\pi)$ due to Rolle's theorem; but this contradicts what has been proved above. From the inequality $p^{(n-2)}(0) > 0$ it follows that $p^{(n-3)}(t) > 0$ on $(0,\pi)$. Proceeding in this way, we shall arrive at the needed conclusion.

The case of the even n is considered similarly.

From the above it follows immediately that if n is odd then

$$\operatorname{sign} p(t) = (-1)^{n - \frac{n-1}{2}} = (-1)^{\frac{n+1}{2}} ,$$

and if n is even and τ is the (unique) zero of the function $p(\cdot)$ on $(0,\pi)$, then

$$\operatorname{sign} p(t) = \begin{cases} (-1)^{\frac{n}{2}} , & 0 < t < \tau , \\ (-1)^{\frac{n-2}{2}} , & \tau < t < \pi . \end{cases}$$

Now check that $\tau = \frac{\pi}{2}$. Indeed, since $\hat{x}^{(n)}(t) = A \operatorname{sign} p(t)$ and $\hat{x}^{(n-1)}(0) = \hat{x}^{(n-1)}(\pi) = 0$, we have:

$$\hat{x}^{(n-1)}(t) = \begin{cases} A(-1)^{\frac{n}{2}} t , & 0 \le t \le \tau , \\ A(-1)^{\frac{n-2}{2}} (t-\pi) , & \tau \le t \le \pi . \end{cases}$$

It follows from the continuity $\hat{x}(\cdot)$ at the point τ that $\tau = \frac{\pi}{2}$ and so (13) is proved.

Thus,

$$\hat{x}^{(n)}(t) = A \operatorname{sign} \sin \left(t + \frac{n+1}{2}\pi\right) .$$

From this it follows that $\hat{x}(\cdot)$ coincides with $A\phi_n\left(t + \frac{n+1}{2}\pi\right)$ up to an additive constant, and since $\hat{x}(0) = 0$ then

$$\hat{x}(t) = A\left(\phi_n\left(t + \frac{n+1}{2}\pi\right) - \mathcal{X}_n\right) .$$

Thus, the necessary condition for maximum in the problem (10) for $K(\cdot) > 0$ is satisfied by the unique function $A\left(\phi_n\left(t + \frac{n+1}{2}\pi\right) - \mathcal{X}_k\right)$, and hence is the solution of this problem.

We emphasize the fact that this solution does not depend on $K(\cdot)$ if $K(\cdot) > 0$.

Let $\bar{K}(\cdot) \in M$ be an even function. It is clear that the function

$$K_\varepsilon(t) = \frac{1}{1+\pi 2\varepsilon}(\bar{K}(t) + \varepsilon) , \qquad \varepsilon > 0 ,$$

belongs to M, $K_\varepsilon(\cdot) > 0$ and $K_\varepsilon(\cdot)$ uniformly converges to $\bar{K}(\cdot)$ as $\varepsilon \to 0$. According to what has been proved above, $\hat{x}(\cdot)$ is the solution of the problem (10) when $K(\cdot) = K_\varepsilon(\cdot)$. Let $\bar{x}(\cdot)$ be the solution of this problem when $K(\cdot) = \bar{K}(\cdot)$. Then we have the inequality

$$-2\int_0^\pi K_\varepsilon(t)\,\hat{x}(t)\,dt \geq -2\int_0^\pi K_\varepsilon(t)\,\bar{x}(t)\,dt .$$

Passing here to the limit as $\varepsilon \to 0$, we obtain that the upper bound in (10) when $K(\cdot) = \bar{K}(\cdot)$ is not greater than

$-2 \int_0^\pi \bar{K}(t)\,\hat{x}(t)\,dt$. But this means that $\hat{x}(\cdot)$ is a solution of the problem (10) when $K(\cdot) = \bar{K}(\cdot)$. ///

From what has been noted above, it follows that the problem (1) (and hence (4)) is equivalent to the following convex programming problem:

$$\left(2A \int_0^\pi K(\tau)\,\psi_n(\tau)\,d\tau\right)^2 + 2\sigma^2 \int_0^\pi K^2(\tau)\,d\tau \to \inf \;,$$
$$\int_0^\pi K(\tau)\,d\tau = \tfrac{1}{2}\;, \qquad \max_\tau\,(-K(\cdot)) \le 0\;. \tag{14}$$

In order that the function $\tilde{K}(\cdot) \in C([0,\pi])$ satisfying the restrictions in the problem (14) be the solution of this problem, it is sufficient (due to the Kuhn-Tucker theorem, [6]) that there exist constants $\lambda_0 > 0$, $\lambda_1, \lambda_2 \ge 0$ and a probability Borel measure $d\mu$ on $[0,\pi]$, whose support is the set of the minimum points $\tilde{K}(\cdot)$ such that

$$\int_0^\pi \left[\lambda_0 8A^2\left(\int_0^\pi \tilde{K}(\tau)\,\psi_n(\tau)\,d\tau\right)\psi_n(t) + \lambda_0 4\sigma^2\tilde{K}(t) + \lambda_1\right]h(t)\,dt$$
$$- \lambda_2 \int_0^\pi h(\tau)\,d\mu(\tau) = 0 \tag{15}$$

for all $h(\cdot) \in C([0,\pi])$ and

$$\lambda_2 \max_\tau\,(-\tilde{K}(\tau)) = 0\;. \tag{16}$$

Let us use this statement to prove that the function $\hat{K}(\cdot)$ of the Theorem is the solution of the problem (1). First let $\sigma^2 < A^2\left(2\pi\chi_n^2 - \|\phi_n\|_2^2\right)$. Check that the corresponding function

$\hat{K}(\cdot)$ satisfies the restriction in (14). Consider the function

$$f(t) = \tfrac{1}{2} + \frac{A^2}{\sigma^2}\left[\int_0^t \psi_n^2(\tau)\,d\tau - \psi_n(t)\int_0^t \psi_n(\tau)\,d\tau\right], \quad t \in [0,2\pi].$$

It is clear that $f(0) = \tfrac{1}{2}$ and $\psi_n(\pi) = -2\chi_n$, therefore

$$f(\pi) = \tfrac{1}{2} + \frac{A^2}{\sigma^2}\left(\tfrac{1}{2}\,\|\phi_n\|_2^2 - \pi\,\chi_n^2\right) .$$

Due to the above assumption, $f(\pi) < 0$. The derivative being negative on $(0,\pi)$, $f(\cdot)$ decreases monotonically and hence it has a unique zero on $(0,\pi)$. This is obviously the unique root of the equation (2).

From the Cauchy-Schwarz inequality it follows that for any $t_0 \in (0,\pi]$ the determinant Δ of the system (3) is positive and therefore α and β are determined uniquely. Having found the explicit expressions we have that $\hat{K}(t) = f(t)/\Delta$ on $[0,t_0]$, and hence $\hat{K}(\cdot)$ is a continuous nonnegative function. The second equality in (3) means that $\int_0^\pi \hat{K}(\tau)d\tau = \tfrac{1}{2}$. Thus $\hat{K}(\cdot)$ is an admissible function in the problem (14).

Now let

$$\lambda_0 = 1, \qquad \lambda_1 = -4\sigma^2\beta ,$$

$$\lambda_2 = -\frac{4\sigma^2}{\Delta}\int_{t_0}^{\pi} f(t)\,dt .$$

As $d\mu$ we shall take the measure which is absolutely continuous relative to Lebesgue measure and has the density

$$g(t) = \begin{cases} 0 , & 0 \le t \le t_0 , \\ \dfrac{f(t)}{\int_{t_0}^{\pi} f(\tau)\, d\tau} , & t_0 \le t \le \pi . \end{cases}$$

Taking into account the first equation in (3), direct calculations show that under these conditions

$$\lambda_0 8A^2 \left[\int_0^{\pi} \hat{K}(\tau)\, \psi_n(\tau)\, d\tau \right] \psi_n(t) + \lambda_0 4\sigma^2 \hat{K}(t) + \lambda_1 - \lambda_2 g(t) \equiv 0,$$

i.e., the identity (15) is true. The equality (16) is true obviously.

Therefore $\hat{K}(\cdot)$ is the solution of the problem (1) when $\sigma^2 < A^2(2\pi X_n^2 - \|\phi_n\|_2^2)$, and since the minimized functional in (14) is strictly convex, this solution is unique. The equality $J = \sigma\sqrt{\beta}$ may be verified readily.

Next let $\sigma^2 \ge A^2(2\pi X_n^2 - \|\phi_n\|_2^2)$. In that case the function $f(\cdot)$ has no zeros on $(0,\pi)$. Set

$$\hat{K}(t) = \alpha\phi_n\left(t + \frac{n+1}{2}\pi\right) + \beta - \alpha X_n = \frac{f(t)}{\Delta} , \quad t \in [0,\pi] ,$$

where α and β are found from (3) for $t_0 = \pi$. If we set $\lambda_0 = 1$, $\lambda_1 = -4\sigma^2\beta$, $\lambda_2 = 0$ and as $d\mu$ we take the probability measure concentrated at the point π, then as before, one may check that $\hat{K}(\cdot)$ is admissible and the relations (15) and (16) are true. Consequently, in this case $\hat{K}(\cdot)$ is the solution of the problem (1) and it is easy to calculate that $J = \sigma\sqrt{\beta}$. It remains to note that α and β for $t_0 = \pi$ have simple explicit expressions as given in the formulation of the Theorem. ///

REFERENCES

[1] Ibragimov, I.A., and Has'minskii, R.Z. *Statistical Estimation: Asymptotic Theory*. Berlin Heidelberg New York: Springer-Verlag Inc., 1981.

[2] Dodynekova, R.D. "Minimax Estimation in Problems with Partially Observed Data." *Uspekhi matem. nauk*, vol.39, no.1 (1984): 133-134.

[3] Berkovitz, L.D., and Pollard, H. "A Nonclassical Variational Problem Arising from an Optimal Filter Problem." I. *Archive Rat. Mech. Analysis*, 26 (1967): 281-301.

[4] Kornijchuk, N.P. *Ekstremal'nye zadachi teorii priblizheniya* (Extremal Problem of Approximation Theory). Moskva: Nauka, 1976.

[5] Tikhomirov, V.M. *Nekotorye zadachi teorii priblizheniya* (Some Problems of Approximation Theory). Moskva: Nauka, 1976.

[6] Alekseev, V.M., Tikhomirov, V.M., and Fomin, S.V. *Optimal'noe upravlenie* (Optimal Control). Moskva: Nauka, 1976.

E.A. FAINBERG and I.M. SONIN

PERSISTENTLY NEARLY OPTIMAL STRATEGIES IN STOCHASTIC DYNAMIC PROGRAMMING

1. INTRODUCTION

This paper considers a discrete time stochastic dynamic programming model with a countable state space. The authors have recently [7] proved a theorem on the existence of stationary strategies which are uniformly nearly optimal with respect to a value function of the class of stationary strategies. This paper deals with a generalization of this result and its extension to some classes of nonstationary strategies, including Markov strategies.

Consider a stochastic dynamic programming model $\mathcal{N} = (X,A,A(\cdot),p,r)$, where

- (i) X is a countable state space,
- (ii) A is an action space,
- (iii) $A(h) \subseteq A$ is a set of actions admissible for each history $h \in H = \bigcup_{n=0}^{\infty} H_n$, $H_n = (X\times A)^n \times X$, $n = 0,1,\ldots,$
- (iv) $p(\cdot \mid ha)$ is a transition function,
- (v) $r(ha)$ is a reward function, $a \in A(h)$.

To simplify notation we will write further $\mathcal{N} = (X,A(\cdot),p,r)$. It is assumed that the set A is endowed with a σ-field $\mathcal{A}$

containing all one-point sets. σ-fields 2^X and A generate σ-fields F_n, F and F_∞ on H_n, H and $H_\infty = (X\times A)^\infty$, respectively. The functions p and r are assumed to be $(F \times A)$-measurable, $-\infty \le r(ha) < +\infty$, $0 \le p(z \mid ha) \le 1$ and $\sum_{z\in X} p(z \mid ha) = 1$.

Let Π be the class of all (possibly randomized) strategies $\pi = \{\pi_n(\cdot \mid h_n),\ n = 0,1,\dots\}$ satisfying the usual measurability conditions and the condition $\pi_n(A(h_n) \mid h_n) = 1$. Using Tulcea's theorem [25, Theorem 1.1], for each history $h = (x_o\, a_o \cdots x_m) \in H$ and the strategy $\pi \in \Pi$, we define the unique probability measure P_h^π on (H_∞, F_∞) such that

$$P_h^\pi(x_o\ a_o\ \cdots\ x_m) = 1\ , \tag{1}$$

$$P_h^\pi(da_n \mid x_o a_o\ \dots\ x_n) = \pi_n(da_n \mid x_o a_o\ \dots\ x_n)\ ,\qquad n \ge m\ , \tag{2}$$

$$P_h^\pi(x_{n+1} \mid x_o a_o\ \dots\ x_n a_n) = p(x_{n+1} \mid x_o a_o\ \dots\ x_n a_n)\ ,\qquad n \ge m\ . \tag{3}$$

Expectations with respect to P_h^π will be denoted by E_h^π.

The total expected reward for a history $h \in H$ and strategy π is denoted by

$$w^\pi(h) = E_h^\pi \sum_{i=n}^{\infty} r(h_i a_i)\ ,\qquad h \in H_n,\quad n = 0,1,\dots\ . \tag{4}$$

To guarantee convergence in (4), we assume that for any h and π

$$E_h^\pi \sum_{i=n}^{\infty} r^+(h_i a_i) < \infty\ ,\qquad h \in H_n,\quad n = 0,1,\dots\ , \tag{5}$$

where $g^+ = \max\{0,g\}$ for any number g, $g^- = g - g^+$.

For $\Delta \subseteq \Pi$

$$v_\Delta(h) = \sup_{\pi\in\Delta} w^\pi(h), \qquad v(h) = v_\Pi(h).$$

Let $\ell: H \to [0,\infty)$ be a measurable function. In this paper we study the existence of persistently $\varepsilon\ell$-optimal strategies for various classes Δ, i.e., strategies $\pi \in \Delta$ such that

$$w^\pi(h) \geq v_\Delta(h) - \varepsilon\ell(h) \qquad \text{for all } h \in H.$$

Let Q be the set of all Markov times $\tau = \tau(h) \leq \infty$ with respect to σ-fields F_n and let $Q^b = \cup_{n=0}^{\infty}\{\tau \in Q: \tau \leq n\}$ be the set of all bounded Markov times.

We denote $H_\Delta = \{h \in H: w^\pi(h) = v_\Delta(h)$ for some $\pi \in \Delta\}$ and for $h \in H_n$, $n = 0,1,\ldots$. Define

$$d_\Delta(h) = \sup_{\pi\in\Pi} \inf_{\substack{\tau\geq n \\ \tau\in Q^b}} E_h^\pi v_\Delta(h_\tau).$$

For $f: H \to [-\infty,+\infty)$, $h \in H$, $a \in A(h)$ we define the operators

$$P^a f(h) = \sum_{z\in X} p(z \mid ha) f(haz),$$

$$Pf(h) = \sup_{a\in A(h)} P^a f(h),$$

$$T^a f(h) = r(ha) + P^a f(h),$$

$$Tf(h) = \sup_{a\in A(h)} T^a f(h),$$

where it is assumed that $Pf^+(h) < \infty$.

Let $f: H \to [-\infty, \infty)$, $H' \subseteq H$. Denote by $L_o(f, H')$ the set of nonnegative measurable functions ℓ on H such that

- (i) $\ell(h) = 0$ for $h \in H'$,
- (ii) $\ell(h) > 0$, $\ell(h) \geq \max\{P\ell(h), f(h)\}$ for $h \in H \setminus H'$.

The model is called Markov if it is defined by a 4-tuple $\mu = (X, A(\cdot), p, r)$ such that $A(h_n) = A(x_n)$, $p(\cdot \mid h_n a) = p(\cdot \mid x_n a)$, $r(h_n a) = r(x_n a)$ for any $h_n = (x_o a_o \cdots x_n) \in H$. For Markov models, the condition (5) transforms to the so-called General Convergence Condition

$$u^\pi(x) = E_x^\pi \sum_{i=0}^{\infty} r^+(x_i a_i) < \infty, \quad x \in X, \quad \pi \in \Pi. \quad (6)$$

The condition (6) implies $u^*(x) = \sup_{\pi \in \Pi} u^\pi(x) < \infty$, [21, Theorem 2.3], [3, p. 108].

Throughout the Introduction and Sections 2 and 3 the Markov model is considered. If $f(h_n) = f(x_n)$, $h_n \in H$, then the operators $P^a f(h)$, $T^a f(h)$, $Pf(h)$, $Tf(h)$ transform to the operators $P^a f(x)$, $T^a f(x)$, $Pf(x)$, $Tf(x)$. Similarly to H_Δ we define $X_\Delta = \{x \in X\colon w^\pi(x) = v_\Delta(x)$ for some $\pi \in \Delta\}$.

For $f: X \to [0, \infty)$ and $X' \subseteq X$ consider the set $L_o(f, X')$ of nonnegative functions ℓ on X, such that

- (i) $\ell(x) = 0$ for $x \in X'$,
- (ii) $\ell(x) > 0$, $\ell(x) \geq \max\{P\ell(x), f(x)\}$ for $x \in X \setminus X'$.

Denote $L(f) = L_o(f, \emptyset)$ and $L = L(0)$.

Let S be the set of all (non-randomized) stationary strategies and let M be the set of all (non-randomized) Markov strategies. Also, let

$$s(x) = v_S(x) \quad , \qquad d(x) = d_S(x) = \sup_{\phi \in S} \inf_{\tau \in Q^b} E_x^\phi s(x_\tau) \quad .$$

The following two results have been proved in [7].

PROPOSITION 1 ([7, Theorem 2.1]). For any $\varepsilon > 0$ and any $\ell \in L_o(s, X_S)$ there exists a stationary strategy ϕ such that $w^\phi \geqslant s - \varepsilon\ell$.

PROPOSITION 2 ([7, Theorem 2.2]). $s = Ts$.

Proposition 1 implies the well-known result of Ornstein [14] and Frid [8] on the existence of stationary εv-optimal strategies in the case $r \geq 0$ and allows us to generalize the extension of the Ornstein-Frid theorem suggested by van der Wal [22] (see [7, Corollaries 2.4, 2.5]).

In Section 3, the following generalization of Proposition 1 is proved.

THEOREM 1. For any $\varepsilon > 0$ and any $\ell \in L_o(d, X_S)$ there exists a stationary strategy ϕ such that $w^\phi \geq s - \varepsilon\ell$.

The example given by Blackwell in [1] shows that if the function s is not bounded from above, then the strategy uniformly ε-optimal in class S may not exist for some positive ε. Proposition 1 implies the existence of such strategies when s is bounded from above ([7, Corollary 2.3]). Theorem 1 implies a more general condition for the existence of uniformly ε-optimal strategies in class S. This condition is the boundedness from above of the function d. For example, in contracting dynamic programming models in the sense of van Nunen and Wessels [13], $d \leq 0$, but s may be unbounded. The substitution of d for the function s in classes $L_o(\cdot, X_S)$ was prompted by the paper of van Dawen and Schäl [2], where the function

$$V(x) = \sup_{\pi \in \Pi} \lim_{n \to \infty} \sup_{n \le \tau \le \infty} E^{\pi} v(x_{\tau}) \, I\{\tau < \infty\} \geq d(x)$$

had been considered. Note that Theorem 1 implies the result reported by Van Dawen and Schäl [2] (see Corollary 1 below).

We also prove that $s = v_{RS}$, where RS is a set of all randomized stationary strategies (Theorem 3).

In Section 4, Theorem 1, Proposition 2 and Theorem 3 are extended to non-Markov models and to various classes of non-stationary strategies by using the evolute (Theorems 4 and 5). These classes are described by using the construction of (f,B)-generated strategies introduced in [6] and by using the transitivity condition ([16, Chapter II, Section 15]). Theorem 4 implies results concerning the existence of Markov uniformly nearly optimal strategies [23], [7, Theorem 2.6], [17].

Note that the classes $L_o(s, X_S)$ and $L_o(d, X_S)$ from Proposition 1 and Theorem 1 can be described in terms of the value functions for some optimal stopping time problems. This points to the existence of a deep and yet unclear relationship between our results and the optimal stopping theory [16].

It should be added that one of the first results on the existence of stationary optimal strategies is due to Viskov and Shiryaev [20]. This paper as well as the paper [15] of Shiryaev influenced significantly the development of the Markov decision theory.

2. PRELIMINARY RESULTS

For $f\colon X \to [-\infty,\infty)$, $\tau \in Q$, $x \in X$ and $\pi \in \Pi$ we write $E_x^\pi f(x_\tau)$ instead of $E_x^\pi I\{\tau < \infty\}\, f(x_\tau)$.

LEMMA 1. ([16, Chapter II, Lemma 3]). For any $\tau \in Q$, $\pi \in \Pi$ and any nonnegative excessive function f (i.e., $f \geq Pf$) on X

$$f(x) \geq E_x^\pi f(x_\tau) \quad , \qquad x \in X \quad . \quad ///$$

For $f\colon X \to [-\infty,\infty)$, $\tau \in Q$ and $\pi \in \Pi$ denote

$$u^\pi(x,\tau,f) = E_x^\pi \left\{ \sum_{i=0}^{\tau-1} r^+(x_i a_i) + f^+(x_\tau) \right\}$$

and if $u^\pi(x,\tau,f) < \infty$ define

$$w^\pi(x,\tau,f) = E_x^\pi \left\{ \sum_{i=0}^{\tau-1} r(x_i a_i) + f(x_\tau) \right\} \quad .$$

LEMMA 2 ([7, Lemma 3.4]). For any $x \in X$, $\pi \in \Pi$ and $\tau \in Q$

$$u^\pi(x,\tau,u^*) < \infty \quad . \quad ///$$

Lemma 2 shows that if $f \leq u^*$ then functions $w^\pi(x,\tau,f)$ are well defined.

LEMMA 3. Let τ, $\tau_n \in Q$, $n = 1,2,\ldots$, $\tau_n \leq \tau$ and $\tau_n \to \tau$. Then for any $\pi \in \Pi$

$$w^\pi(x,\tau,0) = \lim_{n\to\infty} w^\pi(x,\tau_n,0) \quad .$$

The proof of Lemma 3 is almost obvious and is similar to the proof of Lemma 3.5 in [7].

LEMMA 4 (cp. [7, Lemma 3.6]). Let $\tau_n \in Q$, $n = 1,2,\ldots$ and

$\tau_n \to \infty$. If for some $\pi \in \Pi$, $x \in X$, $C < \infty$, $\delta > -\infty$ and for some sequence of functions $f_n: X \to [-\infty,\infty)$ the following conditions are satisfied

- (a) $\liminf_{n\to\infty} w^{\pi}(x,\tau_n,f_n) \geq C$,
- (b) $\liminf_{n\to\infty} E_x^{\pi} f_n(x_{\tau_n}) \leq \delta$,

then

$$w^{\pi}(x) \geq C - \delta .$$

Proof.

$$\begin{aligned} w^{\pi}(x) &= \lim_{n\to\infty} w^{\pi}(x,\tau_n,0) \\ &= \lim_{n\to\infty} \left[w^{\pi}(x,\tau_n,f_n) - E_x^{\pi} f_n(x_{\tau_n}) \right] \\ &\geq \liminf_{n\to\infty} w^{\pi}(x,\tau_n,f_n) - \liminf_{n\to\infty} E_x^{\pi} f_n(x_{\tau_n}) \\ &\geq C - \delta \end{aligned}$$

(the first equality follows from Lemma 3, the last inequality follows from (a) and (b)). ///

LEMMA 5. Let $x \in X$, $\tau \in Q^b$, $\tau_n \to \tau$, $n = 1,2,\ldots$, $\pi \in \Pi$, $f \leq u^*$ and

$$\limsup_{n\to\infty} w^{\pi}(x,\tau_n,f) \geq C .$$

Then

$$w^{\pi}(x,\tau,f) \geq C . \tag{7}$$

Proof. Let $f \geq K > -\infty$, $K < 0$. Then

$$w^{\pi}(x,\tau,f) = w^{\pi}(x,\tau_n,f) + \{w^{\pi}(x,\tau,0) - w^{\pi}(x,\tau_n,0)\} + E_x^{\pi} \sum_{i=0}^{N} f(x_i)(I\{\tau=i\} - I\{\tau_n=i\}) , \tag{8}$$

where $\tau \le N < \infty$.

Since $K \le f(x_i)I\{\tau_n=i\} \le u^*(x_i)$, $P_x^{\pi}\{\tau_n=i\} \to P_x^{\pi}\{\tau=i\}$ and $E_x^{\pi}f^+(x_i) < \infty$ for each $i = 1,2,\ldots,N$, Lebesgue's theorem on majorized convergence implies that the last term in (8) tends to zero. The second term tends to zero by Lemma 3. Therefore the conditions of Lemma 5 imply (7).

When f is unbounded from below, we define $f^K(x) = \max\{f(x), K\}$. Then

$$\limsup_{n\to\infty} w^{\pi}(x,\tau_n,f^K) \ge \limsup_{n\to\infty} w^{\pi}(x,\tau_n,f) \ge C .$$

Since Lemma 5 is proved for functions bounded from below, $w^{\pi}(x,\tau,f^K) \ge C$. Therefore,

$$w^{\pi}(x,\tau,f) = \lim_{K\to-\infty} w^{\pi}(x,\tau,f^K) \ge C . \quad ///$$

3. STATIONARY NEARLY OPTIMAL STRATEGIES

Without loss of generality we assume that $s > -\infty$. Indeed, consider the sets $X_{-\infty} = \{x: s(x) = -\infty\}$ and $\tilde{A}(x) = \{a \in A(x): p(z \mid xa) = 0$ for any $z \in X_{-\infty}\}$. One may easily show that $\tilde{A}(x) \ne \emptyset$ for any $x \in X \setminus X_{-\infty}$ and $s(x) = \tilde{s}(x)$ for $x \in X \setminus X_{-\infty}$, where $\tilde{s}$ is a value function of the class of stationary strategies in the model $(X\setminus X_{-\infty}, \tilde{A}(\cdot), p, r)$.

We prove Theorem 1 only for $\ell \in L(d)$. The extension to $\ell \in L_o(d,X_S)$ is similar to [7, p. 127] and is based on Lemma 4.4 in [7] on the existence of $\psi \in S$ such that

(a) $w^{\psi}(x) = s(x)$ for $x \in X_S$ and

(b) $p(y \mid x, \psi(x)) = 0$ for $x \in X_S$, $y \in X\backslash X_S$.

Let $\ell : X \to [0,\infty)$, $G \subseteq X$. Denote

$$s^{\ell,G} = \begin{cases} \ell(x) , & x \in G , \\ s^{+}(x), & x \notin G . \end{cases}$$

LEMMA 6. Let $\varepsilon > 0$, $\ell \in L$, $1 \le N < \infty$ and Y be a finite subset of X. Then there exists a stationary strategy ϕ, integer $n > N$, finite sets D, $D \supseteq Y$, and G, $G \cap D = \emptyset$, such that

(a) $w^{\phi}(y) \ge s(y) - \varepsilon\ell(y)$, $y \in D$,

(b) $E_y^{\phi}\, s^{\ell,G}(x_\tau) < \varepsilon$, $y \in Y$,

where $\tau = \min\,(n, \min\{k: x_k \in X\backslash D\})$.

Proof. Denote $c = \min\,(1, \min\,\{\ell(y);\ y \in Y\})$. Since Y is finite and $\ell > 0$, we have $c > 0$. Let $\varepsilon < 1$ and $\varepsilon' = \varepsilon^2 c/3 \le \varepsilon^2/3 < \varepsilon/3$.

By Lemma 3.12 from [7] there exists $\phi \in S$ and an integer $n > N$ such that

$$w^{\phi}(y) \ \underline{>}\ s(y) - \varepsilon' , \qquad y \in Y , \tag{9}$$

$$E_y^{\phi}\, s^{+}(x_n) \ < \ \varepsilon' , \qquad y \in Y . \tag{10}$$

Let $\rho = \min\,\{k: x_k \in \tilde{G}\}$, where $\tilde{G} = \{z: w^{\phi}(z) < s(z) - \varepsilon\ell(z)\}$. According to [7, p. 121]

$$E_y^{\phi}\ell(x_\rho) \ < \ \varepsilon'/\varepsilon , \qquad y \in Y . \tag{11}$$

Let us construct D and G. Let $D^o = Y$, $G^o = \emptyset$. The inequality (9) and the definitions of $\tilde{G}$ and ε' imply $D^o \cap \tilde{G} = \emptyset$. If for some $i = 1,2,\ldots,n-1$ the finite sets D^{i-1} and G^{i-1} such that $D^{i-1} \cap \tilde{G} = \emptyset$, $G^{i-1} \subseteq \tilde{G}$ are constructed, then we choose finite sets D^i and G^i such that $D^i \supseteq D^{i-1}$, $D^i \cap \tilde{G} = \emptyset$, $\tilde{G} \supseteq G^i \supseteq G^{i-1}$ and

$$E_y^{\phi} I\{x_1 \notin D^i \cup G^i\}\ s^+(x_1) < \varepsilon/3(n-1)\ , \qquad y \in D^{i-1}\ . \tag{12}$$

Denote $D = D^{n-1}$, $G = G^{n-1}$ and $\tau = \min\,(n, \min\,\{k : x_k \in X\backslash D\})$.

By the inequality (12)

$$E_y^{\phi} I\{\tau < n,\ x_\tau \notin G\}\ s^+(x_\tau) < \varepsilon/3\ , \qquad y \in Y\ . \tag{13}$$

Let us check the fulfillment of the inequalities (a) and (b) from Lemma 6. Since $D \cap \tilde{G} = \emptyset$, the inequality (a) is valid. To check (b) we consider the function $f = s^{\ell,G}$. Then

$$E_y^{\phi} f(x_\tau) = E_y^{\phi} I\{\tau = n,\ x_\tau \notin G\}\ s^+(x_n) + E_y^{\phi} I\{\tau < n,\ x_\tau \notin G\}\ s^+(x_\tau)$$
$$+ E_y^{\phi} I\{\tau \le n,\ x_\tau \in G\}\ \ell(x_\tau)\ . \tag{14}$$

We use the inequalities (10) and $s^+ \geq 0$ to estimate above the first summand in (14) and the inequality (13) to estimate the second summand and, finally the relations (11), $I\{\tau \le n,\ x_\tau \in G\} = I\{\tau = \rho \le n,\ x_\rho \in G\}$ and $G \subseteq \tilde{G}$ to estimate the third summand. Then we have

$$E_y^{\phi} f(x_\tau) \leq \varepsilon' + \varepsilon/3 + \varepsilon'/\varepsilon < \varepsilon\ ,$$

i.e., inequality (b) from Lemma 6 holds. ///

For $\phi \in S$ and $Y \subseteq X$ we define

$$S(\phi,Y) = \{\psi \in S: \psi(x) = \phi(x),\ x \in Y\} ,$$

$$s^{\phi,Y}(x) = \sup \{w^{\psi}(x): \psi \in S(\phi,Y)\} , \qquad x \in X .$$

LEMMA 7 ([7, Lemma 4.2]). Let a set Y, $Y \subseteq X$, a constant $\varepsilon > 0$, a function $\ell: X \to [0,\infty)$, $\ell \geq P\ell$, and a strategy $\phi \in S$ be such that

$$w^{\phi}(y) \geq s(y) - \varepsilon\ell(y) , \qquad y \in Y .$$

Then

$$s^{\phi,Y}(x) \geq s(x) - \varepsilon\ell(x) , \qquad x \in X .$$

Proof of Theorem 1 for the functions $\ell \in L(d)$. Let $\varepsilon > 0$, $\ell \in L(d)$, $X = \{x^1, x^2, \ldots\}$ and ε_i, $i = 1,2,\ldots$, be a sequence of positive numbers such that $\sum_i \varepsilon_i < \varepsilon/2$.

Denoting $Y_1 = \{x^1\}$ and using Lemma 6, we choose $\phi_1 \in S$, n_1 and finite sets D_1, $D_1 \supseteq Y_1$, and G_1, $G_1 \cap D_1 = \emptyset$, such that

$$w^{\phi_1}(x) \geq s(x) - \varepsilon_1\ell(x) , \qquad x \in D_1 , \tag{15}$$

$$E_y^{\phi_1} f_1(x_{\tau_1}) < \varepsilon_1 , \qquad y \in Y_1 ,$$

where $\tau_1 = \min(n_1, \min\{k: x_k \in X\setminus D_1\})$, $f_1 = s^{\ell,G_1}$.

Denote $s_1(x) = s^{\phi_1,D_1}(x)$, $x \in X$. By Lemma 7 (for $Y = D_1$)

$$s_1(x) \geq s(x) - \varepsilon_1\ell(x) , \qquad x \in X . \tag{16}$$

For some $k = 2, 3, \ldots$, let sets $D_1, \ldots, D_{k-1}$, stationary strategies $\phi_1, \ldots, \phi_{k-1}$ and positive integers $n_1 < n_2 < \ldots$ $\ldots < n_{k-1}$ such that

- (i) $D_1 \subseteq \dots \subseteq D_{k-1}$,
- (ii) $x^{k-1} \in D_{k-1}$,
- (iii) $\phi_i(x) = \phi_{i+1}(x) = \dots = \phi_{k-1}(x)$, where $x \in D_i$, $i = 1,\dots,k-1$,
- (iv) $w^{\phi_{k-1}}(x) \geq s(x) - (\varepsilon_1 + \dots + \varepsilon_{k-1})\ell(x)$, $x \in D_{k-1}$,

be constructed.

Denote $s_{k-1}(x) = s^{\phi_{k-1}, D_{k-1}}(x)$. By Lemma 7 we have

$$s_{k-1}(x) \geq s(x) - (\varepsilon_1 + \dots + \varepsilon_{k-1})\ell(x) , \qquad x \in X . \tag{17}$$

For $k = 2,3,\dots$, we consider the model $\mu_{k-1} = \{X, A_{k-1}(\cdot), p, r\}$, where $A_{k-1}(x) = \{\phi_{k-1}(x)\}$ for $x \in D_{k-1}$ and $A_{k-1}(x) = A(x)$ for $x \in X \setminus D_{k-1}$. Let S_{k-1} be the set of stationary strategies in the model μ_{k-1}. Then $S_{k-1} = S(\phi_{k-1}, D_{k-1})$ and s_{k-1} be the value of the class of stationary strategies in the model. Let

$$d_{k-1}(x) = \sup_{\phi \in S_{k-1}} \inf_{\tau \in Q^b} E_x^{\phi} s_{k-1}(x) ,$$

and let $L_{k-1}(d_{k-1})$ denote the set of positive excessive majorants of the function d_{k-1} in the model μ_{k-1}. By $A_{k-1}(\cdot) \subseteq A(\cdot)$ we have $s_{k-1} \leq s$, $d_{k-1} \leq d$ and $L_{k-1}(d_{k-1}) \supseteq L(d)$. Hence, $\ell \in L_{k-1}(d_{k-1})$.

Define $Y_k = D_{k-1} \cup \{x^k\}$. Applying Lemma 6 to the model μ_{k-1}, the set Y_k, the constant ε_k, the integer n_{k-1} and the function ℓ, we obtain the existence of a stationary strategy ϕ_k, integer $n_k > n_{k-1}$ and finite sets D_k, G_k such that $D_k \supseteq Y_k \supseteq D_{k-1}$, $D_k \cap G_k = \emptyset$ and

$$w^{\phi_k}(x) \geq s_{k-1}(x) - \varepsilon_k \ell(x) , \qquad x \in D_k , \tag{18}$$

$$E_x^{\phi_k} f_k(x_{\tau_k}) \leq \varepsilon_k , \qquad x \in Y_k , \tag{19}$$

where $\tau_k = \min(n_k, \min\{i: x_i \in X \setminus D_k\})$,

$$f_k = s_{k-1}^{\ell, G_k} = \begin{cases} \ell(x) , & x \in G_k , \\ s_{k-1}^+(x) , & x \notin G_k . \end{cases} \tag{20}$$

In addition, the definition of μ_{k-1} implies that $\phi_k(x) = \phi_{k-1}(x)$ for $x \in D_{k-1}$.

Applying Lemma 7 to the model μ_{k-1}, the set D_k, the strategy ϕ_k, the constant ε_k and the function ℓ, we have

$$s_k(x) = s^{\phi_k, D_k}(x) \geq s_{k-1}(x) - \varepsilon_k \ell(x) , \qquad x \in X .$$

By the last inequality and (17) we have

$$s_k(x) \geq s(x) - (\varepsilon_1 + \cdots + \varepsilon_k)\, \ell(x) , \qquad x \in X . \tag{21}$$

Now we define the stationary strategy ϕ, setting $\phi(x) = \phi_k(x)$ for $x \in D_k$, $k = 1,2,\ldots$. This definition is correct because $\phi_{k+1}(x) = \phi_k(x)$ for $x \in D_k$, $D_k \subseteq D_{k+1}$, $k = 1,2,\ldots$, and $\bigcup_{1 \leq k < \infty} D_k = \bigcup_{1 \leq k < \infty} \{x^k\} = X$.

Let us fix some $x \in X$ and prove that

$$w^{\phi}(x) \geq s(x) - \varepsilon \ell(x) . \tag{22}$$

To prove (22), we check conditions (a) and (b) from Lemma 4 for $C = s(x) - \varepsilon \ell(x)$, $\delta = 0$, functions $f_k(x)$ and the Markov times

τ_k defined above. The relations $D_k \subseteq D_{k+1}$, $n_{k+1} > n_k$, where $k = 1,2,\ldots$ and $X = \bigcup_k D_k$ imply $\tau_k(h) \not\nearrow \infty$ as $k \to \infty$.

Choose i such that $x \in D_i$. Then for $k \geq i$ we have $x \in D_k$ and by the definition of the strategy ϕ and (19)

$$E_x^{\phi} f_k(x_{\tau_k}) = E_x^{\phi_k} f_k(x_{\tau_k}) \leq \varepsilon_k \quad . \tag{23}$$

Hence condition (b) from Lemma 4 is satisfied.

To verify condition (a) from Lemma 4, we show first that for any $n \geq i$

$$w^{\phi}(x,\tau_n,s) \geq s(x) - \varepsilon\ell(x)/2 \quad . \tag{24}$$

By the definitions of ϕ_n, τ_n and the inequalities (18) (21) one may write

$$\begin{aligned} w^{\phi}(x,\tau_n,s) &= w^{\phi_n}(x,\tau_n,s) \geq w^{\phi_n}(x) \\ &\geq s_{n-1}(x) - \varepsilon_n\ell(x) \\ &\geq s(x) - (\varepsilon_1 + \cdots + \varepsilon_{n-1} + \varepsilon_n)\ell(x) \\ &\geq s(x) - \varepsilon\ell(x)/2 \quad . \end{aligned}$$

Note that if $\ell \in L(s) \subseteq L(d)$ and therefore $\ell \geq s^+ \geq s_{k-1}^+$, then (23) implies

$$E_x^{\phi} s_{k-1}^+(x_{\tau_k}) \leq \varepsilon_k \quad . \tag{25}$$

If $\ell \in L(s)$ then (25) and (21) imply the validity of condition (b) from Lemma 4 and (24) implies condition (a) (for $C = s(x) - \varepsilon\ell(x)/2$, $\delta = \varepsilon\ell(x)/2$ and $f_k = \ell$), i.e., we obtain Proposition 1. As has been proved in [7], Proposition 1 implies

Lemma 8 (i) (we will continue the proof of Theorem 1 after Lemmas 8 - 10).

LEMMA 8. Let $\tau \in Q$. Then

- (i) ([7, Lemma 4.3]) $w^{\phi}(x,\tau,s) \le s(x)$ for any $\phi \in S$, $x \in X$,
- (ii) $w^{\pi}(x,\tau,v) \le v(x)$ for any $\pi \in \Pi$, $x \in X$.

Proof. (ii). Fix $\pi \in \Pi$, $\tau \in Q$ and $x \in X$. For an arbitrary $\varepsilon > 0$ consider a uniformly ε-optimal strategy σ. Define the strategy γ by

$$\gamma_n(\cdot \mid h_n) = \begin{cases} \pi_n(\cdot \mid h_n), & n < \tau, \\ \sigma_{n-\tau}(\cdot \mid x_\tau a_\tau \cdots x_n), & n \ge \tau. \end{cases}$$

Then $v(x) \ge w^{\gamma}(x) = w^{\pi}(x,\tau,w^{\sigma}) \ge w^{\pi}(x,\tau,v-\varepsilon) \ge w^{\pi}(x,\tau,v) - \varepsilon$. Since $\varepsilon > 0$ is arbitrary, the lemma is proved. ///

We will need the following two lemmas.

LEMMA 9. Let $R \subseteq X \times \{0,1,\dots\}$, $\tau,\rho \in Q$, $\tau = \min\{i:(x_i,i) \in R\}$ and $\rho \le \tau$. Then

- (i) $w^{\phi}(x,\tau,s) \le w^{\phi}(x,\rho,s)$ for any $\phi \in S$, $x \in X$,
- (ii) $w^{\pi}(x,\tau,v) \le w^{\pi}(x,\rho,v)$ for any $\pi \in \Pi$, $x \in X$.

Proof. (i). For $k = 0,1,\dots,$ we define Markov times ${}^k\tau = \min\{i: (x_i, i+k) \in R\}$. Then

$$\begin{aligned} w^{\phi}(x,\tau,s) &= w^{\phi}(x,\rho,0) + E_x^{\phi} \sum_{k=0}^{\tau} I\{\rho=k\} \left[\sum_{n=k}^{\tau-1} r(x_n a_n) + s(x_\tau) \right] \\ &= w^{\phi}(x,\rho,0) + E_x^{\phi} \sum_{k=0}^{\tau} I\{\rho=k\}\, w^{\phi}(x_k, {}^k\tau, s) \\ &\le w^{\phi}(x,\rho,0) + E_x^{\phi} \sum_{k=0}^{\tau} I\{\rho=k\}\, s(x_k) = w^{\phi}(x,\rho,s) \end{aligned}$$

(the inequality follows from Lemma 8). The proof of (ii) is similar to (i) and uses the shifts of the strategy π defined following Corollary 1.

LEMMA 10. Let $R_n \subseteq X \times \{0,1,2,...\}$, $n = 0,1,...,$ $\tau_n = \min\{i: (x_i, i) \in R_n\}$, $\tau \in Q^b$ and $\tau_n \to \infty$. If for some $x \in X$ and for $f = s$, or $f = v$

$$\limsup_{n\to\infty} w^\pi(x,\tau_n,f) \geq C ,$$

where $\pi \in S$ when $f = s$ and $\pi \in \Pi$ when $f = v$, then

$$w^\pi(x,\tau,f) \geq C . \tag{26}$$

Proof. Denote $\tilde{\tau}_n = \min(\tau,\tau_n)$. Lemma 9 implies

$$\limsup_{n\to\infty} w^\pi(x,\tilde{\tau}_n,f) \geq \limsup_{n\to\infty} w^\pi(x,\tau_n,f) \geq C . \tag{27}$$

Also, (27) and Lemma 5 imply (26). ///

We continue the proof of Theorem 1. Recall that we have to check condition (a) from Lemma 4 for ϕ, τ_n, f_n defined above and $C = s(x) - \varepsilon\ell(x)$.

Let ε' be an arbitrary positive constant. By the definition of the function d we have that there exist Markov times $\beta_k \in Q^b$ such that

$$\ell(y) \geq d(y) \geq E_y^\phi s(x_{\beta_k}) - \varepsilon' , \quad y \in G_k , \quad k = 1,2,... \tag{28}$$

where the finite sets G_k are defined above.

Let $\tilde{\tau}_k$, $k = 1,2,...,$ be a Markov time defined by

$$\tilde{\tau}_k(h) = \begin{cases} \tau_k(h) & x_{\tau_k} \notin G_k , \\ \tau_k(h) + \beta_k(x_{\tau_k} a_{\tau_k} x_{\tau_k+1} \dots) , & x_{\tau_k} \in G_k . \end{cases}$$

Since $\tau_k \leq n_k$ and $\beta_k \in Q^b$, we have $\tilde{\tau}_k \in Q^b$. By (24) and Lemma 10 applied to the sequence $\tau_n \nearrow \infty$ we have

$$w^{\phi}(x, \tilde{\tau}_k, s) \geq s(x) - \varepsilon \ell(x)/2 , \qquad k = 1,2,\dots . \tag{29}$$

By the definitions of f_k, τ_k and $\tilde{\tau}_k$ we have

$$\begin{aligned} w^{\phi}(x, \tau_k, f_k) &= w^{\phi}(x, \tau_k, 0) + E_x^{\phi} I\{x_{\tau_k} \notin G_k\} s^+_{k-1}(x_{\tau_k}) \\ &\quad + E_x^{\phi} I\{x_{\tau_k} \in G_k\} \ell(x_{\tau_k}) \\ &\geq w^{\phi}(x, \tau_k, 0) + E_x^{\phi} I\{x_{\tau_k} \notin G_k\} s^+(x_{\tau_k}) - \varepsilon \ell(x)/2 \\ &\quad + E_x^{\phi} I\{x_{\tau_k} \in G_k\} s(x_{\tilde{\tau}_k}) - \varepsilon' \end{aligned}$$

(the inequality follows from (21), (28) and Lemma 1).

Therefore

$$\begin{aligned} w^{\phi}(x, \tau_k, f_k) &+ E_x^{\phi} \sum_{i=\tau_k}^{\tilde{\tau}_k - 1} r^+(x_i a_i) \\ &\geq w^{\phi}(x, \tau_k, 0) + E_x^{\phi} I\{x_{\tau_k} \notin G_k\} s^+(x_{\tau_k}) \\ &\quad + E_x^{\phi} I\{x_{\tau_k} \in G_k\} \left(\sum_{i=\tau_k}^{\tilde{\tau}_k - 1} r(x_i a_i) + s(x_{\tilde{\tau}_k}) \right) - \varepsilon \ell(x)/2 - \varepsilon' \\ &= w^{\phi}(x, \tilde{\tau}_k, s) - \varepsilon \ell(x)/2 - \varepsilon' . \end{aligned} \tag{30}$$

Since $\tau_k \nearrow \infty$, one may choose k_o such that for $k > k_o$

$$E_x^{\phi} \sum_{i=\tau_k}^{\tilde{\tau}_k - 1} r^+(x_i a_i) \leq E_x^{\phi} \sum_{i=\tau_k}^{\infty} r^+(x_i a_i) < \varepsilon' . \quad (31)$$

By (29) - (31) we have for $k > k_o$

$$w^{\phi}(x, \tau_k, f_k) > s(x) - \varepsilon \ell(x) - 2\varepsilon' .$$

Since $\varepsilon' > 0$ is arbitrary, we have that condition (a) from Lemma 4 is proved. Thus (22) is also proved. ///

Hordijk [11] has proved that the stationary strategy ϕ is optimal if and only if $T^{\phi(x)}v(x) = v(x)$, $x \in X$ and

$$\limsup_{n \to \infty} E_x^{\phi} v(x_n) \leq 0 , \qquad x \in X . \quad (32)$$

The following theorem generalizes this assertion.

THEOREM 2. A stationary strategy ϕ is optimal if and only if $T^{\phi(x)}v(x) = v(x)$, $x \in X$ and

$$\inf_{\tau \in Q^b} E_x^{\phi} v(x_{\tau}) \leq 0 , \qquad x \in X . \quad (33)$$

Proof. Necessity. (32) implies (33).

Sufficiency. For an arbitrary n we have $w^{\phi}(x,n,v) = (T^{\phi})^n v(x) = v(x)$. Let $\varepsilon_n \searrow 0$. Fix $x \in X$. For each n we choose a finite set $Y_n \subseteq X$ such that $E_x^{\phi} v^+(x_n) I\{x_n \notin Y_n\} \leq \varepsilon_n$. Choose $\gamma_n \in Q^b$ such that $E_x^{\phi} v(x_{\partial n}) \leqslant \varepsilon_n$, $x \in Y_n$.

Define $\tau_n \in Q^b$ by

$$\tau_n = \begin{cases} n , & x_n \notin Y_n , \\ n + \gamma_n(x_n a_n x_{n+1} a_{n+1} \cdots) , & x_n \in Y_n . \end{cases}$$

For any $\rho \in Q^b$ Lemmas 10 and 8 (ii) imply $w^\phi(x,\rho,v) = v(x)$. Thus we have $w^\phi(x,\tau_n,v) = v(x)$. In addition,

$$E_x^\phi v(x_{\tau_n}) = E_x^\phi v(x_n) I\{x_n \notin Y_n\} + E_x^\phi I\{x_n \in Y_n\} E_{x_n}^\phi v(x_{\gamma_n})$$
$$\leq 2\varepsilon_n .$$

Applying Lemma 4 to the sequence $\tau_n \to \infty$, we get $w^\phi(x) = v(x)$. ///

Note that if we replace v by s in the formulation of Theorem 2 then we shall get a necessary and sufficient condition for $w^\phi = s$.

<u>COROLLARY 1</u> (cp. van Dawen, Schäl [2]). Let $\ell: X \to [0,\infty)$ and $\ell \geq \max(d, P\ell)$. Suppose that $s = v$ and for each $x \in X^\ell = \{x \in X: \ell(x) = 0\}$ there exists $a \in A(x)$ such that $T^a v(x) = Tv(x)$. Then for any $\varepsilon > 0$ there exists $\phi \in S$ such that

$$w^\phi(x) = v(x) \qquad \text{for all } x \in X_\Pi ,$$

$$w^\phi(x) \geq v(x) - \varepsilon\ell(x) \qquad \text{for all } x \in X \setminus X_\Pi .$$

<u>Proof</u>. First we show that $X_S = X_\Pi$. If there exists an optimal strategy, then there exists a stationary optimal strategy [22]. Slightly modifying the proofs in [22, Theorem 1.2], one may get an existence of a stationary strategy ϕ such that $w^\phi(x) = v(x)$, $x \in X_\Pi$. The assumption $v = s$ implies $X_S = X_\Pi$.

If it is shown that $X^\ell \subseteq X_\Pi$, then Theorem 1 implies Corollary 1. When $x \in X^\ell$ (i.e., $\ell(x) = 0$), then $\ell \geq 0$, $\ell \geq P\ell$ imply $p(z \mid xa) = 0$ for any $z \in X \setminus X^\ell$ and $a \in A(x)$. Consequently the relation $X^\ell \subseteq X_\Pi$ follows from Theorem 2 applied to the model $\{X^\ell, A(\cdot), p, r\}$. ///

THEOREM 3. $s = v_{RS}$.

LEMMA 11. If $v_{RS} \leq 0$ then $s = v_{RS}$.

Proof. Let $\sigma \in RS$. Then $w^\sigma \leq 0$. For $f \leq u^*$ we define

$$T^\sigma f(x) = \int_{A(x)} T^a f(x)\, \sigma(da \mid x) \quad , \qquad x \in X \quad .$$

Then $w^\sigma = T^\sigma w^\sigma$. There exists $\phi \in S$ such that $T^\phi w^\sigma \geq T^\sigma w^\sigma = w^\sigma$ [3, p. 41]. If $(T^\phi)^i w^\sigma \geq w^\sigma$ for some $i = 1,2,\ldots,$ then $T^\phi (T^\phi)^i w^\sigma \geq T^\phi w^\sigma \geq w^\sigma$. Consequently for any $n = 1,2,\ldots$

$$w^\phi(x,n,w^\sigma) = (T^\phi)^n w^\sigma(x) \geq w^\sigma(x) \quad , \qquad x \in X \quad .$$

Let $n \to \infty$. Then by Lemma 3, we have for $x \in X$

$$w^\phi(x) = \lim_{n\to\infty} w^\phi(x,n,0) \geq \lim_{n\to\infty} w^\phi(x,n,w^\sigma) \geq w^\sigma(x) \text{ . ///}$$

Proof of Theorem 3. Fix arbitrary $\varepsilon > 0$ and $x \in X$. Consider $\sigma \in RS$ such that

$$w^\sigma(x) \geq v_{RS}(x) - \varepsilon/2 \quad . \tag{34}$$

Choose $n = 1,2,\ldots$ such that

$$E_x^\sigma \sum_{i=n}^{\infty} r^+(x_i a_i) < \varepsilon/2 \quad . \tag{35}$$

Consider the model $\mu_- = \{X, A(\cdot), p, r^-\}$. Let w_-^π be the total expected reward when r is replaced by r^-, and let $s_-(\cdot) = \sup_{\phi \in S} w_-^{\phi}(\cdot)$. By Lemma 11 we have $s \geq s_- \geq w_-^{\sigma}$. By the last inequalities and (35) we have

$$w^{\sigma}(x) = w^{\sigma}(x,n,w^{\sigma}) \leq w^{\sigma}(x,n,w_-^{\sigma}) + \varepsilon/2 \leq w^{\sigma}(x,n,s) + \varepsilon/2. \tag{36}$$

Show that

$$w^{\sigma}(z,n,s) \leq s(z) , \qquad z \in X . \tag{37}$$

From the definition of T^{σ} and Proposition 2 we have

$$w^{\sigma}(z,1,s) = T^{\sigma}s(z) \leq Ts(z) = s(z) , \qquad z \in X .$$

For some $i = 1,2,\ldots,$ let

$$w^{\sigma}(z,i,s) = (T^{\sigma})^i s(z) \leq s(z) , \qquad z \in X .$$

Then

$$w^{\sigma}(z,i+1,s) = T^{\sigma}(T^{\sigma})^i s(z) \leq T^{\sigma}s(z) \leq s(z) , \qquad z \in X .$$

Thus (37) is proved. From (37), (36) and (34) we have

$$s(x) \geq w^{\sigma}(x,n,s) \geq w^{\sigma}(x) - \varepsilon/2 \geq v_{RS}(x) - \varepsilon .$$

Since $\varepsilon > 0$ is arbitrary, Theorem 3 is proved. ///

Note that Theorem 3 and Lemma 8 (i) imply that Lemma 8 (i) remains valid for $\phi \in RS$.

4. PERSISTENTLY NEARLY OPTIMAL (f,B)-GENERATED STRATEGIES

In this section we extend Theorem 1 and Proposition 2 to some classes of nonstationary strategies in general (non-Markov) models.

Let B be a countable (or finite) set and let $f:H \to B$ be a measurable function. Denote $H^{x,b} = \{h_n \in H : f(h_n) = b,\ x_n = x,\ n = 0,1,\ldots\}$, where $(x,b) \in X \times B$, and $Z = \{(x,b) \in X \times B : H^{x,b} \neq \emptyset\}$.

<u>DEFINITION 1</u> ([6]). A non-randomized strategy ϕ is called (f,B)-generated if there exists a function $\tilde{\phi}: Z \to A$ such that $\phi(h_n) = \tilde{\phi}(x_n, f(h_n))$ for any $h_n \in H$. A strategy is called randomized (f,B)-generated if there exist measures $\tilde{\pi}(\cdot \mid x,b)$, $(x,b) \in Z$, on A such that $\pi(\cdot \mid h_n) = \tilde{\pi}(\cdot \mid x_n, f(h_n))$ for any $h_n \in H$.

For given f and B we denote B^f and RB^f the sets of all (f,B)-generated and randomized (f,B)-generated strategies, respectively. Note that if $\Delta = B^f$ for some $\Delta \subseteq \Pi$, then we will sometimes write $R\Delta$ instead of RB^f.

<u>Condition (A)</u>. (f and B generate a transitive statistic, cf. Shiryaev [16, Chapter II, Section 15]). If $x_n = x'_m$ and $f(h_n) = f(h'_m)$ for some h_n, $h'_m \in H$, $n,m = 0,1,\ldots,$ then

$$A(h_n) = A(h'_m) \ , \tag{38}$$

$$f(h_n az) = f(h'_m az) \ , \qquad a \in A(h_n) \ , \qquad z \in X \ , \tag{39}$$

$$p(z \mid h_n a) = p(z \mid h'_m a) \ , \qquad a \in A(h_n) \ , \qquad z \in X \ , \tag{40}$$

$$r(h_n a) = r(h'_m a) \ , \qquad a \in A(h_n) \tag{41}$$

(Note that for Markov models the equalities (38), (40) and (41) hold when $x_n = x'_m$).

Let us consider some examples of (f,B)-generated strategies satisfying condition (A).

EXAMPLE 1. Let A be countable. Setting $B = H$ and $f(h) = h$, we have that B^f is a set of all non-randomized strategies.

In the following examples, the model is assumed to be Markov and condition (A) is reduced to (39).

EXAMPLE 2. Stationary strategies. Let $B = \emptyset$ and $f(h) = \emptyset$ for any h. Then $S = B^f$.

EXAMPLE 3. Markov strategies. Let $B = \{0,1,\ldots\}$ and $f(h_n) = n$ for any $h_n \in H_n$, $n = 0,1,\ldots$. Then $M = B^f$.

EXAMPLE 4. Let $Y \subseteq X$. Consider the class Δ of all non-randomzied strategies ϕ such that $\phi(h_n) = \tilde{\phi}(x_n, m_Y(h_n))$, where $m_Y(h_n) = \sum_{i=0}^{n} I\{x_i \in Y\}$, $h_n \in H$, $n = 0,1,\ldots$. Let $B = \{0,1,\ldots\}$ and $f(h_n) = m_Y(h_n)$. Then $\Delta = B^f$.

EXAMPLE 5. Non-randomized strategies of the form $\phi(h_n) = \phi(x_o\cdots x_n)$. In this case $B = \bigcup_{n=1}^{\infty} X^n$ and $f(h_n) = (x_o x_1 \cdots x_n)$.

Tracking strategies (see [10], [24], [6], [18]) are an example of (f,B)-generated strategies which do not satisfy the condition (A).

THEOREM 4. Let $\Delta = B^f$, where f and B satisfy condition (A). Then for any $\varepsilon > 0$ and any function $\ell \in L_o(d_\Delta, H_\Delta)$ satisfying for any $(x,b) \in Z$ the condition (we set $\inf\{\emptyset\} > 0$)

$$\inf\{\ell(h) : h \in H^{x,b} \setminus H_\Delta\} > 0 \quad , \tag{42}$$

there exists a strategy $\phi \in \Delta$ such that for any $h \in H$

$$w^\phi(h) \geq v_\Delta(h) - \varepsilon\ell(h) \quad .$$

<u>THEOREM 5</u>. Let $\Delta = B^f$, where f and B satisfy condition (A). Then

• (i) $v_\Delta(h) = Tv_\Delta(h)$, $h \in H$,

• (ii) $v_\Delta = v_{R\Delta}$.

To prove Theorems 4 and 5 we consider the Markov model $\tilde{\mu} = \{\tilde{X}, \tilde{A}(\cdot), \tilde{p}, \tilde{r}\}$, where

• (i) $\tilde{X} = Z$,

• (ii) $\tilde{A}(x,b) = A(h)$ for $h \in H^{x,b}$,

• (iii) $\tilde{p}(x',b' \mid x,b,a) = \tilde{p}(x' \mid x,b,a) I_{\{b'\}} (\tilde{f}(x,b,a,x'))$,

$$(43)$$

where $\tilde{f}(x,b,a,z) = f(haz)$, $\tilde{p}(z \mid xba) = p(z \mid ha)$, $h \in H^{x,b}$,

• (iv) $r(xba) = r(ha)$, $h \in H^{x,b}$, where $(x,b), (x',b') \in \tilde{X}$, $a \in \tilde{A}(x,b)$.

The model $\tilde{\mu}$ is the <u>evolute</u> of the initial model N. As a rule, all the objects in the model $\tilde{\mu}$ will be labeled with ~. For example, $\tilde{S}$ is the set of stationary strategies in the model $\tilde{\mu}$.

Any strategy $\pi \in RB^f$ is defined by the conditional distributions $\tilde{\pi}(\cdot|x,b)$, where $(x,b) \in \tilde{X}$. So there are one-to-one correspondences between RB^f and $R\tilde{S}$, B^f and $\tilde{S}$.

For any $\pi \in RB^f$ we denote by $\tilde{\pi}$ the strategy from $\tilde{S}$ defined by transition probabilities $\tilde{\pi}(\cdot \mid \cdot)$. The strategy $\tilde{\pi} \in \tilde{S}$ defines measures $\tilde{P}^{\tilde{\pi}}_{x,b}$ on $(\tilde{H}_\infty, \tilde{F}_\infty) = ((\tilde{X} \times A)^\infty, (2^{\tilde{X}} \times \mathcal{A})^\infty)$. These measures are defined by the following relations:

$$\tilde{P}^{\tilde{\pi}}_{x,b}(\tilde{x}_o = x,\ \tilde{b}_o = b) = 1 \ , \tag{44}$$

$$\tilde{P}^{\tilde{\pi}}_{x,b}(\tilde{a}_n \in A' \mid \tilde{x}_o \tilde{b}_o \tilde{a}_o \cdots \tilde{x}_n \tilde{b}_n) = \tilde{\pi}(A' \mid \tilde{x}_n \tilde{b}_n) \ , \tag{45}$$

$$\tilde{P}^{\tilde{\pi}}_{x,b}(\tilde{x}_{n+1} = x',\ \tilde{b}_{n+1} = b' \mid \tilde{x}_o\tilde{b}_o\tilde{a}_o\cdots\tilde{x}_n\tilde{b}_n\tilde{a}_n)$$
$$= \tilde{p}(x' \mid \tilde{x}_n\tilde{b}_n\tilde{a}_n)\ I_{\{b'\}}\ (\tilde{f}(\tilde{x}_n\tilde{b}_n\tilde{a}_n x')) \quad . \tag{46}$$

For any $h' \in H$ and $\pi \in \Pi$ we extend measures $P^{\pi}_{h'}$ from $(H_\infty, \mathcal{F}_\infty)$ to $(\tilde{H}_\infty, \tilde{\mathcal{F}}_\infty)$ by using the transition probabilities

$$P^{\pi}_{h'}(b_n \mid x_o b_o a_o x_1 \cdots x_n) = I_{\{b_n\}}(f(x_o a_o x_1 \cdots x_n)) \ , \tag{47}$$
$$n = 0,1,\ldots \ .$$

<u>LEMMA 12</u>. Let $\pi \in RB^f$, where f and B satisfy condition (A). Then for any $h' = (x'_o a'_o \cdots x'_m) \in H_m$, $m = 0,1,\ldots$ and $\tilde{C} \in \tilde{\mathcal{F}}$

$$P^{\pi}_{h'}(x_m b_m a_m x_{m+1} \cdots \in \tilde{C}) = \tilde{P}^{\tilde{\pi}}_{x,b}(\tilde{x}_o \tilde{b}_o \tilde{a}_o \tilde{x}_1 \cdots \in \tilde{C}) \ , \tag{48}$$

where $x = x'_m$, $b = f(h'_m)$.

<u>Proof</u>. The measure $\tilde{P}^{\tilde{\pi}}_{x,b}$ is defined by the relations (44) - (46). Check that the measure from the left-hand side of (48) is defined by the same initial and transition probabilities. By (1) and (47) we have $P^{\pi}_{h'}(x_m = x,\ b_m = b) = 1$. By (2) and (45) we have ($P^{\pi}_{h'}$ - a.s.) for $A' \subseteq \tilde{A} = A$:

$$P^{\pi}_{h'}(a_{m+n} \in A' \mid x_o b_o a_o \cdots x_{m+n} b_{m+n})$$
$$= \tilde{\pi}(A' \mid x_{m+n},\ f(x_o a_o \cdots x_{m+n}))$$
$$= \tilde{\pi}(A' \mid x_{m+n},\ b_{m+n}) \quad .$$

Therefore

$$P^{\pi}_{h'}(a_{m+n} \in A' \mid x_m b_m a_m \cdots x_{m+n} b_{m+n}) = \tilde{\pi}(A' \mid x_{m+n},\ b_{m+n}) \ .$$

By (47) and (3) we have (P^{π}_{h}' - a.s.)

$$P_{h'}^{\pi}(x_{m+n+1} = x',\ b_{m+n+1} = b' \mid x_o b_o a_o \cdots x_{m+n} b_{m+n} a_{m+n})$$

$$= P_{h'}^{\pi}(x_{m+n+1} = x' \mid x_o a_o \cdots x_{m+n} a_{m+n}) I_{\{b'\}}(\tilde{f}(x_{m+n} b_{m+n} a_{m+n} x'))$$

$$= p(x' \mid x_{m+n} b_{m+n} a_{m+n})\ I_{\{b'\}}(\tilde{f}(x_{m+n} b_{m+n} a_{m+n} x')).$$

Therefore

$$P_{h'}^{\pi}(x_{m+n+1} = x',\ b_{m+n+1} = b' \mid x_m b_m a_m \cdots x_{m+n} b_{m+n} a_{m+n})$$

$$= \tilde{p}(x' \mid x_{m+n} b_{m+n} a_{m+n})\ I_{\{b'\}}\ (\tilde{f}(x_{m+n} b_{m+n} a_{m+n} x'))\quad . \quad ///$$

<u>LEMMA 13</u>. If f and B satisfy condition (A), then for any $h_m \in H$, $m = 0,1,\ldots$

- (i) $w^{\pi}(h_m) = \tilde{w}^{\tilde{\pi}}(x_m,\ f(h_m)),\quad \pi \in RB^f,$
- (ii) $v_{B^f}(h_m) = \tilde{s}(x_m,\ f(h_m)),$
- (iii) $v_{RB^f}(h_m) = \tilde{v}_{R\tilde{S}}(x_m,\ f(h_m)),$
- (iv) $d_{B^f}(h_m) = \tilde{d}_{\tilde{S}}(x_m, f(h_m)),$
- (v) $d_{RB^f}(h_m) = \tilde{d}_{R\tilde{S}}(x_m, f(h_m)).$

<u>Proof</u>. Lemma 12 implies (i). By (i) and the one-to-one correspondence between B^f and $\tilde{S}$, RB^f and $R\tilde{S}$ we get (ii) and (iii). If $\tilde{\tau}$ is a Markov time defined in $(\tilde{H}_{\infty}, \tilde{\mathcal{F}}_{\infty})$, then for

$$\tau(x_o a_o x_1 a_1 \cdots) = \tilde{\tau}(x_o\ f(x_o)\ a_o\ x_1\ f(x_o f(x_o) a_o x_1)\ a_1 \ldots)$$

we have $\tau = \tilde{\tau}$ $(P_h^{\pi}$ - a.s.) for any $h \in H$ and $\pi \in RB^f$. By this remark, (ii), (iii) and Lemma 12 we have (iv), (v).

<u>LEMMA 14</u>. Let f and B satisfy condition (A) and let $\Delta = B^f$. If $\ell \in L_o(d_{\Delta}, H_{\Delta})$, then $\tilde{\ell} \in \tilde{L}_o(\tilde{d}_{\tilde{S}}, \tilde{H}_{\tilde{S}})$, where

$$\tilde{\ell}(x,b) \;=\; \inf\,\{\ell(h):\ h \in H^{x,b}\}\,, \qquad (x,b) \in \tilde{X}\,.$$

Proof. Lemma 13 (i, ii) implies $H_\Delta = \bigcup_{(x,b)\in\tilde{X}_{\tilde{S}}} H^{x,b}$. Fix some $(x,b) \in \tilde{X} \setminus \tilde{X}_{\tilde{S}}$. Then Lemma 13 (iv) implies $\tilde{\ell}(x,b) \geq \tilde{d}_{\tilde{S}}(x,b)$. The condition (42) implies $\tilde{\ell}(x,b) > 0$.

It remains to show that $\tilde{\ell}(x,b) \geq \tilde{P}\tilde{\ell}(x,b)$. For $\varepsilon > 0$ we choose $h \in H^{x,b}$ such that $\tilde{\ell}(x,b) \geq \ell(h) - \varepsilon$. Then by the definitions of $\tilde{\ell}$ and $\tilde{P}$ we have

$$\tilde{\ell}(x,b) \;\geq\; \ell(h) - \varepsilon \;\geq\; P\ell(h) - \varepsilon \;\geq\; \tilde{P}\tilde{\ell}(x,b) - \varepsilon\,.$$

Since $\varepsilon > 0$ is arbitrary, $\tilde{\ell}(x,b) \geq \tilde{P}\tilde{\ell}(x,b)$. ///

Proof of Theorem 4. Theorem 1 and Lemma 14 imply the existence of $\tilde{\phi} \in \tilde{S}$ such that $\tilde{w}^{\tilde{\phi}}(x,b) \geq \tilde{s}(x,b) - \varepsilon\tilde{\ell}(x,b)$ for $(x,b) \in \tilde{X}$ and $\tilde{w}^{\tilde{\phi}}(x,b) = \tilde{s}(x,b)$ for $(x,b) \in \tilde{X}_{\tilde{S}}$. By Lemmas 13 and 14 we have $w^{\phi}(h_n) = v_{B^f}(h_n)$ for $h_n \in H_\Delta$ and for $h_n \in H \setminus H_\Delta$

$$\begin{aligned} w^{\phi}(h_n) \;&=\; \tilde{w}^{\tilde{\phi}}(x_n,\, f(h_n)) \;\geq\; \tilde{s}(x_n,\, f(h_n)) - \varepsilon\tilde{\ell}(x_n,\, f(h_n)) \\ &\geq\; v_{B^f}(h_n) - \varepsilon\ell(h_n)\,. \end{aligned}$$ ///

Proof of Theorem 5. •(i). Proposition 2 implies $\tilde{s} = \tilde{T}\tilde{s}$. By Lemma 13 (ii) and condition (A) we have for any $h_n \in H$, $n = 0,1,\ldots$

$$\begin{aligned} v_\Delta(h_n) \;&=\; \tilde{s}(x_n,\, f(h_n)) \;=\; \tilde{T}\tilde{s}(x_n,\, f(h_n)) \\ &=\; Tv_\Delta(h_n)\,. \end{aligned}$$

•(ii) follows from Lemma 13 (ii, iii) and Theorem 3. ///

Consider some corollaries from Theorem 4. Let $N\Pi$ be the set of all non-randomized strategies. If $(A,\mathcal{A})$ is a standard

Borel space, then for each $h \in H$ and $\pi \in \Pi$ there exists $\phi \in N\Pi$ such that $w^{\phi}(h) \geq w^{\pi}(h)$ ([12]; [9, Theorem 2,1], [4] - [5]). Hence $H_{N\Pi} = H_{\Pi}$ and $v_{N\Pi} = v$. By Theorem 4 and Example 1 we have the following corollary.

COROLLARY 2. If A is countable, then for any $\varepsilon > 0$ and any $\ell \in L_o(d_{N\Pi}, H_{\Pi})$ there exists $\phi \in N\Pi$ such that

$$w^{\phi}(h) > v(h) - \varepsilon\ell(h), \qquad h \in H .$$

In the sequel, we will consider the Markov model μ.

Let $\pi \in \Pi$ and $h'_n = (x'_o a'_o \cdots x'_n) \in H$, $n = 0,1,\ldots$. Let $h'_n[\pi]$ denote the strategy σ defined by

$$\sigma_i(\cdot \mid h_i) = \pi_{n+i}(\cdot \mid x'_o a'_o \cdots x'_{n-1} a'_{n-1} x_o a_o \cdots x_i)$$

for any $h_i = x_o a_o \cdots x_i \in H_i$, $i = 0,1,\ldots$. Note that

$$w^{\pi}(h_n) = w^{h_n[\pi]}(x_n) , \qquad h_n \in H , \quad n = 0,1,\ldots .$$

For $\Delta \subseteq \Pi$ we consider the following condition.

Condition (B) (Invariance condition). $h[\Delta] = \Delta$ for any $h \in H$.

Note that the classes of strategies described in Examples 2 - 5 satisfy condition (B).

If some class $\Delta \subseteq \Pi$ satisfies condition (B), then $v_{\Delta}(h_n) = \sup_{\Delta} w^{\pi}(x_n) = v_{\Delta}(x_n)$. If $\ell \in L_o(d_{\Delta}, X_{\Delta})$, then the function $\ell(h_n) = \ell(x_n)$ belongs to $L_o(d_{\Delta}, H_{\Delta})$ and satisfies the condition (42). Hence for Markov models Theorem 4.1 implies the following result.

COROLLARY 3. Let f and B satisfy condition (A) and let $\Delta = B^f$ satisfy condition (B). Then for any $\varepsilon > 0$ and any

$\ell \in L_o(d_\Delta, X_\Delta)$ there exists $\phi \in \Delta$ such that for any $h \in H_n$, $i = 0,1,\ldots$

$$w^\phi(h_n) \geq v_\Delta(x_n) - \varepsilon\ell(x_n) \quad .$$

Let $\phi = \{\phi_o, \phi_1, \ldots\} \in M$. Denote $n\phi = \{\phi_n, \phi_{n+1}, \ldots\}$, $n = 0,1,\ldots$. For each $h_n \in H$, $n = 0,1,\ldots$ we have $h_n[\phi] = n\phi$ and $w^\phi(h_n) = w^{n\phi}(x_n)$. For Markov strategies, Corollary 3 and the equalities $v_M = v$ (see [4], [6]) imply the following result generalizing [22], [7, Theorem 2.6]; [17].

COROLLARY 4. For any $\varepsilon > 0$ and any $\ell \in L_o(d_M, X_M)$ there exists a Markov strategy ϕ such that $W^{n\phi}(x) > v(x) - \varepsilon\ell(x)$ for any $n = 0,1,\ldots,$ $x \in X$.

In conclusion, we give some examples (models to be considered are Markov).

EXAMPLE 6. (The condition (42) is violated and the statement of Theorem 4 is not valid.) Let $\Delta = M$, $X = \{x^1, x^2, \ldots\}$, $x^1 \notin X_M$ and $v(x) \leq 0$ for any $x \in X$. Let $\ell(h_n) = \ell(x_o, x_n) = 1/i$ when $x_o = x^i$. Since $v \leq 0$, $\ell(h) = \max(P\ell(h), v(h))$, $h \in H$. But ℓ does not satisfy the condition (42). Fix $n = 1,2,\ldots$ and $\varepsilon = 1$. If the statement of Theorem 4 is valid, then $w^{n\phi}(x^1) \geq v(x^1) - 1/i$ for some $\phi \in M$ and any $n, i = 1,2,\ldots$. Therefore $w^{n\phi}(x^1) = v(x^1)$, $n = 1,2,\ldots$. This contradicts $x^1 \notin X_M$. ///

EXAMPLE 7. (f and B satisfy condition (A), $\Delta = B^f$ does not satisfy condition (B) and $v_\Delta(h_n) \neq v_\Delta(x_n)$.) Let $X = X^1 \cup X^2$ and let Δ be the class of strategies being Markov if $x_o \in X^1$ and stationary if $x_o \in X^2$. Let $B = \{-1,0,1,2,\ldots\}$. Consider

$f(h_n) = n$ if $x_o \in X^1$ and $f(h_n) = -1$ if $x_o \in X^2$. Then f and B satisfy condition (A) and $\Delta = B^f$ does not satisfy condition (B).

Let $v(x) \neq s(x)$ for some $x \in X$ (see [24, Example 3.5]). Then for $h_n \in H$, $n = 1,2,\ldots$ such that $x_n = x$, we have $v_\Delta(h_n) = v(x)$ if $x_o \in X^1$ and $v_\Delta(h_n) = s(x)$ if $x_o \in X^2$. ///

If $v \leq 0$ then for any sequence of positive numbers ε_i there exists $\phi \in M$ such that $w^{n\phi}(x) \geq v(x) - \varepsilon_n$ for any $n = 0,1,\ldots$ and $x \in X$. (The proof of this result is similar to the proof of Theorem 8.1 in [19]). The following example shows that when $r \geqslant 0$ the similar result does not hold. Hence in Corollary 4 the constant ε cannot be replaced by an arbitrary sequence of positive numbers.

EXAMPLE 8. Let $X = \{x^0, x^1, \ldots\}$ and $A = \{1,2\}$. The state x^0 is absorbing and $A(x^i) = A$, $i = 1,2,\ldots$. Let $r(x^i,1) = 0$, $r(x^i,2) = 1 - 1/i$ and $p(x^{i+1} \mid x^i 1) = p(x^0 \mid x^i 2) = 1$, $i = 1,2,\ldots$ Then $v(x^i) = 1$ for $i = 1,2,\ldots,$ but there is no $\phi \in M$ such that $w^{n\phi}(x^n) \geq 1 - 1/2n$, $n = 1,2,\ldots$.

REFERENCES

[1] Blackwell, D. "Positive Dynamic Programming." *Proc. 5th Berkeley Symposium Mathem. Statist. and Probab.*, vol.1 (1967): 719-726.

[2] Van der Dawen, R., and Schäl, M. "On the Existence of Stationary Optimal Policies in Markov Decision Models." *Z. Angew. Math. Mech.*, vol.63, no.5 (1983): T403-T404.

[3] Dynkin, E.B., and Yushkevich. A.A. *Controlled Markov Processes*. Berlin Heidelberg New York: Springer-Verlag Inc., 1979.

[4] Fainberg, E.A. "Nonrandomized Markov and Semi-Markov Strategies in Dynamic Programming." *Theory Probab. Applications*. vol.27, no.1 (1982): 116-126.

[5] Fainberg, E.A. "Controlled Markov Processes with Arbitrary Numerical Criteria." *Theory Probab. Applications*, vol.27, no.3 (1982): 486-503.

[6] Fainberg, E.A. "On Some Classes of Policies in Dynamic Programming." *Theory Probab. Applications* (to appear).

[7] Fainberg, E.A., and Sonin, I.M. "Stationary and Markov Policies in Countable State Dynamic Programming." *Lecture Notes in Math.*, pp. 111-129, vol.1021. Berlin Heidelberg New York: Springer-Verlag Inc., 1983.

[8] Frid, E.B. "On a Problem of D. Blackwell from the Theory of Dynamic Programming." *Theory Probab. Applications*, vol.15, no.3 (1970): 719-722.

[9] Gihman, I.I., and Skorokhod, A.V. *Controlled Stochastic Processes*. Berlin Heidelberg New York: Springer-Verlag Inc., 1979.

[10] Hill, Th.P. "On the Existence of Good Markov Strategies." *Transactions Amer. Math. Soc.*,vol.247 (1979): 157-176.

[11] Hordijk, A. *Dynamic Programming and Markov Potential Theory*. Mathl. Centre tracts, vol.51. Amsterdam: Mathematisch Centrum, 1974.

[12] Krylov, N.V. "Construction of an Optimal Strategy for a Finite Controlled Chain." *Theory Probab. Applications*, vol.10, no.1 (1965): 45-54.

[13] Van Nunen, J., and Wessels, J. *Markov Decision Processes with Unbounded Rewards. Markov Decision Theory*. Mathl. Centre tracts, vol.93. Amsterdam: Mathematisch Centrum, 1976.

[14] Ornstein, D. "On the Existence of Stationary Optimal Strategies." *Proc. Amer. Math. Soc.*, vol.20 (1969): 563-569.

[15] Shiryaev, A.N. "Some New Results in the Theory of Controlled Random Processes." *Selected Translations in Mathematical Statistics and Probability*, vol.8 (1970): 49-130. (Translated from *Trans. 4th Prague Conf. on Information Theory*, 1965.)

[16] Shiryaev, A.N. *Optimal Stopping Rules*. Berlin Heidelberg New York: Springer-Verlag Inc., 1978.

[17] Sonin, I.M. "The Existence of Persistently Nearly Optimal Markov Policies in Markov Decision Chains with Countable State Space." In *Modeli i metody v stokhasticheskoi optimizatsii* (Models and Methods in Stochastic Optimization). Moskva: Tsentr. Econ.-Matem. Institut Akademii Nauk SSSR, 1984.

[18] Sonin, I.M., and Fainberg, E.A. "Sufficient Classes of Strategies in Controllable Countable Markov Chains with Total Criterion." *Soviet Math. Doklady*, vol.29, no.2 (1984): 308-311.

[19] Strauch, R. "Negative Dynamic Programming." *Ann. Math. Stat.*, vol.37 (1966): 871-889.

[20] Viskov, O.V., and Shiryaev, A.N. "On Controls Leading to Optimal Stationary Policies." *Trudy Matemat. Instit. imeni Steklova Akademii Nauk SSSR*, vol.71 (1964): 35-45.

[21] Van der Wal, J. *Stochastic Dynamic Programming*. Mathl. Centre tracts, vol.139. Amsterdam: Mathematisch Centrum, 1974.

[22] Van der Wal, J. *On Stationary Strategies*.Eindhoven, the Netherlands: University of Technology. Memorandum-COSOR 81-14, 1981.

[23] Van der Wal, J. *On Uniformly Nearly Optimal Markov Strategies*. Eindhoven, the Netherlands: University of Technology. Memorandum-COSOR 81-16, 1981.

[24] Van der Wal, J., and Wessels, J. *On the Use of Information in Markov Decision Processes*. Eindhoven, the Netherlands: University of Technology. Memorandum-COSOR 81-20, 1981.

[25] Yushkevich, A.A., and Chitashvili, R.Ya. "Controlled Stochastic Sequence and Markov Chains." *Russian Math. Surveys*, vol.37, no.6 (1982): 239-274.

L.I. GAL'CHUK

GAUSSIAN SEMIMARTINGALES

INTRODUCTION

Gaussian processes and semimartingales are two classes of well-studied processes. Therefore, an investigation of necessary and sufficient conditions in terms of a mean and covariance for a Gaussian process to be a quasimartingale, a semimartingale, or a process of bounded variation is an interesting and natural problem. Apparently, Professor A.N. Shiryaev was the first who formulated this problem. The problem has been studied also by N.C. Jain and D. Monrad in [1] and by M. Emery in [2]. Some related results have been obtained by C. Stricker in [3]. [1] presents a necessary and sufficient condition for a Gaussian quasimartingale to be decomposed into a sum of a martingale and a process of bounded variation, which comprise a joint Gaussian process. [2] presents a necessary and sufficient condition for a Gaussian process to be a quasimartingale or a semimartingale.

Unlike [2] - [3], we do not assume that the processes under consideration are right continuous. We prove that the class of Gaussian semimartingales coincides with the class of Gaussian quasimartingales. We also give a new necessary and sufficient condition for a Gaussian process to be a semimartingale.

1. AUXILIARY RESULTS

Let (Ω, F, P) be a complete probability space with filtration (F_t), $t \in R_+$. By M_{loc}, V, A_{loc} we denote the space of local martingales of processes with finite variation and processes with locally integrable variation, respectively. By T and T_+ we also denote the families of stopping times with respect to (F_t) and (F_{t+}). For the notions used, see [4].

DEFINITION 1. An (F_t)-optional process $X = (X_t)$, $t \in R_+$, is called a semimartingale (we write $X \in S$) if there exists a decomposition

$$X = M + A , \qquad M \in M_{loc} , \qquad A \in V . \quad (1)$$

For a process X we define

$$V_t(X, \pi_n) = \sum_{i=1}^{n} \left| E\left[X_{t_i} - X_{t_{i-1}} \mid F_{t_{i-1}}\right]\right| + \left|X_{t_0}\right| ,$$

$$V_t(X) = \sup_{\pi_n} E\, V_t(X, \pi_n) ,$$

where $\pi_n = (0 = t_0 < t_1 < \cdots < t_n = t)$ is a partition of $[0,t]$ and sup is over all finite partitions.

DEFINITION 2. An (F_t)-optional process $X = (X_t)$, $t \in R_+$ is called a quasimartingale $(X \in Q)$ if $V_t(X) < \infty$ for any $t \in R_+$.

Suppose $X \in S$. There exist two sequences $(S_n) \in T$, $(U_n) \in T_+$ involving the jumps of X and $P(U_n = T < \infty) = 0$ for any $n \in \mathbb{N}$, $T \in \mathcal{T}$ (see [4]). There also exists a sequence $(T_n) \in \mathcal{T}$ which is a set of optional separability of X (see [5]). We define a sequence of partitions (σ_n), $n \in \mathbb{N}$,

$$\sigma_n = (0 = \tau_0 \le \tau_1 \le \cdots \le \tau_{2n+1} = t) ,$$

where τ_k, $k \le 2n$ are stopping times obtained from $T_1 \wedge t, \ldots, T_n \wedge t$, $S_1 \wedge t, \ldots, S_n \wedge t$ by rearranging the order

$$[\![\tau_0]\!] = [\![0]\!] ,$$

$$[\![\tau_{k+1}]\!] = \min\left([\![T_1 \wedge t]\!] \setminus \bigcup_{i=1}^{k} [\![\tau_i]\!], \ldots, [\![T_n \wedge t]\!] \setminus \bigcup_{i=1}^{k} [\![\tau_i]\!] ,\right.$$
$$\left. [\![S_1 \wedge t]\!] \setminus \bigcup_{i=1}^{k} [\![\tau_i]\!], \ldots, [\![S_n \wedge t]\!] \setminus \bigcup_{i=1}^{k} [\![\tau_i]\!]\right) .$$

Denote

$$\rho_{\sigma_n}(X) = \sum_{k=1}^{2n+1} \left(X_{\tau_k} - X_{\tau_{k-1}}\right)^2 . \tag{2}$$

THEOREM 1. Suppose $X \in \mathcal{S}$. Then

$$\rho_{\sigma_n}(X) \overset{P}{\to} [X,X]_t , \qquad n \to \infty ,$$

where

$$[X,X]_t = \langle X^c, X^c \rangle_t + \sum_{0<s\le t} (\Delta X_s)^2 + \sum_{0\le s<t} (\Delta^+ X_s)^2 ,$$

$$\Delta X_s = X_s - X_{s-} , \qquad \Delta^+ X_s = X_{s+} - X_s .$$

At first we prove an auxiliary result.

LEMMA 1. Suppose $X \in \mathcal{S}$, $X = M + A$, $M \in \mathcal{M}^2$, $A \in \mathcal{A}_{loc}$, $E[\operatorname{var}_{[0,t]} A]^2 < \infty$,

$$\operatorname*{var}_{[0,t]} A = \int_0^t |dA^r|_s + \sum_{0 \le s < t} |\Delta^+ A_s| ,$$

$$A_t^r = A_t - \sum_{0 \le s < t} \Delta^+ A_s .$$

Then $\rho_{\sigma_n}(X)$ are uniformly integrable with respect to n and converge to $[X,X]_t$ in L_1.

Proof. Omitting the index σ_n, we obtain $\rho(X) \le 2(\rho(M) + \rho(A))$ from (2). For $\rho(A)$ we have

$$\rho(A) \le \left(\max_k \left|A_{\tau_k} - A_{\tau_{k-1}}\right|\right) \cdot \sum_k \left|A_{\tau_k} - A_{\tau_{k-1}}\right| \le (\operatorname*{var}_{[0,t]} A)^2 .$$

Hence the family $\rho_{\sigma_n}(A)$ is uniformly integrable.

We denote the optional modification of the martingale $\left(E[M_\infty I_{|M_\infty| \le \lambda} \mid F_s]\right)$, $s \in R_+$ by $U = (U_s)$, $s \in R_+$, and assume that $V = M - U$. Then $\rho(M) \le 2(\rho(U) + \rho(V))$. For $\rho(V)$ we have

$$E\rho(V) \le EM_t^2 \le E|M_\infty|^2 \, I_{|M_\infty| > \lambda} .$$

Choosing λ sufficiently large, we obtain the inequality $E\rho(V) < \varepsilon$ for a given $\varepsilon > 0$. Now we shall prove that $\rho(U)$ is bounded in L_2. We have

$$E\rho^2(U) = E\left(\sum_{k=1}^{2n+1} \left(U_{\tau_k} - U_{\tau_{k-1}}\right)^2\right)^2$$

$$= E\left[\sum_{k=1}^{2n+1} \left(U_{\tau_k} - U_{\tau_{k+1}}\right)^4 + \right.$$

$$\left. + \; 2 \sum_{k=1}^{2n+1} \left(U_{\tau_k} - U_{\tau_{k-1}}\right)^2 \sum_{j>k} \left(U_{\tau_j} - U_{\tau_{j-1}}\right)^2 \right] .$$

Since $|U| \leq \lambda$, the first sum in the right-hand term is majorated by $4\lambda^2\ \Sigma(\cdot)^2$, and since

$$E\left|\sum_{j>k}\left(U_{\tau_j} - U_{\tau_{j-1}}\right)^2 \mid F_{\tau_k}\right| = E\left|\left(U_t - U_{\tau_{k+1}}\right)^2 \mid F_{\tau_k}\right| ,$$

the second sum is majorated by the same value. Hence

$$E\rho^2(U) \leq 12\lambda^2 E \sum_{k=1}^{2n+1}\left(U_{\tau_k} - U_{\tau_{k-1}}\right)^2 \leq 12E\lambda^2|M_\infty|^2\, I_{|M_\infty|\leq\lambda} < \infty .$$

Since L_1-norm of $\rho(V)$ may be arbitrarily small and L_2-norm of $\rho(U)$ is bounded, the family $\rho_{\sigma_n}(M)$ is uniformly integrable. For L_1-convergence of $\rho_{\sigma_n}(X)$ to $[X,X]_t$ it is sufficient to prove the convergence in probability by virtue of uniform integrability. We write X in the form

$$X = m + A + Z = Y + Z , \qquad Y = m + A ,$$

where the local martingale m consists of the continuous part and the sum of the first n compensated jumps of M. We have

$$\rho(X) - [X,X]_t = \rho(X) - \rho(Y) + [Y,Y]_t - [X,X]_t + \rho(Y) - [Y,Y]_t .$$

From the Kunita-Watanaba inequality (see [4]) we obtain

$$[X,X]_t - [Y,Y]_t = [2Y+Z,\ Z]_t = [2X-Z,\ Z]_t$$
$$\leq [2X-Z,\ 2X-Z]_t^{\frac{1}{2}} \cdot [Z,Z]_t^{\frac{1}{2}} .$$

In the right-hand term the first factor is bounded and the second one converges to zero in probability. By analogy we prove that $\rho(X) - \rho(Y) \to 0$ in probability.

We have to prove now that $\rho(Y) - [Y,Y]_t \to 0$. Since $Y = m + A$ and m has a finite number of jumps, we may include them in A and assume that m is continuous. We have

$$\rho(Y) - [Y,Y]_t = \rho(m) - [m,m]_t + 2\sum_k \left(m_{\tau_k} - m_{\tau_{k-1}}\right)\left(A_{\tau_k} - A_{\tau_{k-1}}\right)$$

$$+ \rho(A) - [A,A]_t \quad .$$

It is proved in [6] that $\rho_{\sigma_n}(m) - [m,m]_t \overset{P}{\to} 0$ for a continuous martingale m.

Since the process A has a finite variation, we see that $\rho_{\sigma_n}(A) \overset{P}{\to} [A,A]_t$. Further,

$$\left|\sum_k \left(m_{\tau_k} - m_{\tau_{k-1}}\right)\left(A_{\tau_k} - A_{\tau_{k-1}}\right)\right| \leq \max_k \left|m_{\tau_k} - m_{\tau_{k-1}}\right| \operatorname*{var}_{[0,t]} A \quad .$$

The right-hand term converges to zero in probability because $\operatorname{var} A < \infty$, $\max \left|m_{\tau_k} - m_{\tau_{k-1}}\right| \overset{P}{\to} 0$.

Proof of Theorem 1. Denote

$$T_\lambda = \inf\,(s: |\Delta X_s| \geq \lambda) \qquad S_\lambda = \inf\,(s: |\Delta^+ X_s| \geq \lambda) \quad ,$$

$$T_\lambda = \infty \quad , \qquad\qquad S_\lambda = \infty \quad ,$$

if the sets $(\cdot)$ are empty. It is easy to see that $T_\lambda \in \mathcal{T}$, $S_\lambda \in \mathcal{T}_+$. Taking into account the decomposition (see [4])

$$X = X^c + X^d + X^g = X^r + X^g \, , \qquad X^r = X^c + X^d \quad ,$$

we denote

$$\tilde{X} = X^r I_{[\![0,T_\lambda[\![} + X^r_{T_\lambda -} I_{[\![T_\lambda,\infty[\![} + X^g I_{[\![0,S_\lambda]\!]} + X^g_{S_\lambda} I_{]\!]S_\lambda,\infty[\![} \quad .$$

For large λ the processes X and $\tilde{X}$ differ by a small probability, so it is sufficient to prove the Theorem for the process $\tilde{X}$. The process $\tilde{X}$ is a special semimartingale because its jumps are bounded (see [7]). It has the decomposition $\tilde{X} = N + B$, $B \in A_{loc}$, B is strictly predictable, i.e., $B_+ = (B_{t+})$ is a (F_t)-optional process and B is predictable. In the equalities $\Delta\tilde{X}_T = \Delta N_T + \Delta B_T$, $T \in T$, T is predictable, $\Delta^+\tilde{X}_T = \Delta^+ N_T + \Delta^+ B_T$, $T \in T$, taking the conditional expectations with respect to F_{T-} and F_T respectively we obtain $E[\Delta\tilde{X}_T \mid F_{T-}] = \Delta B_T$ for a predictable stopping time T and $E[\Delta^+\tilde{X}_T \mid F_T] = \Delta^+ B_T$, $T \in T$. Hence the jumps of B are bounded. Then the jumps of N are bounded and $N \in M^2_{loc}$ (see [4]). Denote

$$T^r_\delta = \inf\left\{s: \int_0^s |dB^r|_u \geq \delta\right\} ,$$

$$T^g_\delta = \inf\left\{s: \sum_{0 \leq u < s} |\Delta^+ B_u| \geq \delta\right\} ,$$

$T_\delta = T^r_\delta \wedge T^g \in T$. We have $\operatorname*{var}_{[0,\infty[} B^{T_\delta} \leq \lambda + 2\delta$. Since $N^{T_\delta} \in M^2_{loc}$ there exist sequences $(R_k) \in T_+$ and $(Y^{(k)}) \in M^2$ such that in the stochastic interval $[\![0,R_k]\!]$ $N^{T_\delta} = Y^{(k)}$ (see [4]). Taking k so large that $P(R_k < t) < \varepsilon$ we obtain $\tilde{X}^{T_\delta} = Y^{(k)} + A^{T_\delta}$ in $[\![0,R_k]\!]$. This decomposition and Lemma yield the necessary result. ///

Let a Gaussian process $X = (X_t)$, $t \in R_+$, $(X \in G)$ be given. Further, (F_t) will be a filtration generated by X, i.e., $F_t = \sigma\{X_s, s \leq t\}$.

<u>LEMMA 2</u>. Let $X = (X_t) \in G$ be a separable measurable process having left-hand limits for $t > 0$ and right-hand limits for

$t \geq 0$ (làglàd). Then there exists a non-random countable set $D \subset R_+$ which contains all discontinuity times of X. At each point $t \in D$ at least one of the variable ΔX_t, $\Delta^+ X_t$ differs from zero a.s.

Proof. Denote $D = \{t: \lim_{s \to t} E(X_t - X_s)^2 \neq 0\}$. If D is empty the result follows from Itô and Nicio [8]. We assume that D is not empty and $t \in D$. The process X is a random variable with values in the space $(R^{R_+}, \mathcal{B}^{R_+})$ which is a space of all functions over R_+ with σ-algebra of cylindrical sets. In this space F_t will be a set of all functions $x = (x(s))$ làglàd for which $\Delta x(t)$ or $\Delta^+ x(t)$ differs from zero. F_t is a subspace. By virtue of [1] the mean $m = (m(s))$ belongs to F_t where $m(s) = EX_s$. Due to the 0-1 Law [9] $P(x - m \in F_t) = 1$ or 0. Hence at a point t the process X has jumps with probability 1. Since X is làglàd, D is at most countable. The function m has no discontinuity on the complement of D; further we may assume that X has a zero mean.

Denote

$$J = \{\omega, t : X_t(\omega) \neq X_{t-}(\omega)\} ,$$

$$J^+ = \{\omega, t : X_t(\omega) \neq X_{t+}(\omega)\} .$$

The sets J and J^+ are $\mathcal{B}(R_+) \times \mathcal{F}_\infty$ - measurable and sparse, i.e., each ω-section is at most countable. There exists at most a countable family of $\mathcal{F}_\infty$-measurable functions $f_i(\omega)$, $i \in \mathbb{N}$ the sum of whose graphics is equal to J. We shall assume that $P(f_i(\omega) = t_k,\ t_k \in D) = 0$ for any $i \in \mathbb{N}$. By virtue of Doob's theorem we may suppose that f_i is a function over $(R^{R_+}, \mathcal{B}^{R_+})$, i.e., $f_i(\omega) = f_i(X(\omega))$. We may assume that $P(\Delta X_{f_i(X)} \neq 0) > 0$ for some i. In $(R^{R_+}, \mathcal{B}^{R_+})$ we consider a collection F_{f_i} of functions $x = (x(t))$ for which $\Delta x_{f_i(x)} \neq 0$. F_{f_i} is a subspace. According to the 0-1 Law [9] $P(X \in F_{f_i(X)}) = 1$. There are times $s \leq t$, $s,t \notin D$ such that $P(X \in F_{f_i(X) \vee s \wedge t}) = 1$. This implies the contradiction because either $P(X \in F_s) = 1$ or $P(X \in F_t) = 1$ but $s,t \notin D$. Consequently $P(\Delta X_{f_i(X)} \neq 0) = 0$ for any i. In this manner for the set J^+ we obtain that X has a.s. no jumps into random times that are disjoint a.s. with the set D. ///

Let X be from the preceding lemma. $\tilde{X}$ will be an $(\mathcal{F}_t)$-optional projection of X (it exists and is a Gaussian process).

Suppose $t \in R_+$. In the interval $(0,t)$ we denote a set of times of the jumps of X by $D = \{t_k\}$, $k \in \mathbb{N}$ and a countable everywhere dense set without any point of D by $S = \{s_k\}$, $k \in \mathbb{N}$.

We introduce a sequence (π_n) of the partitions of the segment $[0,t]$ putting $\pi_n = (0 < t_1 < \cdots < t_n < t;\ 0 < s_1 < \cdots < s_N < t$, $t_i \in D$, $s_i \in S$, $i \in \mathbb{N})$. The number N of points s_i, $i = 1,\ldots,N$ provides at least one point of s_i between any points t_k, t_{k-1}, as well as on the left of t_1 and on the right of t_n. The points of $s_i \in \pi_n$ nearest to t_k on the left will be denoted by $s(i(k))$.

LEMMA 3. Nearly all trajectories of $\tilde{X}$ have limits on the right and on the left and jumps only in the set D.

Proof. Denote

$$X_u^{(n)} = X_0 I(u=0) + X_{0+} I(0<u\le s_1) + \sum_{m=1}^{i(1)-1} X_{s_m} I(s_m \le u < s_{m+1})$$
$$+ \sum_{k=1}^{n-1} \sum_{m=i(k)+1}^{i(k+1)-1} X_{s_m} I(s_m \le u < s_{m+1}) + \sum_{i=1}^{n} X_{s(i)} I(s(i) \le u < t_i)$$
$$+ \sum_{i=1}^{n} X_{t_i} I(u=t_i) + \sum_{i=1}^{n} X_{t_{i+}} I(t_i < u \le s(i)+1)$$
$$+ \sum_{m=i(n)+1}^{N} X_{s_m} I(s_m \le u < s_N) + X_{s_N} I(s_N \le u < t) + X_t I(u=t)$$
$$0 \le u \le t .$$

The process $\{X_u^{(n)}\}$, $u \in [0,t]$ is progressive-measurable and the sequences $X_u^{(n)}(\omega)$ converge to $X_u(\omega)$ for any ω, u. Consequently, X is a progressive-measurable process.

It follows from the definition of optional projection and X_T being F_T-measurable for any $T \in T$ for a progressive-measurable process X that

$$\tilde{X}_T = E[X_T \mid F_T] = X_T \quad \text{a.s. on } (T < \infty) . \tag{3}$$

The optional process $\tilde{X}$ is optional separable, i.e., there is a sequence $(T_n) \in T$ such that the graphic of $\tilde{X}$ is in the closure of a set of the values of random quantities $\tilde{X}_{T_n}$ (see [5]). It follows from (3) that outside some set Ω_0, $P(\Omega_0) = 0$

$$\tilde{X}_{T_n(\omega)}(\omega)\, I\,(T_n(\omega) < \infty) = X_{T_n(\omega)}(\omega)\, I\,(T_n(\omega) < \infty)$$

for all $n \in \mathbb{N}$, $\omega \notin \Omega_0$. This implies that outside Ω_0 the process $\tilde{X}$ has a limit on the left for $t > 0$ and a limit on the right for $t \ge 0$, which are coincident with the corresponding limits of X and the jumps of $\tilde{X}$ take place only in the set D. ///

THEOREM 2. Suppose $X \in GS$.

•a. There is $\varepsilon > 0$ such that

$$E \exp \{\varepsilon [X,X]_t\} < \infty .$$

•b. X is a special semimartingale with decomposition $X = m + A$, m is a square integrable martingale, A is strictly predictable with variation integrable along any segment, the pair (m,A) is a Gaussian process. Moreover

$$m_t^d = \sum_{t_i \le t} \left(\Delta X_{t_i} - E\left[\Delta X_{t_i} \mid \mathcal{F}_{t_i -}\right]\right) ,$$

$$m_t^g = \sum_{t_i < t} \left(\Delta^+ X_{t_i} - E\left[\Delta^+ X_{t_i} \mid \mathcal{F}_{t_i}\right]\right) ,$$

$$A_t^d = \sum_{t_i \le t} E\left[\Delta X_{t_i} \mid \mathcal{F}_{t_i -}\right] ,$$

$$A_t^g = \sum_{t_i < t} E\left[\Delta^+ X_{t_i} \mid \mathcal{F}_{t_i}\right] ,$$

where the last two series converge a.s. absolutely and the first two converge a.s.

•c. There is $\varepsilon > 0$ such that

$$E \exp \left\{\varepsilon [X,X]_t + \left(\operatorname*{var}_{[0,t]} A\right)^2\right\} < \infty .$$

<u>Proof</u>. •a. Let $(R^{[0,t]}, \mathcal{B}^{[0,t]})$ be a space of all functions on the segment $[0,t]$ with σ-algebra of cylindrical sets. Let (π_n) be a sequence of the partitions of the segment $[0,t]$ introduced before Lemma 3. By Theorem 2

$$\rho_{\pi_n}(X) = \sum_i \left(X_{\tau_i} - X_{\tau_{i-1}}\right)^2 \overset{P}{\to} [X,X]_t .$$

For some subsequence convergence with probability 1 will hold.

We denote $N^2(X) = \sup_{\pi_n} \rho_{\pi_n}(X)$. N is a function measurable with respect to $\mathcal{B}^{[0,t]}$. Now in $(R^{[0,t]}, \mathcal{B}^{[0,t]})$ we consider a collection F of functions $x = (x(s))$ with limits on the right and on the left for which $N(X) < \infty$. F is a subspace. According to [1] the mean $m = (m(s))$ of X belongs to F. Since $P(N(X-m) < \infty) > 0$, then according to [9] there is $\varepsilon' > 0$ such that $E \exp\{\varepsilon' N^2(X-m)\} < \infty$. Then $E \exp\{\varepsilon' [X-m, X-m]_t\} < \infty$. This implies $E \exp\{\varepsilon [X,X]_t\} < \infty$ for some $\varepsilon > 0$.

• b. As a result of the above discourse, $E[X,X]_t < \infty$ and consequently X is a special semimartingale (see [7]) with a unique decomposition $X = m + A$, where A is strictly predictable. It follows from the inequality $E[m,m]_t \le E[X,X]_t$ that m is a square integrable martingale.

Orthogonality of Gaussian martingales

$$\left\{\Delta X_{t_i} - E\left[\Delta X_{t_i} \mid \mathcal{F}_{t_i -}\right]\right\} I(t_i \le t) \quad , \qquad i = 1,2,\dots$$

and the last inequality gives convergence of the series in mean square

$$m_t^d = \sum_i \left\{\Delta X_{t_i} - E\left[\Delta X_{t_i} \mid \mathcal{F}_{t_i -}\right]\right\} I(t_i \le t) \quad .$$

Convergence also takes place with probability 1 because the summands are independent Gaussian quantities. This is true about the series

$$m_t^g = \sum_i \left\{\Delta^+ X_{t_i} - E\left[\Delta^+ X_{t_i} \mid \mathcal{F}_{t_i}\right]\right\} I(t_i \le t) \quad .$$

The processes m^d, m^g are square integrable martingales; m^d (m^g) is right continuous (left continuous) and has a limit on the left (right). Then since $\Delta A_{t_i} = E[\Delta X_{t_i} \mid F_{t_i -}]$, $\Delta^+ A_{t_i} = E[\Delta^+ X_{t_i} \mid F_{t_i}]$ the series

$$A_t^d = \sum_i E\left|\Delta X_{t_i} \mid F_{t_i -}\right| I(t_i \le t)$$

$$A_t^g = \sum_i E\left|\Delta^+ X_{t_i} \mid F_{t_i}\right| I(t_i < t)$$

converge a.s. absolutely. The process A^d (A^g) is right continuous (left continuous) and has a limit on the left (right). The process $X^c = X - (m^d + A^d) - (m^g + A^g)$ is a continuous Gaussian semimartingale with decomposition $X^c = m^c + A^c$, m^c is a continuous martingale, A^c is a continuous process of finite variation. The process $A^c + A^d + A^g$ is strictly predictable. It coincides with A because the special martingale has a unique decomposition. Then $m = m^c + m^d + m^g$. The collection $(X, m^d, m^g, A^d, A^g, m^c{+}A^c)$ is a Gaussian process by construction. It is established in [3] that the quantities

$$\sum_{\tau_i \in \pi_n} E\left|X^c_{\tau_i} - X^c_{\tau_{i-1}} \mid F_{\tau_{i-1}}\right|$$

converge in mean square to A_t^c. This implies that the collection $(X, m^c, m^d, m^g, A^c, A^d, A^g)$ is Gaussian at the partition points π_n. The existence of limits on the right and on the left implies that they are a Gaussian process.

• c. The inequality $E \exp\{\varepsilon(\operatorname{var}_{[0,t]} A)^2\} < \infty$ is established in [1]. We deduce the required inequality from this one and take first part of the theorem by the Cauchy-Buniakowski inequality.

COROLLARY 1. If $X \in G\mathcal{S}$, then $X \in G\mathcal{Q}$. This follows immediately from the Definition 2 if X is replaced by its decomposition $X = m + A$, where m is a martingale, A is a process with variation integrable along any segment.

2. THE MAIN RESULT

Denote for $s_i \le s_j$, $t_i \le t_j$

$$\delta_{i,j} = R(s_i,t_i) + R(s_j,t_j) - R(s_i,t_j) - R(s_j,t_i) ,$$

$$C_{r-1} = (\delta_{i,j})^{-1}_{0\le i,j\le r-1} ,$$

$$\delta_r = \begin{bmatrix} \delta_{r,0} \\ \delta_{r,1} \\ \vdots \\ \delta_{r,r-1} \end{bmatrix} ,$$

where $R(s,t)$ is a covariance function of X.

THEOREM 3. $X \in G\mathcal{Q} \iff$ •1) $m \in \mathcal{V}$;

•2) $\sup\limits_{\pi_n} \sum\limits_{r\le n} [\delta'_r C_{r-1} \delta_r]^{\frac{1}{2}} < \infty$,

where δ'_r is the transpose of δ_r.

Proof. $\Rightarrow$ •1) Suppose $X \in G\mathcal{Q}$, $\tilde{X}$ is an independent version of X. Then the process $1/\sqrt{2}\,(X-\tilde{X})$ and $X - m$ have the same distribution. Then $X - m \in G\mathcal{Q}$. Consequently $m = X - (X-m) \in G\mathcal{Q}$.

Now the definition of quasimartingale yields $m \in \mathcal{V}$.

•2) Assume that $m = 0$. Let $\pi_n = (0 = t_0 < \cdots < t_n = t)$ be a fixed partition. Denote $\Delta_i = X_{t_{i+1}} - X_{t_i}$, $i \geq 0$, $\Delta_{-1} = X_{t_0}$. By the normal correlation theorem (see [10]) we have

$$E[\Delta_r \mid \Delta_{-1}, \ldots, \Delta_{r-1}] = \delta_r' C_{r-1} \begin{pmatrix} \Delta_{-1} \\ \vdots \\ \Delta_{r-1} \end{pmatrix} \sim N(0, \sigma_r^2) ,$$

$$\sigma_r = \left(\frac{2}{\pi}\right)^{\frac{1}{2}} [\delta_r' C_{r-1} \delta_r]^{\frac{1}{2}} .$$

Further,

$$E\,|E[\Delta_r \mid \Delta_{-1}, \ldots, \Delta_{r-1}]| = \{E\,|E[\Delta_r \mid \Delta_{-1}, \ldots, \Delta_{r-1}]|^2\}^{\frac{1}{2}}$$
$$= [\delta_r' C_{r-1} \delta_r]^{\frac{1}{2}} .$$

Therefore

$$\sup_{\pi_n} \sum_{r \leq n} [\delta_r' C_{r-1} \delta_r]^{\frac{1}{2}} = \sup_{\pi_n} \sum_{r \leq n} E|E[\Delta_r \mid \Delta_{-1}, \ldots, \Delta_{r-1}]|$$

$$= \sup_{\pi_n} \sum_{r \leq n} E\left|E\left[M_{t_{r+1}} - M_{t_r} + A_{t_{r+1}} - A_{t_r} \mid \Delta_{-1}, \ldots, \Delta_{r-1}\right]\right|$$

$$= \sup_{\pi_n} \sum_{r \leq n} E\left|E\left[A_{t_{r+1}} - A_{t_r} \mid \Delta_{-1}, \ldots, \Delta_{r-1}\right]\right|$$

$$\leq \sup_{\pi_n} E \sum_{r \leq n} \left|A_{t_{r+1}} - A_{t_r}\right| < \infty .$$

$\Leftarrow$) Let $\pi_n = \{0 = t_n^0 \leq \cdots \leq t_n^n = t\}$ be some partition of $[0,t]$. Denote $X_t^n = X(t_k^n)$, $t_k^n \leq t < t_{k+1}^n$, $F^n(t) = \sigma\{X(t_j^n), t_j^n \leq t\}$.

Processes $X^n = ((X_t^n), (F_t^n))$ are semimartingales for any n. For each X^n we have a decomposition

$$X^n = M^n - A^n , \qquad M^n \in M_{loc} , \qquad A^n \in V,$$

where

$$M_t^n = \sum_{t_k^n \le t} \{X(t_k^n) - E[X(t_k^n) \mid \mathcal{F}^n(t_{k-1}^n)]\} ,$$

$$A_t^n = \sum_{t_k^n \le t} E[X(t_{k-1}^n) - X(t_k^n) \mid \mathcal{F}^n(t_{k-1}^n)] .$$

For fixed n for the variation of A^n we have

$$E \operatorname*{var}_{[0,t]} A^n = E \sum_{t_k^n \le t} |E[X(t_k^n) - X(t_{k-1}^n) \mid \mathcal{F}^n(t_{k-1}^n)]| .$$

Under condition 2) we have

$$E \operatorname*{var}_{[0,t]} A^n = \sup_{\pi_n} \sum_{r \le n} [\delta_r' C_{r-1} \delta_r]^{\frac{1}{2}} < \infty .$$

Consequently the sequence $\{A_t^n\}$, $n \in \mathbb{N}$, is bounded in L_1 for any t. Since the terms of the sequence are Gaussian, then the sequence is bounded in L_2. Then it converges weakly in L_2. Denote

$$A_t = \lim A_t^n \quad \text{(weakly in } L_2) .$$

As a result of weak convergence, A is Gaussian. From weak convergence and Fatou's lemma we have

$$E|A_t - A_s| = \liminf_{n\to\infty} E|A_t^n - A_s^n| \le B_t - B_s , \qquad s \le t ,$$

where

$$B_t = \sup_{\pi_n} E \sum_{t_k^n \le t} |E[X(t_k^n) - X(t_{k-1}^n) \mid \mathcal{F}^n(t_{k-1}^n)]| .$$

Consequently the process A has an integrable variation on any segment. Then A has right and left-hand limits (lådlåg).

Each process (A_t^n), $t \in R_+$, possesses the following properties: the variable A_t^n is $\mathcal{F}_{t-}$-measurable and A_{t+}^n is $\mathcal{F}_t$-measurable for any t. The same is true about the limit process A.

Denote

$$A_t^d = \sum_{0<t_i\le t} \Delta A_{t_i}, \qquad A_t^g = \sum_{0\le t_i<t} \Delta^+ A_{t_i},$$

where $\{t_i\}$, $i \in \mathbb{N}$ is a sequence of non-random times which include all jumps of A. The series on the right-hand side converge a.s. absolutely for any t. The process A^d is càdlàg, predictable, A^g is cáglád strictly predictable. The process $A^c = A - A^d - A^g$ is continuous. Hence the process A is strictly predictable.

Denote

$$M = X + A .$$

The process M is optional.

For any points $t_k^{n_o}$, $t_{k+1}^{n_o}$ from a partition π_{n_o} and for any bounded $\mathcal{F}^{n_o}\left(t_k^{n_o}\right)$ - measurable random variable ϕ we have due to weak convergence A^n to A that

$$E\left[M\left(t_{t+1}^{n_o}\right) - M\left(t_k^{n_o}\right)\right]\phi = E\left[X\left(t_{k+1}^{n_o}\right) - X\left(t_k^{n_o}\right) + A\left(t_{k+1}^{n_o}\right) - A\left(t_k^{n_o}\right)\right]\phi$$

$$= \lim_{n\to\infty} E\left[X\left(t_{k+1}^{n_o}\right) - X\left(t_k^{n_o}\right) + A^n\left(t_{k+1}^{n_o}\right) - A^n\left(t_k^{n_o}\right)\right]\phi$$

$$= \lim_{n\to\infty} E\left\{\left[X\left(t_{k+1}^{n_o}\right) - X\left(t_k^{n_o}\right)\right]\right.$$

$$\left. + \sum_{t_k^{n_o}\le t_j^n<t_{k+1}^{n_o}} \left(X(t_j^n) - E[X(t_{j+1}^n) \mid \mathcal{F}^n(t_j^n)]\right)\phi\right\}$$

$$= \lim_{n\to\infty} E \sum_{t_k^{n_o}\le t_j^n<t_{k+1}^{n_o}} [M^n(t_{j+1}^n) - M^n(t_j^n)]\phi = 0 .$$

By the theorem on monotone classes this equality holds for any $F(t_k^{n_0})$-measurable bounded variable ϕ.

Therefore M is a martingale over $\bigcup_n \pi_n$. Since M has limits on the left and on the right and it is sufficient to pass to the limit over $\bigcup_n \pi_n$, then it is easy to see that

$E(M_t - M_s)\phi = 0, \quad s \le t, \quad \phi \in F_s$. ///

Condition 2) of the Theorem is simpler when X is a Gauss-Markov process. In fact by virtue of the Markov property

$$[\delta_r' C_{r-1} \delta_r]^{\frac{1}{2}} = E\left|\bar{X}(t_{r-1}^n) - E[\bar{X}(t_r^n) \mid \bar{X}(t_{r-1}^n)]\right| ,$$

where $\bar{X}(t) = X_t - m_t$.

By the normal correlation theorem we have

$$[\delta_r' C_{r-1} \delta_r]^{\frac{1}{2}} = E\left|\left(1 - R\left(t_r^n, t_{r-1}^n\right) R^{-1}\left(t_{r-1}^n, t_{r-1}^n\right) \bar{X}(t_{r-1}^n)\right)\right|$$
$$= \left(\frac{2}{\pi}\right)^{\frac{1}{2}} \left|R\left(t_{r-1}^n, t_{r-1}^n\right) - R\left(t_r^n, t_r^n\right)\right| R^{-\frac{1}{2}}\left(t_{r-1}^n, t_{r-1}^n\right).$$

We formulate the last theorem for this case.

THEOREM 4. Let $X = (X_t)$, $t \in R_+$, be an optional Gauss-Markov process with mean m and covariance function $R(s,t)$. Then

$X \in Q \iff$ 1') $m \in V$;

2') $\sup_{\pi_n} \sum_r \left|R\left(t_{r-1}^n, t_{r-1}^n\right) - R\left(t_r^n, t_r^n\right)\right| R^{-\frac{1}{2}}\left(t_{r-1}^n, t_{r-1}^n\right) < \infty.$

A covariance function is called triangular if it is of the form

$$R(s,t) = \begin{cases} \phi(s)\psi(t), & s \le t , \\ \psi(s)\phi(t), & s \ge t , \end{cases}$$

where ϕ, ψ are functions distinct from zero, ϕ/ψ is nondecreasing.

A Gaussian process has a triangular covariance function iff it is Markovian (see [11]). For a Markov process with a triangular covariance function, condition 2') takes the form

$$\sup_{\pi_n} \sum_r |\psi(t_r) - \Psi(t_{r-1})| \sqrt{\frac{\phi}{\psi}(t_{t-1})} < \infty \quad .$$

Since ϕ/ψ is a non-decreasing function, the last condition is equivalent to

$$\sup_{\pi_n} \sum_r |\psi(t_r) - \psi(t_{t-1})| < \infty \quad .$$

According to [11] a Gauss-Markov process X has a triangular covariance function iff it has the form $X_t = \psi(t)W\left\{\frac{\phi}{\psi}(t)\right\}$, where W is a Wiener process. It follows from all this that $X \in Q$ iff $\psi \in V$.

NOTE. An (F_t)-optimal process X is a Gaussian (F_t)-martingale iff there exists an increasing deterministic process B and an (F_t)-Wiener process W such that $X_t = W_{B(t)}$ for any t. Indeed, suppose that X is a Gaussian (F_t)-martingale. Then X is Gaussian martingale with independent increments and the process $B = \langle X, X\rangle = EX^2$ is a deterministic process. Let $\widetilde{W}$ be a Wiener process independent of X. The process $\widetilde{W}_B$ has the same covariance as X. Therefore, $X = W_B$, where W is a Wiener process with respect to the family (F_t). Hence each (F_t)-optimal Gaussian semimartingale X has the form

$$X = W_B + A \quad ,$$

where W is an (F_t)-Wiener process, B is an increasing deterministic process, A is a process with integrable variation, and the pair (W_B, A) is a Gaussian process.

REFERENCES

[1] Jain, N.C., and Monrad, D. "Gaussian Quasimartingales." *Z. Wahrscheinlichkeitstheorie und Verw. Gebiete*, vol.59 (1982): 139-159.

[2] Emery, M. "Covariance des semimartingales gaussiennes." *C.r. Acad. sci.*, ser.1, vol.295, no.12 (1982): 703-705.

[3] Stricker, C. "Semimartingales gaussiennes: Applications au problem de l'innovation." *Z. Wahrscheinlichkeitstheorie und Verw. Gebiete*, vol.64 (1983): 303-312.

[4] Gal'čuk, L. "Optional Martingales." *Math. USSR Sbornik*, vol. 40, no.4 (1981): 435-468.

[5] Doob, J.L. "Stochastic Process Measurability Conditions." *Ann. Inst. Fourier*, Grenoble, vol.25, no.3-4 (1975): 163-176.

[6] Meyer, P.-A. Un cours sur les intégrales stochastique. In *Lecture Notes in Math.*, pp. 245-500, vol.511. Berlin Heidelberg New York: Springer-Verlag Inc., 1976.

[7] Gal'chuk, L. "Stochastic Integrals with Respect to Optional Semimartingales." *Theory Probab. Applications*, vol.29, no.1 (1984): 93-107.

[8] Itô, K, and Nisio, M. "On the Oscillation Function of Gaussian Processes." *Math. Scand.*, 22 (1968): 209-223.

[9] Fernique, X. "Regularité des trajectoires des fonctions aléatoires gaussiennes." In *Lecture Notes in Math.*, pp. 1-96, vol.480. Berlin Heidelberg New York: Springer-Verlag Inc., 1974.

[10] Liptser, R.Sh., and Shiryayev, A.N. *Statistics of Random Processes*, I, II. New York Berlin Heidelberg: Springer-Verlag Inc., 1977.

[11] Neveu, J. "Processus aléatoires guassiennes." *Le fascicule de Laboratoire de calcul des probabilités de l'Universités Paris*. VI, Paris, 1972.

I.I. GIKHMAN

ON THE STRUCTURE OF TWO-PARAMETER DIFFUSION FIELDS

In this paper, we give a definition of one class of two-parameter fields, which we call diffusion fields, and show that these fields can be represented as a sum of boundary values and two double integrals. The first summand is an ordinary integral and the second summand is a stochastic integral with respect to a Wiener measure. Furthermore, the integrads are functionals of the field values in the "strict past" with respect to the given values of the arguments. The proof is based on an immediate integration over the diffusion field.

The usual definition of stochastic integrals over semimartingales involves the representation of the process as a sum of the function of bounded variation and the martingale. For some cases one can give a direct definition of a stochastic integral without a priori decomposition of the field into the corresponding components. Moreover, the selection of the martingale part of the process follows directly from integration theory. We shall illustrate this procedure, using continuous two-parameter diffusion fields.

Let us introduce some notation. Let R^d be a d-dimensional linear space with inner product $(\cdot,\cdot)$ and let $|a|$ be the norm of a vector or an operator a. If p and q are two vectors in R^d, $p \times q$ designates an operator acting upon the third vector r according to $(p \times q)r = p(q,r)$. Also, $L(R^d, R^m)$ is the space of linear operators mapping R^d into R^m, and $L^+(R^d, R^m)$ is the subspace of $L(R^d, R^m)$ which consists of symmetric positive definite operators. Furthermore, R_+^2 is a set of pairs of real numbers $z = (x,y)$, $u = (s,t)$ (or $z' = (x',y')$, $u' = (s',t')$); $z \le u$ means that $x \le s$, $y \le t$; and $z < u$ means that $x < s$, $y < t$; $(z,z']$ is a rectangle $\{u: z<u\le z'\}$, $[z,z'] = \{u: z\le u\le z'\}$ and $|(z,z']|$ is its area.

The random variables and functions considered below are supposed to be defined on some fixed probability space (Ω, F, P) where filtration of σ-algebras (F_z), $z \in [0,z^*]$, $z^* \in R_+^2$, is defined.

Let $F_z^1 = F_{(x,y^*)}$, $F_z^2 = F_{(x^*,y)}$, $S_z = F_z^1 \vee F_z^2$, $S_u^z = F_{(s,y)} \vee F_{(x,t)}$ for $u < z$. All functions considered below are assumed to be (F_z)-adapted even if it is not always explicitly stated.

We shall call the random function $\xi = \xi(z)$, $z \in [0,z^*]$ a hyperbolic field if $\forall z \le z'$, $z,z' \in [0,z^*]$

$$P[\xi(z') \in B/S_z] = P[\xi(z') \in B/S_z^{z'}] \tag{1}$$

for all Borel sets B from R^d. The term "hyperbolic" shows the character of relationship between the values of the function $\xi(z)$.

For a point $z \in R_+^2$, the word "future" means the first quadrant $\{z': z'\ge z\}$ with vertex at z, and "past" means the third quadrant $\{z': z'\leqslant z\}$. Then (1) means that the distribution $\xi(z')$

(in the "future" with respect to the point z), when all the known events do not belong to the "future," depends on the events belonging to the "past" with respect to the point z' only. Lines which are parallel to the co-ordinate axes are "characteristics" of the field.

Hyperbolic property of the field is connected with the Cairoli-Walsh "commutation" condition (or condition F.4), which is widely used in the theory of two-parameter martingales.

It is as follows: σ-algebras F_z^1 and F_z^2 are conditionally independent of F_z, $\forall\, z \geq 0$. If Cairoli-Walsh's condition is fulfilled, then any (F_z)-adapted field is the hyperbolic field. On the other hand, if $\xi(z)$ is the hyperbolic field and F_z is σ-algebra generated by the variables $\xi(u)$, $u \in [0,z]$, then the filtration (F_z) satisfies the Cairoli-Walsh condition, and for integrable random variable η

$$E[\eta / F_z^1 / F_z^2] = E[\eta / F_z^2 / F_z^1] = E[\eta / F_z] .$$

By a diffusion field we mean the continuous random hyperbolic field $\xi(z)$, $z \in [0,z^*]$ if for all $\varepsilon > 0$

$$P(|\xi(z,z']| \geq \varepsilon) = \rho_\varepsilon^0(z,z']\ |(z,z']| , \tag{2}$$

$$E[I_\varepsilon(\xi(z,z'])\xi(z,z'] / S_z] = (\beta(z) + \rho_\varepsilon'(z,z'))\ |(z,z']| , \tag{3}$$

$$D[I_\varepsilon(\xi(z,z'])(\xi(z,z'],\ p) / S_z]$$
$$= [(\alpha(z)p,\ p) + (\rho_\varepsilon''(z,z')p,\ p)]\ |(z,z']| . \tag{4}$$

Here $I_\varepsilon(p)$ is an indicator of sphere $\{p,\ |p|<\varepsilon\}$, $p \in R^d$; $D[\gamma/S_z]$ is a conditional variation of random variable γ with respect to σ-algebra S_z; $\alpha(z)$, $\beta(z)$ are (F_z)-adapted measurable functions, $\beta(z)$ admits values in R^d and $\alpha(z)$ admits values in $L^+(R^d,R^m)$. The values $\rho^0_\varepsilon(z,z')$, $\rho'_\varepsilon(z,z')$, $\rho'_\varepsilon(z,z')$ tend to zero for $z' \downarrow z$ in the following sense:

$$\lim_{\delta\downarrow 0} \sup_{\substack{z<z' \\ |z'-z|<\delta}} [\rho^0_\varepsilon(z,z') + E(|\rho'_\varepsilon(z,z')| + |\rho''_\varepsilon(z,z')|)] = 0 . \quad (5)$$

We shall call the functions $\alpha(z)$, $\beta(z)$ the infinitesimal coefficients of the diffusion field.

Let $f(z)$, $z \in [0,z^*]$, $z^* \in R^2_+$ be Riemann integrable (F_z)-adapted operator-valued random function and let $f(z)\colon [0,z^*] \to L(R^d,R^m)$. To define the stochastic integral

$$\iint_{[0,z^*]} f(u)\ \xi(du)$$

as a limit in probability of the corresponding integral sums, we consider a certain division λ of the rectangle $[0,z^*]$ by the lines $x = s_k$, $y = t_r$, $k = 0,1,\ldots,n_1$, $r = 0,1,\ldots,n_2$, $0 = s_0 < s_1 < \cdots < s_{n_1} = x^*$, $0 = t_0 < t_1 < \cdots < t_{n_2} = y^*$ on the partial rectangle $\square_{kr} = (u_{kr}, u_{k+1\ r+1}]$, $u_{kr} = (s_k,t_r)$.

Let $\Delta s_k = s_{k+1} - s_k$, $\Delta t_r = t_{r+1} - t_r$, $|\lambda| = \max\limits_{k,r} (\Delta s_k \vee \Delta t_r)$.

Set

$$\iint_{[0,z^*]} f(u)\ \xi(du) \underset{\text{Def.}}{=} P \lim_{|\lambda|\to 0} \sum_{k,r=0}^{n_1-1,\ n_2-1} f(u_{kr})\ \xi(u_{kr}, u_{k+1\ r+1}] . \quad (6)$$

<u>THEOREM 1</u>. If $f(z) \in H_1$ where H_1 is a class of Riemann integrable operator-valued random functions defined on $[0,z^*] \times \Omega$ and if $\xi(z)$ is a diffusion process with continuous infinitesimal

coefficients, then the stochastic integral (6) exists.

To prove Theorem 1 we set

$$\sum_\lambda = \sum_\lambda^0 + \sum_\lambda' + \sum_\lambda'' ,$$

where

$$\sum_\lambda = \sum_{k,r=0}^{n_1-1,\, n_2-1} f(u_{kr})\, \square\xi_{kr} , \qquad \square\xi_{kr} = \xi(u_{kr}, u_{k+1\, r+1}] ,$$

$$\sum_\lambda^0 = \sum_{k,r=0}^{n_1-1,\, n_2-1} f(u_{kr})\, [1 - I_\varepsilon(\square\xi_{kr})]\, \square\xi_{kr} ,$$

$$\sum_\lambda' = \sum_{k,r=0}^{n_1-1,\, n_2-1} f(u_{kr})\, E[I_\varepsilon(\square\xi_{kr})\, \square\xi_{kr}/S_{kr}] , \qquad S_{kr} = S_{u_{kr}} ,$$

$$\sum_\lambda'' = \sum_{k,r=0}^{n_1-1,\, n_2-1} f(u_{kr})\, [I_\varepsilon(\square\xi_{kr})\square\xi_{kr} - E[I_\varepsilon(\square\xi_{kr})\, \square\xi_{kr}/S_{kr}]] .$$

Theorem 1 is a direct result of Lemmas 1, 2, 4.

<u>LEMMA 1</u>. If the random function $\xi(z)$ satisfies the condition

$$\lim_{\delta\downarrow 0} \sup_{\substack{z<z' \\ |z'-z|<\delta}} \frac{1}{|(z,z']|} P(|\xi(z,z']| > \varepsilon) = 0 , \qquad \forall\, \varepsilon > 0 , \tag{7}$$

then

$$P \lim_{|\lambda|\to 0} \sum_\lambda^0 = 0 .$$

Indeed, $P\left\{\left|\sum_\lambda^0\right| > 0\right\} \le \sum_{k,r=0}^{n_1-1,\, n_2-1} P(|\square\xi_{kr}| > \varepsilon)$ and the assertion of Lemma 1 follows from (7).

<u>LEMMA 2</u>. If the function $f(z) \in H_1$, $\xi(z)$ satisfies the condition (3), the function $\beta(z)$ is continuous and

$$\lim_{\delta\downarrow 0} \sup_{\substack{z<z' \\ |z'-z|<\delta}} E|\rho_\varepsilon'(z,z')| = 0 \quad , \tag{8}$$

then

$$P \lim_{|\lambda|\to 0} \sum_\lambda' = \iint_{[0,z^*]} f(u)\ \beta(u)\ ds\ dt \quad . \tag{9}$$

Proof. Let

$$\sum_\lambda' = \sum_{\lambda_1}' + \sum_{\lambda_2}' \quad ,$$

where

$$\sum_{\lambda_1}' = \sum_{k,r=0}^{n_1-1,\ n_2-1} f(u_{kr})\ \beta(u_{kr})\ |\square_{kr}| \quad ,$$

$$\sum_{\lambda_2}' = \sum_{k,r=0}^{n_1-1,\ n_2-1} f(u_{kr})\ \rho_\varepsilon'(k,r)\ |\square_{kr}| \quad ,$$

$|\square_{kr}|$ is the area of the rectangle $\square_{kr}$, $\rho_\varepsilon'(k,r) = \rho_\varepsilon'(u_{kr}, u_{k+1\,r+1})$.
Then

$$\lim \sum_{\lambda_1}' = \iint_{[0,z]} f(u)\ \beta(u)\ ds\ dt \qquad \text{a.s.}$$

and

$$P\left\{\left|\sum_{\lambda_2}'\right| > \varepsilon\right\}$$

$$\leq P\left\{\sup_{u\in[0,z^*]} |f(u)| > N\right\} + P\left\{\sum_{k,r=0}^{n_1-1,\ n_2-1} |\rho_\varepsilon'(k,r)|\cdot|\square_{kr}| > \frac{\varepsilon}{N}\right\} \quad ;$$

thus $P\left\{\left|\sum_{\lambda_2}'\right| > \varepsilon\right\} \to 0$ by continuity of the function $f(u)$ and (8).

Consider the sums $\sum_\lambda''$. Let

$$\gamma_{k\ell} = I_\varepsilon(\square\xi_{k\ell})\square\xi_{k\ell} - E[I_\varepsilon(\square\xi_{k\ell})\ \square\xi_{k\ell}/S_{k\ell}] \quad .$$

It follows from the hyperbolity of the field $\xi(z)$ that the variables $\gamma_{k\ell}$ are $F_{k+1,\ell+1}$-measurable $\left(F_{k\ell} \overset{=}{\mathrm{Def}} F_{u_{k\ell}}\right)$ and mutually orthogonal, i.e., $E\gamma_{ij}\gamma_{k\ell} = 0$ if $(i,j) \neq (k,\ell)$. For example, if $i < k$ then the variables γ_{ij} are $S_{k\ell}$-measurable and

$$E\gamma_{ij}\gamma_{k\ell} = E\gamma_{ij}\, E[\gamma_{k\ell} / S_{k\ell}] = 0 .$$

Moreover, if $(i,j) \le (k,\ell)$, $(i,j) \le (r,s)$ and $(k,\ell) \neq (r,s)$, then analogous use of hyperbolity shows that

$$E[(f(u_{k\ell})\gamma_{k\ell}) \times (f(u_{rs})\gamma_{rs}) / S_{ij}] = 0 . \quad (10)$$

Note that $|\gamma_{ij}| \le 2\varepsilon$. Introduce the field $\sum_\lambda''(i,j)$ with discrete argument $(i,j) = (0,0), \ldots, (n_1,n_2)$

$$\sum_\lambda''(i,j) = \sum_{\substack{(k,\ell) \\ (0,0)\le(k,\ell)\le(i-1,j-1)}} f(u_{k\ell})\,\gamma_{k\ell} .$$

Let $E|f(u)| < \infty$ $\forall\, u \in [0,z^*]$. Then $\sum_\lambda''(i,j)$ is a strong (F_{ij})-martingale. If, in addition, $E|f(u)|^p < \infty$ for all $u \in [0,z^*]$ and $p > 1$, the Doob-Cairoli inequality holds, i.e.,

$$E \max_{(0,0)\le(i,j)\le(n_1,n_2)} \left|\sum_\lambda''(i,j)\right|^p \le q^{2p}\, E\left|\sum_\lambda''(n_1,n_2)\right|^p ,$$

where $q = \frac{p}{p-1}$.

Set

$$\sigma_\lambda(i,j) = \sum_{\substack{0\le k\le i-1 \\ 0\le p\le j-1}} f(u_{k\ell})\, E[\gamma_{k\ell} \times \gamma_{k\ell} / S_{k\ell}]\, f(u_{k\ell}) .$$

It follows from (10) that

$$E\left[\Box\sum_\lambda''(i,j) \times \Box\sum_\lambda''(i,j) \Big/ S_{ij}\right] = E[\Box\sigma_\lambda(i,j) / S_{ij}] ,$$

where

$$\Box\sum_\lambda''(i,j) = \sum_\lambda''(r,s) - \sum_\lambda''(r,j) - \sum_\lambda''(i,s) + \sum_\lambda''(i,j) \quad ,$$

and the variable $\Box\sigma_\lambda(i,j)$ has an analogous sense. In particular,

$$E\left[\left|\Box\sum_\lambda''(i,j)\right|^2 / S_{ij}\right] = E[\Box sp\sigma_\lambda(i,j) / S_{ij}] \quad .$$

Now prove the existence of the limit of the sums $\sum_\lambda''$ for $|\lambda| \to 0$.

LEMMA 3. Assume that $\xi(z)$ is a hyperbolic field satisfying the condition (4), where

$$\lim_{\delta \downarrow 0} \sup_{\substack{z<z' \\ |z'-z|<\delta}} E|\rho_\delta''(z,z')| = 0 \quad , \tag{11}$$

and $\alpha(u)$ is a continuous random function.

Then for each $\varepsilon > 0$, $N > 0$

$$\overline{\lim_{|\lambda|\to 0}} P\left\{\left|\sum_\lambda''\right| > \varepsilon\right\} \leq \frac{N}{\varepsilon^2} + P\left(sp \iint_{[0,z^*]} f(u)\, \alpha(u)\, f^T(u)\, ds\, dt > N\right). \tag{11a}$$

Proof. The sequence $\sum_\lambda''(i,n_2)$, $i = 0,1,\ldots,n_1$ is an (F_i^1)-square integrable martingale, $\left(F_i^1 = F_{u_{in_2}}^1\right)$ and if $\Delta\sum_\lambda''(i,n_2) = \sum_\lambda''(i+1,n_2) - \sum_\lambda''(i,n_2)$ then

$$\delta_i \overset{=}{\text{Def}} E\left[\left|\Delta\sum_\lambda''(i,n_2)\right|^2 / F_i^1\right] = E\left[\left|\sum_{j=0}^{n_2-1} f(u_{ij})\gamma_{ij}\right|^2 / F_i^1\right]$$

$$= E\left[\sum_{j=0}^{n_1-1} sp\ f(u_{ij})\, E[\gamma_{ij} \times \gamma_{ij} / S_{ij}]\, f^T(u_{ij}) / F_i^1\right] .$$

Let $\pi_i^N = 1$ if $\sum_{k=0}^{i} \delta_i \leq N$, and $\pi_i^N = 0$ otherwise. Obviously,

$$P\left\{\left|\Sigma''_\lambda\right| < \varepsilon\right\} \le P\left(\pi^N_{n_1-1} = 0\right) + P\left\{\left|\sum_{i=0}^{n_1-1} \pi^N_i \, \Delta\Sigma''_\lambda(i,n_2)\right| > \varepsilon\right\}$$

and $P\left(\pi^N_{n_1-1} = 0\right) = P\left\{\sum_{i=0}^{n_1-1} \delta_i > N\right\}$,

$$P\left\{\left|\sum_{i=0}^{n_1-1} \pi^N_i \, \Delta\Sigma''_\lambda(i,n_2)\right| > \varepsilon\right\} \le \frac{1}{\varepsilon^2} E\left|\sum_{i=0}^{n_1-1} \pi^N_i \, \Delta\Sigma''_\lambda(i,n_2)\right|^2$$

The variables π^N_i are F^1_i-measurable, and the members of the sequence $\pi^N_i \, \Delta\Sigma''_\lambda(i,n_2)$, $i = 0,\ldots,n_1-1$ are mutually orthogonal. Therefore

$$E\left|\sum_{i=0}^{n_1-1} \pi^N_i \, \Delta\Sigma''_\lambda(i,n_2)\right|^2 = E \sum_{i=0}^{n_1-1} \pi^N_i \, \delta_i \le N$$

It follows from (4) that

$$\sum_{i=0}^{n_1-1} \delta_i = \sum_{i,j} \operatorname{sp}\left(f(u_{ij})\, \alpha(u_{ij})\, f^T(u_{ij})\right) |\square_{ij}|$$

$$+ \sum_{i,j} E\left[\operatorname{sp}\left(f(u_{ij})\, \rho''_\varepsilon(i,j)\, f^T(u_{ij})\right) |\square_{ij}| \,/\, F^1_i\right] ,$$

where $\rho''_\varepsilon(i,j) = \rho''_\varepsilon(u_{ij}, u_{i+1\, j+1})$. Thus, for each $a > 0$

$$P\left\{\left|\Sigma''_\lambda\right| > \varepsilon\right\} \le \frac{N}{\varepsilon^2} + P\left\{\sum_{i,j} \operatorname{sp}(f\circ\alpha\circ f)^T(u_{ij})|\square_{ij}| > N-a\right\} + b_\lambda(a) , \quad (12)$$

where

$$b_\lambda(a) = P\left\{\sum_{ij} E\left[\operatorname{sp} f(u_{ij})\, \rho''_\varepsilon(i,j)\, f^T(u_{ij})|\square_{ij}| \,/\, F^1_i\right] > a\right\} .$$

For each $N_1 > 0$

$$b_\lambda(a) \le P\left\{\sup_{u\in[0,z^*]} |f(u)| > N_1\right\} + P\left\{\sum_{ij} E\left[sp\rho''_\varepsilon(i,j) / F^1_i\right] > \frac{a}{N_1^2}\right\}$$

$$\le P\left\{\sup_{u\in[0,z^*]} |f(u)| > N_1\right\} + \frac{N_1^2}{a} \sup_{|z'-z|<|\lambda|} E|\rho''_\varepsilon(z,z')| \quad . \qquad (13)$$

It follows from the boundedness of $f(u)$ and (11) that as $|\lambda| \to 0$, $b_\lambda(a) \to 0$ for all $a > 0$. Passing to the limit in the inequality (12) at first as $|\lambda| \to 0$ and then as $a \to 0$, we obtain (11a). ///

<u>REMARK 1</u>. The same reasoning (using Doob's inequality for one-parameter martingales) shows that

$$\overline{\lim_{|\lambda|\to 0}} P\left\{\sup_i \left|\sum''_\lambda(i,n_2)\right| > \varepsilon\right\}$$

$$\le \frac{4N}{\varepsilon^2} + P\left\{sp \iint_{(0,z^*]} f(u)\ \alpha(u)\ f^T(u)\ ds\ dt > N\right\} , \qquad (14)$$

$$\forall\ \varepsilon > 0\ ,\quad N > 0\ .$$

<u>LEMMA 4</u>. If the conditions of Lemmas 2 and 3 hold, then

$$P \lim_{|\lambda|\to 0} \sum''_\lambda \ \overset{=}{\text{Def}} \iint_{[0,z]} f(u)\ \check{\xi}(ds,dt) \quad .$$

<u>Proof</u>. Let λ' and λ'' be two arbitrary decompositions of the rectangle $[0,z^*]$ and let $\bar{\lambda} = \lambda' \wedge \lambda''$ be the division of the rectangle $[0,z^*]$ obtained by applying the divisions λ' and λ''. Thus we have

$$P\left\{\left|\sum''_{\lambda'} - \sum''_{\lambda''}\right| > \delta\right\} \le P\left\{\left|\sum''_{\lambda'} - \sum''_{\bar\lambda}\right| > \frac{\delta}{2}\right\} + P\left\{\left|\sum''_{\bar\lambda} - \sum''_{\lambda''}\right| > \frac{\delta}{2}\right\} \quad .$$

The difference $\sum''_{\lambda'} - \sum''_{\bar\lambda}$ can be represented as

$$\sum\nolimits''_{\lambda'} - \sum\nolimits''_{\bar{\lambda}} = S_1 + S_2 + S_3 \quad ,$$

where

$$S_1 = \sum_{ij} \sum_{k\ell} [f(u_{ij}) - f(u_{ij,k\ell})]\gamma_{ij,k\ell} \quad ,$$

$$S_2 = \sum_{ij} f(u_{ij}) \sum_{k\ell} (I_\varepsilon(\Box\xi_{ij}) - I_\varepsilon(\Box\xi_{ij,k\ell}))\, \Box\xi_{ij,k\ell} \quad ,$$

$$S_3 = \sum_{ij} f(u_{ij})\{E[I_\varepsilon(\Box\xi_{ij})\Box\xi_{ij} / S_{ij}]$$

$$- \sum_{k\ell} E[I_\varepsilon(\Box\xi_{ij,k\ell})\Box\xi_{ij,k\ell} / S_{ij,k\ell}]\} \quad .$$

The second pair of indices (k,ℓ) in the notation $u_{ij,k\ell}$, $\Box\xi_{ij,k\ell}$, $\chi_{ij,k\ell}$, $S_{ij,k\ell}$ corresponds to indexing rectangles into which the rectangle $[u_{ij}, u_{i+1,j+1}]$ is divided when we use the division $\bar{\lambda}$.

To estimate the sum S_1 we utilize the inequalities (12) and (13). Here we replace f by $f - f_{\lambda'}$, where $f_{\lambda'}(u)$ is defined by the equalities $f_{\lambda'}(u) = f(u_{ij})$, if $u \in [u_{ij}, u_{i+1,j+1})$, $i = 0,\dots,n_1-1$, $j = 0,\dots,n_2-1$ and λ implies $\bar{\lambda}$. We obtain

$$P(|S_1| > \varepsilon)$$

$$\leq \varepsilon + P\left(\sup_{u\in[0,z^*]} |\alpha(u)| \sum_{ij} \sup_{u\in\Box_{ij}} |f(u) - f_{\lambda'}(u)| \cdot |\Box_{ij}| > \frac{\varepsilon^3}{2} \right) + b_\lambda\left(\frac{\varepsilon^3}{2}\right) \quad ,$$

from which one can see that $P \lim_{|\lambda'| \vee |\lambda''| \to 0} S_1 = 0$. Furthermore, for almost all ω one can find $\delta_0 = \delta_0(\delta, \varepsilon)$ such that we obtain $\max_{ij,k\ell} |\Box\xi_{ij,k\ell}| < \varepsilon$, $\max |\Box\xi_{ij}| < \varepsilon$ for $|\lambda'| \vee |\lambda''| < \delta_0$. Hence,

$S_2 = 0$ for these λ' and λ''. Therefore, $\lim S_2 = 0$ a.s. as $|\lambda'| \vee |\lambda''| \to 0$. Next we utilize (3) and continuity of the function $\beta(u)$ and obtain, as in the proof of Lemma 2, that $\operatorname{Plim} S_3 = 0$ as $|\lambda'| \vee |\lambda''| \to 0$. ///

It follows from Lemmas 1, 2 and 4 that there exists the integral (6) and the equality

$$\iint_{[0,z^*]} f(u)\, \xi(ds,dt) = \iint_{[0,z^*]} f(u)\, \beta(u)\, ds\, dt + \iint_{[0,z^*]} f(u)\, \check{\xi}(ds,dt) . \tag{14}$$

In this formula the first term on the right-hand side is a usual Riemann integral and the second one is the only notation of the limit of the sums $\sum_{\lambda}''$ as $|\lambda| \to 0$.

The given construction of the integrals can be applied to arbitrary rectangle $[z,z']$ for which the conditions of Theorem 1 are fulfilled.

Consequently, we can introduce the function $\mathcal{T}(z)$ (or $\mathcal{T}(z,f)$)

$$\mathcal{T}(z,f) = \mathcal{T}(z) \underset{\text{Def}}{=} \iint_{[0,z]} f(u)\, \check{\xi}(ds,dt)$$

a.s. for all $z \in [0,z^*]$. Obviously, the function $\mathcal{T}(z)$ is adapted to (F_z) and

$$\mathcal{T}(z,z'] = \iint_{(z,z']} f(u)\, \check{\xi}(ds,dt) .$$

Note that the definition of integral $\mathcal{T}(z)$ contains $\varepsilon > 0$. As the formula (14) shows, the random function $\mathcal{T}(z)$ does not depend on ε. This can be shown directly.

It follows from (12) that

$$P(|\mathcal{T}(z)| > \varepsilon) \le \frac{N}{\varepsilon^2} + P\left\{ \operatorname{sp} \iint_{[0,z]} f(u)\, \alpha(u)\, f^T(u)\, ds\, dt > N \right\} . \tag{15}$$

Now we consider the properties of the function $\mathcal{T}(z)$ for additional conditions. Suppose that

there exists a non-random constant c_o such that

$$|\rho''_\varepsilon(z,z')| \leq c_o \qquad \text{for all} \quad \varepsilon \in (0,\varepsilon_o],\ \varepsilon_o > 0 \ . \tag{16}$$

We denote by H_o the class of adapted operator-valued functions $f(z)$, $z \in [0,z^*]$, $f: R^d \to R^d$ satisfying the following conditions:

- 1. $f(z)$, $z \in [0,z^*]$ is a Riemann integrable function,
- 2. there exists a non-random constant $c = c(f)$ such that

$$|f(z)| + |(f\alpha f^T)(z)| \leq c \ , \qquad \forall\ z \in [0,z^*] \ . \tag{17}$$

THEOREM 2. Suppose that the random function $\xi(z)$, $z \geq 0$ satisfies the conditions of Theorem 1, the inequality (16), and $f \in H_o$.

Then $\mathcal{T}(z)$ is the limit of the sum $\sum''_\lambda$ in the sense of m.s. convergence. The random function $\mathcal{T}(z)$ of the argument z, $z \in [0,z^*]$ has a continuous modification and is a strong square integrable martingale with the strong matrix characteristic

$$\langle \mathcal{T},\mathcal{T}\rangle(z) = \iint_{[0,z]} f(u)\ \alpha(u)\ f^T(u)\, ds\, dt \ . \tag{18}$$

Proof. First we show that when the conditions of Theorem 2 are fulfilled the set of variables $\left\{\left|\sum''_\lambda\right|^2\right\}$ is uniformly integrable (with respect to λ). In order to show this we estimate the fourth moments of the sum $\sum''_\lambda$. Recall that by Burkhölder's inequality (see, e.g., [3]) there exists a constant c (depending on a dimension of R^d only) such that

$$E\left|\sum_{\lambda}''\right|^4 \leq cE\left(\sum_{ij} |(f\gamma)_{ij}|^2\right)^2 ,$$

where $(f\gamma)_{ij} = f(u_{ij})\gamma_{ij}$. Thus

$$E\left|\sum_{\lambda}''\right|^4 \leq cE\left(\sum_{ij} sp(f(\gamma\times\gamma)f^T)_{ij}\right)^2$$

$$= cE \sum_{(ij)(k\ell)} sp(f(\gamma\times\gamma)f^T)_{ij}\ sp(f(\gamma\times\gamma)f^T)_{k\ell} .$$

Let $\mu_{(ij),(k\ell)} \overset{\text{Def}}{=} E\ sp(f(\gamma\times\gamma)f^T)_{ij}\ sp(f(\gamma\times\gamma)f^T)_{k\ell}$. If $k \neq i$, e.g., $i > k$, then

$$\mu_{(ij)(k\ell)} = E\ E[sp(f(\gamma\times\gamma)f^T)_{ij} / S_{ij}]\ sp(f(\gamma\times\gamma)f^T)_{k\ell}$$

$$\leq E\ sp\left(f_{ij}E[(\gamma\times\gamma)_{ij}/S_{ij}]f_{ij}^T\right)\ sp(f(\gamma\times\gamma)f^T)_{k\ell}$$

$$\leq E\left(sp(f\,\alpha\,f^T)_{ij} + sp(f\,\rho_{\varepsilon}''\,f^T)_{ij}\ sp(f(\gamma\times\gamma)f^T)_{k\ell}\right)$$

$$\leq E\left(c_1 + c_2^2c_o\right)|\square_{ij}|\ sp(f(\gamma\times\gamma)f^T)_{k\ell}$$

$$\leq \left(c_1 + c_2^2c_0\right)^2|\square_{ij}|\cdot|\square_{k\ell}| .$$

For $j \neq \ell$ we obtain the same inequality. Utilizing (18) for $i = k$, $j = \ell$, we obtain

$$\mu(ij)(ij) \leq 4\varepsilon^2c_2^2\ E\ sp(f(\gamma\times\gamma)f^T)_{ij}$$

$$\leq 4\varepsilon^2c_2^2\left(c_1 + c_2^2c_o\right)|\square_{ij}| .$$

Thus

$$E\left|\sum_{\lambda}''\right|^4 \leq c\sum_{(ij)(k\ell)}\mu_{(ij)(k\ell)}$$

$$\leq c\left(c_1 + c_2^2c_o\right)^2(x^*y^*)^2 + 4c\varepsilon^2c_2^2\left(c_1 + c_2^2c_o\right)x^*y^* .$$

The uniform integrability of the variables $\left|\sum_\lambda''\right|^2$ is proved. Therefore we can rewrite the equality

$$E|\mathcal{T}(z^*)|^2 = E\left(P\lim\left|\sum_{ij} f(u_{ij})\gamma_{ij}\right|^2\right)$$

in the form

$$E|\mathcal{T}(z^*)|^2 = \lim_{|\lambda|\to 0} E\sum_{ij} sp(f(\gamma\times\gamma)f^T)_{ij}$$

$$= \lim_{|\lambda|\to 0} E\sum_{ij} sp(fE(\gamma\times\gamma)f^T)_{ij}$$

$$= \lim_{|\lambda|\to 0} E\sum_{ij} sp(f\,(\alpha+\rho_\varepsilon'')\,f^T)_{ij}|\Box_{ij}| \quad .$$

Taking into consideration the conditions (5) and (17), we have

$$E|\mathcal{T}(z^*)|^2 = E\iint_{[0,z^*]} sp(f(u)\;\alpha(u)\;f^T(u))\,ds\,dt \quad . \tag{19}$$

Analogously,

$$\lim_{|\lambda|\to 0} E\left|\mathcal{T}(z^*) - \sum_\lambda''\right|^2 = E\,P\lim_{|\lambda|\to 0}\left|\mathcal{T}(z^*) - \sum_\lambda''\right|^2 = 0 \quad ,$$

i.e., the sums $\sum_\lambda''$ converge to $\mathcal{T}(z^*)$ in m.s. Thus the random function $\mathcal{T}(z)$ is square integrable and is defined for each $z \in [0,z^*]$ a.s. and is adapted to (F_z). We obtain easily

$$E[\mathcal{T}(z,z']\,/\,S_z] = 0 \quad ,$$

i.e., $\mathcal{T}(z)$ is a strong martingale. Next let $\mathcal{T}(z)$ designate its separable modification. Analogously to (19) we obtain

$$E[\mathcal{T}(z,z'] \times \mathcal{T}(z,z']\,/\,S_z] = E\left[\iint_{]z,z^*]} (f(u)\;\alpha(u)\;f^T(u)\,ds\,dt\,/\,S_z\right]$$

for $0 \le z \le z'$. This equality implies that the matrix characteristic of the strong martingale $\mathcal{T}(z)$ is given by (18). Analogously to the formulas (18) and (19) one can obtain the inequality

$$E|\mathcal{T}(z,z']|^4 \le c_3|(z,z']|^2 + 4\varepsilon^2 c_3|(z,z']| ,$$

where c is a constant independent of ε. The left-hand side of this inequality does not depend on ε. Passing to the limit as $\varepsilon \to 0$, we obtain

$$E|\mathcal{T}(z,z']|^4 \le c_3|(z,z']|^2 . \tag{20}$$

Fix some y and y', $0 \le y < y' \le y^*$. We shall consider the square integrable martingale $\psi(x) = \mathcal{T}(x,y_2) - \mathcal{T}(x,y_1)$ with an argument x. It follows from (20) that $E|\psi(x+\Delta x) - \psi(x)|^4 \le c_3|y_2 - y_1|^2\Delta x^2$. Therefore, the martingale $\psi(x)$ has a continuous modification. The matrix characteristic of $\psi(x)$ is bounded by a non-random constant, and $\psi(x)$ has finite moments of any order. Using Burkhölder's inequality, we obtain for any $p > 1$

$$\begin{aligned} E|\mathcal{T}(z,z']|^p &= E|\psi(x') - \psi(x)|^p \\ &\le c_p E\left[\mathrm{sp} \iint_{[z,z']} (f\,\alpha\, f^T)\, ds\, dt\right]^{p/2} \\ &\le c_p c_4 [(x'-x)(y'-y)]^{p/2} , \end{aligned}$$

where c_p is an absolute constant. It follows from the last inequality that the strong martingale $\mathcal{T}(z)$ has a continuous modification (see e.g., [4]).

Now we generalize the definition of stochastic integral for

wider classes of random functions. Also, we suppose (if it is not mentioned otherwise) that $\xi(z)$ is a diffusion field satisfying (16).

We denote by H_2 the class of operator-valued adapted Borel functions $f(z)$, $z \in [0,z^*]$ satisfying the following conditions:

•1. the realizations of a function f are a.s. bounded on $[0,z^*]$, i.e.,

$$\sup \{|f(z)|, z \in [0,z^*]\} < \infty \quad \text{a.s.} ,$$

•2. $\|f\|^2 \underset{\text{Def}}{=} E \iint_{(0,z^*]} sp(f \alpha f^T)(u)\, ds\, dt < \infty .$

It is easy to verify that for each function f from H_2 one can find a sequence of functions $\{f_n, n=1,2,\ldots\}$ possessing the following properties: $f_n \in H_o$ and $\|f - f_n\| \to 0$ as $n \to \infty$.

Define a stochastic integral for any function $f \in H_2$.

For $z \in [0,z^*]$ let

$$\mathcal{T}(z,f) \underset{\text{Def}}{=} \lim \mathcal{T}(z,f_n) ,$$

where $f_n \in H_o$, $\|f - f_n\| \to 0$ as $n \to \infty$.

<u>THEOREM 3</u>. The integral $\mathcal{T}(z,f)$ exists for any function $f \in H_2$, it has a continuous modification, and it is a square integrable strong martingale with the characteristic (18).

<u>Proof</u>. Let f_n be a sequence from H_o such that $\|f - f_n\| \to 0$ as $n \to \infty$. Then $\|f_{n+m} - f_n\| \to 0$. It follows from Doob's inequality that

$$E \sup_{z \in [0,z^*]} |\mathcal{T}(z,f_n) - \mathcal{T}(z,f_{n+m})|^2 \le 16 \|f_{n+m} - f_n\|^2 \to 0 .$$

Choose the subsequence f_{n_k} in such a way that

$16 \left\| f_{n_{k+1}} - f_{n_k} \right\|^2 \le \frac{1}{2^{2k}}$. Then the series

$$\sum_k \sup_{z \in [0,z^*]} \left| \mathcal{T}\left(z, f_{n_{k+1}}\right) - \mathcal{T}\left(z, f_{n_k}\right) \right|^2$$

converges a.s. Therefore the limit $\mathcal{T}'(z,f) = \lim_{k \to \infty} \mathcal{T}(z, f_{n_k})$ exists as $n \to \infty$ uniformly on z a.s. Let $\mathcal{T}'(z,f) \underset{\text{Def}}{=} \mathcal{T}(z,f)$. The function $\mathcal{T}(z,f)$ is continuous on z a.s., $z \in [0,z^*]$, $(\mathcal{F}_z)$-adapted and

$$E \sup_{0 \le z \le z^*} |\mathcal{T}(z,f) - \mathcal{T}(z,f_n)|^2 \le \lim_{k \to \infty} E \sup_{z \in [0,z^*]} \left| \mathcal{T}\left(z, f_{n_k}\right) - \mathcal{T}(z,f_n) \right|^2$$

$$\le \lim_{k \to \infty} 16 \left\| f_{n_k} - f_n \right\|^2 .$$

Thus, $\lim \mathcal{T}(z,f_n) = \mathcal{T}(z,f)$ exists $\forall\ f \in H_2$ and has a continuous modification. Clearly, $\mathcal{T}(z,f)$ is a square integrable martingale and we can pass to the limit as $n \to \infty$ in the equality

$$\int_A \mathcal{T}(z,z'](f_n) \times \mathcal{T}(z,z'](f_n)\, dP = \int_A \iint_{(0,z]} (f_n \,\alpha\, f_n^T)\, ds\, dt\, dP, \quad A \in \mathcal{S}_z.$$

We shall obtain

$$\int_A \mathcal{T}(z,z'](f) \times \mathcal{T}(z,z'](f)\, dP = \int_A \iint_{[0,z]} (f \,\alpha\, f^T)\, ds\, dt\, dP, \quad \forall\ A \in \mathcal{S}_z,$$

and it follows from this equality that a matrix characteristic of the strong martingale $\mathcal{T}(z)$ can be defined by (18). ///

<u>LEMMA 5</u>. (Localization of Stochastic Integrals). If $f_k \in H_2$, $k = 1,2$ and $f_1(z) = f_2(z)$, $\forall\ z \in [0,z^*]$ for $\omega \in A$, then $\mathcal{T}(z,f_1) = \mathcal{T}(z,f_2)$ on A a.s.

<u>Proof</u>. Consider first the case where $f_k \in H_o$, $k = 1,2$. Then the integral $\mathcal{T}(z,f)$ is defined as the limit of "integral sums" and the equality $\mathcal{T}(z,f_1) = \mathcal{T}(z,f_2)$ on A a.s. is obvious. It is then clear how to pass to the proof of the assertion for $f_k \in H_2$. ///

Next we establish equality that allows us to generalize the definition of stochastic integrals for the cases when the existence of finite moments is not assumed.

Let H denote the space of all Borel adapted operator functions $f(z)$, $z \in [0,z^*]$ for which $\sup_{z\in[0,z^*]} sp(f \alpha f^T)(z) < \infty$ a.s. Set $f_N(z) = f(z)$ if $sp(f \alpha f^T)(z) < N$ and $f_N(z) = 0$ otherwise. For any $\varepsilon > 0$, $N > 0$ we have

$$P\left\{\sup_{z\in[0,z^*]} |\mathcal{T}(z,f)| > \varepsilon\right\}$$

$$\leq P\left\{\sup_{0\leq z\leq z^*} sp(f \alpha f^T)(z) > N\right\} + P\left\{\sup_{0\leq z\leq z^*} |\mathcal{T}(z,f_N)| > \varepsilon\right\} .$$

It follows from Chebyshev's and Doob's inequalities that

$$P\left\{\sup_{0\leq z\leq z^*} |\mathcal{T}(z,f_N)| > \varepsilon\right\} \leq \frac{1}{\varepsilon^2} E \sup_{0\leq z\leq z^*} |\mathcal{T}(z,f_N)|^2$$

$$\leq \frac{1}{\varepsilon^2} 16E \iint_{[0,z^*]} sp(f_N \alpha f^T)(u)\, ds\, dt$$

$$\leq \frac{16 N |z^*|}{\varepsilon^2} .$$

Thus

$$P\left\{\sup_{0\leq z\leq z^*} |\mathcal{T}(z,f)| > \varepsilon\right\} \leq \frac{16N|z^*|}{\varepsilon^2} + P\left\{\sup_{0\leq z\leq z^*} sp(f \alpha f^T) > N\right\} . \qquad (21)$$

Let $f \in H$. Then there exists a sequence $\{f_n\}$, $f_n \in H_2$,

$n = 1,2,\ldots$ such that $P\left\{\sup_{0\le z\le z^*} |f(z) - f_n(z)| > \varepsilon\right\} \to 0$ for any $\varepsilon > 0$. Assuming $N = \frac{\varepsilon^3}{16|z^*|}$ in (21), we shall obtain

$$P\left\{\sup_{0\le z\le z^*} \left|\mathcal{T}(z,f_n) - \mathcal{T}(z,f_{n+m})\right| > \varepsilon\right\}$$

$$\le \varepsilon + P\left\{\sup \operatorname{sp}\left[(f_{n+m} - f_n)\,\alpha\,(f_{n+m} - f_n)^T\right] > \frac{\varepsilon^3}{16|z^*|}\right\}$$

$$\le 2\varepsilon$$

for all sufficiently large $n,m > 0$. Thus $P \lim \mathcal{T}(z,f_n)$ exists for any $z \in [0,z^*]$.

Let

$$\mathcal{T}(z,f) \underset{\text{Def}}{=} P \lim \mathcal{T}(z,f_n) \quad .$$

Therefore the function $\mathcal{T}(z,f)$ has a continuous modification.

THEOREM 4. For an arbitrary operator function $f \in H$ the stochastic integral

$$\mathcal{T}(z,f) = \iint_{[0,z]} f(u)\, \check{\xi}(ds,dt) \tag{22}$$

is defined. It is a continuous adapted function and for any $\varepsilon > 0$ and $N > 0$ the inequality (21) holds. If $f \in H_2$ then $\mathcal{T}(z,f)$ is a square integrable strong martingale with the matrix characteristic (18).

Note that besides the inequality (21) for the stochastic integral $\mathcal{T}(z,f)$ the inequality

$$P(|\mathcal{T}(z,f)| > \varepsilon) \le \frac{N}{\varepsilon^2} + P\left\{\iint_{(0,z^*]} \operatorname{sp}(f\,\alpha\,f^T)\,ds\,dt \ge N\right\} \tag{23}$$

also holds.

Indeed, it was initially established for $f \in H_1$ and holds for successive generalizations of the integral. The inequality (23) shows that for successive generalizations of the notion of integral its values are not changed a.s.

That is a corollary from the following limit theorem: if for the functions, f, f_n the integrals $\mathcal{T}(z,f)$, $\mathcal{T}(z,f_n)$ are defined in one of the above meanings and

$$\iint_{(0,z]} (f - f_n)\,\alpha\,(f - f_n)^T \, ds\, dt \;\to\; 0$$

in probability, then

$$\mathcal{T}(z,f) = \operatorname*{P\,lim}_{n\to\infty} \mathcal{T}(z,f_n) \qquad \text{a.s.} \;.$$

The right-hand side of the equality (22) has been only a definition and it has not been identified with a stochastic integral. Moreover, the random function $\check{\xi}$ is not yet defined.

Theorem 4 allows us to define the function $\check{\xi}$ by the equalities

$$\check{\xi}(z) = \mathcal{T}(z,1) \;,$$

where 1 is the identity operator.

The function $\check{\xi}(z)$ is a local strong martingale in the following sense: there exists a sequence of (F_z)-strong continuous square integrable martingales $\check{\xi}_n(z)$, such that

$$\check{\xi}(z) = \check{\xi}_{n_0}(z) = \check{\xi}_{n_0+1}(z) = \cdots \tag{24}$$

starting with a certain number $n_0 = n_0(\omega)$.

To verify this assertion we shall assume that $f_n(z) = 1$ if

$\operatorname{sp} \alpha(z) \le n$, and $f_n(z) = 0$ otherwise. Then $\iint_{(0,z^*)} \operatorname{sp}(f_n \alpha f^T)\, ds\, dt \le n|z^*|^2$ and $\mathcal{T}(z,f_n)$ is a strong square integrable martingale. Moreover, all moments of the variables $\mathcal{T}(z,f_n)$ are finite. It follows from Lemma 5 that $\mathcal{T}(z,f_n) = \mathcal{T}(z,f_{n+1}) = \cdots = \mathcal{T}(z,1)$ on the set $\sup_{z \in [0,z^*]} \alpha(z) \le n$. Assuming $\check{\xi}_n(z) = \mathcal{T}(z,f_n)$, $n = 1,2,\ldots,$ we shall obtain the sequence which satisfies (24).

Note that matrix characteristics of the processes $\check{\xi}_n$ are of the form

$$\langle \check{\xi}_n, \check{\xi}_n \rangle(z) = \iint_{(0,z]} I_n(u)\, \alpha(u)\, ds\, dt$$

and

$$\langle \check{\xi}_n, \check{\xi}_n \rangle(z) = \iint_{(0,z^*]} \alpha(u)\, ds\, dt$$

on the set $\sup \operatorname{sp} \alpha(u) \le n$.

In this sense we say that a local strong martingale $\check{\xi}(z)$ has a matrix characteristic

$$\iint_{(0,z]} \alpha(u)\, ds\, dt \quad .$$

Now one can consider stochastic integrals

$$\iint_{(0,z]} f(u)\, \check{\xi}(ds,dt) \quad .$$

It is easy to verify that this integral coincides with the function $\mathcal{T}(z,f)$. Note the next formula of the composition for two stochastic integrals. If

$$\zeta(z) = \iint_{(0,z]} f(u)\, \mu(ds,dt) , \qquad \eta(z) = \iint_{(0,z]} g(u)\, \zeta(ds,dt)$$

then

$$\eta(z) = \iint_{(0,z]} g(u)\, f(u)\, \mu(ds,dt) ,$$

when all integrals are defined.

THEOREM 5. If $\operatorname{sp} \alpha(z) > 0 \quad \forall z \in [0,z^*]$ then

$$\check{\xi}(z) = \iint_{(0,z]} \alpha^{\frac{1}{2}}(u)\, w(ds,dt) ,$$

where w is an (F_z)-adapted Wiener field.

Proof. Set $f(z) = [\alpha(z)]^{-\frac{1}{2}}$. Then $f(z)$ is a continuous adapted matrix function, $f\,\alpha\, f^T = I$ and the continuous strong martingale $T(z,\alpha^{-\frac{1}{2}})$ has the matrix characteristic Ixy. Consequently, $w(z) \underset{\text{Def}}{=} T(z,\alpha^{-\frac{1}{2}})$ is a two-dimensional Wiener field.

Consider a stochastic integral

$$\zeta(z) = \iint_{(0,z]} \alpha^{\frac{1}{2}}(u)\, w(ds,dt) .$$

Since the field $w(z)$ can be expressed in terms of the stochastic integral

$$w(z) = \iint_{(0,z]} \alpha^{-\frac{1}{2}}(u)\, \check{\xi}(ds,dt) ,$$

then

$$\zeta(z) = \iint_{(0,z]} \alpha^{\frac{1}{2}}(u)\, \alpha^{-\frac{1}{2}}(u)\, \check{\xi}(ds,dt) = \check{\xi}(z) \quad .///$$

COROLLARY 1. The continuous diffusion field $\xi(z)$ satisfying (18) with $\alpha(z) > 0 \quad \forall z \in [0,z^*]$ a.s. allows the following integral representation:

$$\xi(z) = \chi(z) + \iint_{(0,z]} \beta(u)\, ds\, dt + \iint_{(0,z]} \alpha^{\frac{1}{2}}(u)\, w(ds,dt) , \qquad (25)$$

where w is an (F_z)-adapted Wiener field and $\chi(z)$ are boundary values of the field $\xi(z)$, with $\chi(z) = \xi(0,y) + \xi(x,0) - \xi(0,0)$.

If the coefficients $\alpha(u)$, $\beta(u)$ are of the form

$$\alpha^{\frac{1}{2}}(u) = A(u, \xi(u)) , \qquad \beta(u) = B(u, \xi(u)) ,$$

where $A(u,p)$, $B(u,p)$ are continuous non-random functions $(u,p) \in [0,z^*] \times R^d$, then we shall call $\xi(z)$ a two-parameter diffusion (or a two-parameter diffusion field).

COROLLARY 2. If a two-parameter diffusion satisfies (16) and $\alpha(z) > 0 \quad \forall z \in [0,z^*]$ a.s., then it is a solution of Goursat's stochastic problem

$$\begin{cases} \dfrac{\partial^2 \xi}{\partial s \partial t} = B(u,\xi(u)) + A(u,\xi(u)) \dfrac{\partial^2 w}{\partial s \partial t} , & u > 0 \\ \xi(u) = \chi(u) \qquad \text{if} \quad u(s,0) \quad \text{or} \quad u = (0,t) . \end{cases}$$

REMARK. We can do without the assumption $\operatorname{sp} \alpha(z) > 0$ in Theorem 5 and Corollary 2.

Assume that on the probability space $(\Omega, F_0, P, (F_t))$ there is defined a Wiener field $w_1(z)$ which is (F_z)-adapted and independent of the process $\xi(z)$. If it had not been the case, we could expand the initially defined probability space in a proper way.

In that case the proof of the existence of the representation (25) is analogous to the proof of the theorem on the representation of continuous martingales with absolutely continuous characteristics by stochastic integrals on a Wiener process (e.g., [2], p. 109).

REFERENCES

[1] Cairoli, R., and Walsh, J.B. "Stochastic Integrals on the Plane." *Acta Math.*, 134, no.1-2 (1975): 111-183.

[2] Gihman, I.I., and Skorohod, A.V. *The Theory of Stochastic Processes*. Vol.3. Berlin Heidelberg New York: Springer- Verlag Inc., 1979.

[3] Métraux, C. "Quelques inégalités pour martingales à paramètre bidimensionnel." *Lecture Notes in Math.*, pp. 170-179, vol. 649. Berlin Heidelberg New York: Springer-Verlag Inc., 1977.

[4] Yadrenko, M.I. *Spectral Theory of Random Fields*. New York Los Angeles: Optimization Software, Inc., 1983. (Distributed by Springer-Verlag Inc.)

L. GIRAITIS and D. SURGAILIS

A LIMIT THEOREM FOR A TRIANGULAR ARRAY OF SYMMETRIC STATISTICS

SUMMARY

A limit theorem for (infinite order) symmetric statistics

$$D^{(n)} = \sum_{k\le n} \sum_{1\le i_1<\dots<i_k\le n} h_k^{(n)}\left(\xi_k^{(n)}, \dots, \xi_k^{(n)}\right)$$

over a triangular array of i.i.d. random variables $\xi_k^{(n)}$, $1 \le k \le n$, $n \ge 1$ is proved. The limit distribution of $D^{(n)}$ is written in terms of multiple stochastic integrals with respect to Gaussian and Poisson random measures.

0. INTRODUCTION

Let $\xi_k^{(n)}$, $k,n \ge 1$ be a double array of real random variables which are i.i.d. for any fixed n, and $h_k^{(n)}: \mathbb{R}^k \to \mathbb{R}$ be symmetric functions such that

$$E\left\{h_k^{(n)}\left(\xi_1^{(n)}, \dots, \xi_k^{(n)}\right)\right\}^2 < \infty \tag{1}$$

and

$$E\left\{h_k^{(n)}\left(\xi_1^{(n)}, \dots, \xi_k^{(n)}\right) \mid \xi_2^{(n)}, \dots, \xi_k^{(n)}\right\} = 0 . \tag{2}$$

Consider the following <u>symmetric statistics</u>:

$$D^{(n)}(t) = \sum_{k\le[nt]} D_k^{(n)}(t) , \tag{3}$$

where $D_k^{(n)}(t)$ are symmetric statistics of order k:

$$D_k^{(n)}(t) = D_k^{(n)}\left(t;\ h_k^{(n)}\right)$$

$$= \sum_{1\le i_1<\dots<i_k\le[nt]} h_k^{(n)}\left(\xi_{i_1}^{(n)}, \dots, \xi_{i_k}^{(n)}\right) , \tag{4}$$

$k \le [nt]$. Perhaps the most important among symmetric statistics are U-statistics of Hoeffding [4], who also derived the representations (3), (4) for U-statistics, with h_k satisfying (1) and (2). Asymptotical distribution of symmetric statistics was investigated by a number of authors, most of them examined the convergence to the Gaussian law or its refinements (see e.g., Serfling [9]). On the other hand, Rubin and Vitale [8], Dynkin and Mandelbaum [1] (also see Fillipova [2]) have recently discussed the situation when $D^{(n)}(t)$ tend in distribution to a random process expressed as a sum of multiple Itô-Wiener integrals with respect to some Gaussian random measure (in [8], as a sum of products of Hermite polynomials of Gaussian variables). These authors have assumed that $\xi_j^{(n)}$ belong to the domain of attraction of the Gaussian law. In this paper we consider a similar problem when $\sum_{j\le n} \xi_j^{(n)}$ converge to general infinitely divisible law. More precisely, we assume that

$$nP\left\{\xi_1^{(n)} \in dx\right\} \rightarrow \pi(dx) \tag{5}$$

and

$$P\left\{\xi_1^{(n)} \in dx/\sqrt{n}\right\} \rightarrow \nu(dx) \tag{6}$$

weakly as $n \to \infty$, where π and ν are some measures on the real line $\mathbb{R}$. Under some conditions on $h_k^{(n)}$ (condition (A) of the Theorem, which can be described as the uniform approximation by simple functions), finite dimensional distributions of $(D^{(n)}(t))_{t\geq 0}$ tend to the corresponding distributions of the random process

$$
\begin{aligned}
D(t) &= \sum_{k\geq 1} D_k(t) \\
&= \sum_{k\geq 1} \left(\frac{1}{k!}\right) \int_{((0,t]\times\bar{\mathbb{R}})^k} \bar{h}_k(x_1,\ldots,x_k)\, \bar{W}(ds_1,dx_1)\cdots\bar{W}(ds_k,dx_k) \quad ,
\end{aligned}
\tag{7}
$$

where $\bar{\mathbb{R}} = \mathbb{R}' \cup \mathbb{R}_0''$, $\mathbb{R}'$, $\mathbb{R}''$ are two copies of $\mathbb{R}$, $\mathbb{R}_0 = \mathbb{R} \setminus \{0\}$, $\bar{h}_k : \bar{\mathbb{R}}^k \to \mathbb{R}$ are some functions and $\bar{W}(ds,dx)$ is random measure on $[0,\infty) \times \bar{\mathbb{R}}$ with independent values on nonintersecting sets which is Gaussian on $[0,\infty) \times \mathbb{R}'$ and Poisson on $[0,\infty) \times \mathbb{R}_0''$, with variance equal to $ds\, \nu(dx)$ and $ds\, \pi(dx)$, respectively. Condition (6) is rather restrictive and is not necessary in general for the convergence of $\sum_{j\leq n} \xi_j^{(n)}$ (to an infinitely divisible law). It is satisfied, however, in the case $\xi_j^{(n)} = \xi_j / \sqrt{n}$, where ξ_j, $j \geq 1$ are i.i.d. random variables with distribution independent of n, which was considered in [1].

The proof of our Theorem is different from [1, 8] although it seems to be rather natural for establishing the convergence of distributions of multiple stochastic integrals. Namely, under the approximation condition (A), the statistics $D_k^{(n)}(t)$ can be approximated in the mean square by finite sums

$$\sum_{\Delta_1 \cdots \Delta_k} g_k^{\Delta_1 \cdots \Delta_k} \bar{V}^{(n)}((0,t] \times \Delta_1) \cdots \bar{V}^{(n)}((0,t] \times \Delta_k) \quad , \tag{8}$$

where Δ_i are either (1) (fixed) intervals of the real line or (2) $\Delta_i = (a_i/\sqrt{n},\ b_i/\sqrt{n})$; $\bar{V}^{(n)}((0,t] \times \Delta_i)$ is equal to the centered jump counting measure

$$\sum_{j \le [nt]} \left(1\left\{ \xi_j^{(n)} \in \Delta_i \right\} - P\left\{ \xi_j^{(n)} \in \Delta_i \right\} \right)$$

in case (1), while in case (2) this counting measure must be multiplied by $1/\sqrt{n}$. It is easy to show that under conditions (5) and (6), $\bar{V}^{(n)}((0,t] \times \Delta_i)$ tends to Poisson or Gaussian limit depending on case (1) or (2), which implies that (8) converges in distribution to a multiple stochastic integral of the form (7). A shortcoming of such a proof, at least in its present form, is that it requires the continuity of the limiting measures π and ν. In particular, the result of Dynkin and Mandelbaum [1] follows from our Theorem under the additional condition of continuity of the distribution of ξ_j. The continuity of π and ν is implicit in the approximation condition (A), as our definition of a simple function requires, as usual, that it must vanish on "diagonals" (see Definition 4 below). It seems that by modifying this definition it would be possible to extend our results to general π and ν. Finally, we'd like to remark that other methods could probably produce similar or even better results; in particular the general martingale methods of convergence of stochastic processes could be useful as the statistics $D_k^{(n)}(t)$ and their limits $D_k(t)$ are, in fact, martingales.

The headlines of the remaining sections are: 1. Definitions and Notation. 2. The Main Theorem. 3. Some Corollaries and Special Cases. 4. Proof of the Theorem. 5. Appendix: Multiple Stochastic Integrals (definition and basic properties).

1. DEFINITIONS AND NOTATION

Everywhere below $\xi_j^{(n)}$, $j \geq 1$ are i.i.d. real random variables with the common probability distribution $F^{(n)}(dx) = P(\xi_j^{(n)} \in dx)$, $n = 1,2,\ldots$. Set also

$$X_t^{(n)} = \sum_{j \leq [nt]} \xi_j^{(n)} , \qquad X^{(n)} = X_1^{(n)} , \tag{9}$$

$$\pi^{(n)}(dx) = nF^{(n)}(dx) , \tag{10}$$

$$\nu^{(n)}(dx) = F^{(n)}(dx/\sqrt{n}) , \tag{11}$$

$$\mu_\varepsilon^{(n)}(dx) = x^2 \nu^{(n)}(dx)\, 1(|x| \leq \sqrt{n}\,\varepsilon) , \quad \varepsilon > 0 , \tag{12}$$

$$p^{(n)}((0,t]\times A) = \sum_{1 \leq j \leq [nt]:\xi_j^{(n)} \in A} 1 , \tag{13}$$

$$\begin{aligned} q^{(n)}((0,t]\times A) &= p^{(n)}((0,t]\times A) - Ep^{(n)}((0,t]\times A) \\ &= p^{(n)}((0,t]\times A) - ([nt]/n)\, \pi^{(n)}(A) , \end{aligned} \tag{14}$$

$$v^{(n)}((0,t]\times A) = q^{(n)}((0,t]\times n^{-\frac{1}{2}}A)\, n^{-\frac{1}{2}} . \tag{15}$$

Below $\overset{d}{=}$, $\overset{d}{\to}$ will denote the identity and the (weak) convergence of (finite-dimensional) distributions, respectively. All the limits, unless otherwise specified, are taken as $n \to \infty$.

DEFINITION 1. A (positive) measure π on $\mathbb{R}_o = \mathbb{R} \setminus \{0\}$ is called

Lévy measure if

$$\int_{\mathbb{R}_0} \min(x^2,1)\,\pi(dx) < \infty .$$

Let π be a Lévy measure and $\pi^{(n)}$ be of the form (10). We'll write $\pi^{(n)} \Rightarrow \pi|_{\mathbb{R}_0}$ if $\pi^{(n)}(A) \to \pi(A)$ for any Borel set $A \subset \mathbb{R}_0$ such that $\pi(\partial A) = 0$, where ∂A is the boundary of A.

DEFINITION 2. Let be given $a \in \mathbb{R}$, $b > 0$ and a Lévy measure π. We say that a random variable ξ is infinitely divisible with characteristics (a,b,π) if

$$E\exp(it\xi) = \exp\left(ita - bt^2/2 + \int_{\mathbb{R}_0} (e^{itx} - 1 - (itx/(1+x^2)))d\pi\right), \quad t \in \mathbb{R} . \tag{16}$$

PROPOSITION 1 (see e.g., [3]). Let $(X_t)_{t\ge 0}$ be a homogeneous process with independent increments such that X_1 is infinitely divisible with characteristics (a,b,π). The relation $\left(X_t^{(n)}\right)_{t\ge 0} \overset{d}{\to} (X_t)$ is equivalent to

$$\int_{\mathbb{R}} x/(1+x^2)\, d\pi^{(n)} \to a \tag{17}$$

$$\limsup_n \mu_\varepsilon^{(n)}(\mathbb{R}) - b \to 0 \qquad (\varepsilon \to 0) , \tag{18}$$

$$\pi^{(n)} \Longrightarrow \pi|_{\mathbb{R}_0} . \tag{19}$$

Introduce the following condition:

$$\mu_\varepsilon^{(n)} \Rightarrow \mu \qquad \text{(weakly in } \mathbb{R}) \quad \forall\, \varepsilon > 0 , \tag{20}$$

where μ is a finite measure independent of ε. Clearly (20) implies (18) with $b = \mu(\mathbb{R})$, but the converse is not true. Set

$$\nu(dx) = \mu(dx)/X^2 . \tag{21}$$

Note that ν is finite, in fact $\nu(\mathbb{R}) \le 1$. Condition (20) implies also that $\int f\, d\nu^{(n)} \to \int f\, d\nu$ for every continuous function f with compact support.

2. THE MAIN THEOREM

Let $\mathbb{R}'$, $\mathbb{R}''$ be two copies of the real line. Set

$\bar{\mathbb{R}} = \mathbb{R}' \cup \mathbb{R}''_0$,

$$\bar{\nu}^{(n)}(dx) = \begin{cases} \nu^{(n)}(dx) , & x \in \mathbb{R}' , \\ \pi^{(n)}(dx) , & x \in \mathbb{R}''_0 , \end{cases} \tag{22}$$

$$\bar{\nu}(dx) = \begin{cases} \nu(dx) , & x \in \mathbb{R}' , \\ \pi(dx) , & x \in \mathbb{R}''_0 . \end{cases} \tag{23}$$

Here $\nu^{(n)}$, $\pi^{(n)}$, ν, π are the same as in the previous Section.

DEFINITION 3. By the sequence of partitions we mean a sequence $(\Delta)_N$, $N \ge 1$ of countable monotone*/ partitions of $\bar{\mathbb{R}}$ by intervals Δ which are either in $\mathbb{R}'$ or in $\mathbb{R}''_0$, such that

$$\sup\,(\mathrm{diam}\,(\Delta) : \Delta \in (\Delta)_N) \to 0 \qquad N \to \infty \tag{24}$$

and

$$\bar{\nu}(\partial\Delta) = 0 \qquad \forall\, \Delta \in (\Delta)_N , \qquad N \ge 1 . \tag{25}$$

Let $(\Delta)^k_N = \{\Delta_1 \times \cdots \times \Delta_k \subset \bar{\mathbb{R}}^k : \Delta_i \in (\Delta)_N\}$ be the corresponding sequence of partitions of $\bar{\mathbb{R}}^k = \bar{\mathbb{R}} \times \cdots \times \bar{\mathbb{R}}$ (k times).

DEFINITION 4. A symmetric function $\bar{g} : \bar{\mathbb{R}}^k \to \mathbb{R}$ is called simple

---

*/ I.e., $(\Delta)_N \subset (\Delta)_{N+1}$ $\forall\, N \ge 1$.

if (1) it takes constant values $\left(\equiv \bar{g}^{\Delta_1 \cdots \Delta_k}\right)$ on sets $\Delta_1 \times \cdots \times \Delta_k \in (\Delta)_N^k$ for some $N \geq 1$ and vanishes on a finite number of such sets and (2) $\bar{g}$ vanishes on "diagonals": $\bar{g}^{\Delta_1 \cdots \Delta_k} = 0$ if $\Delta_i = \Delta_j$ for some $i \neq j$.

For each $\varepsilon > 0$ and $n \geq 1$ we denote by ${}_{\varepsilon,n}\bar{G}_k$ the set of all simple functions $\bar{g}\colon \bar{\mathbb{R}}^k \to \mathbb{R}$ which vanish on the set

$$\bigcup_{i=1}^{k} \left(\{x_i \in \mathbb{R}' : |x_i| > \varepsilon\sqrt{n}\} \cup \{x_i \in \mathbb{R}''_o : |x_i| \leq \varepsilon\}\right) . \tag{26}$$

Given a symmetric function $h\colon \mathbb{R}^k \to \mathbb{R}$ and $\varepsilon > 0$, $n \geq 1$ we define a new symmetric function ${}_{\varepsilon,n}\bar{h}\colon \bar{\mathbb{R}}^k \to \mathbb{R}$ by

$$\begin{aligned} &{}_{\varepsilon,n}\bar{h}(x_1, \ldots, x_i, \ldots, x_k) \\ &\quad = (\sqrt{n})^i\, h(x_1/\sqrt{n}, \ldots, x_i/\sqrt{n}, x_{i+1}, \ldots, x_k) \end{aligned} \tag{27}$$

if $X_1 \in \mathbb{R}'$, ..., $x_i \in \mathbb{R}'$, $|x_1| \leq \varepsilon\sqrt{n}$, ..., $|x_i| \leq \varepsilon\sqrt{n}$, $x_{i+1} \in \mathbb{R}''_o$, ..., $x_k \in \mathbb{R}''_o$, $|x_{i+1}| > \varepsilon$, ..., $|x_k| > \varepsilon$, $i = 0,1,\ldots,k$; ${}_{\varepsilon,n}\bar{h} = 0$ if otherwise in $\bar{\mathbb{R}}^k$. Note that ${}_{\varepsilon,n}\bar{h}$ vanishes on the set (26) by definition. Now we are ready to formulate our main result.

THEOREM. Assume that conditions (17)-(20) hold. Also, let the symmetric functions $h_k^{(n)}\colon \mathbb{R}^k \to \mathbb{R}$, $n,k \geq 1$ satisfying (1), (2), and $(D^{(n)}(t))_{t\geq 0}$ be defined by (3), (4). If, moreover,

(A) for any $\delta > 0$ there exist $\varepsilon = \varepsilon(\delta) > 0$, $n_o = n_o(\varepsilon,\delta) \geq 1$, $k_o \geq 1$ and simple functions $\bar{g}_k \in {}_{\varepsilon,n_o}\bar{G}_k$, $k \geq 1$, $\bar{g}_k = 0$ if $k \geq k_o$ such that for all $n \geq n_o$

$$\sum_{k=1}^{\infty} \int_{\bar{\mathbb{R}}^k} \left(_{\varepsilon,n}\bar{h}^{(n)} - \bar{g}_k\right)^2 d^k \bar{\nu}^{(n)}/k! \quad < \quad \delta \tag{28}$$

and

$$\sum_{k=1}^{\infty} \int_{\bar{\mathbb{R}}^k} (\bar{h}_k - \bar{g}_k)^2 \, d^k\bar{\nu} / k! \quad < \quad \delta \quad , \tag{29}$$

where $\bar{h}_k \in L^2(\bar{\mathbb{R}}^k; \bar{\nu}^{(k)})$ are symmetric functions independent of δ, ε and n, then

$$(D^{(n)}(t))_{t\geq 0} \overset{d}{\to} (D(t))_{t\geq 0} \quad , \tag{30}$$

where $D(t)$ is given by (7), where

$$\bar{W}(ds,dx) = \begin{cases} W(ds,dx) \ , & x \in \mathbb{R}' \ , \\ q(ds,dx) \equiv p(ds,dx) - ds\pi(dx) \ , & x \in \mathbb{R}''_o \ , \end{cases}$$

and $W(ds,dx)$, $p(ds,dx)$ are mutually independent Gaussian and Poisson random measures on $\mathbb{R}_+ \times \mathbb{R}$ and $\mathbb{R}_+ \times \mathbb{R}_o$, respectively, with independent values on nonintersecting sets such that $E(W(ds,dx))^2 = ds\ \nu(dx)$, $Ep(ds,dx) = ds\ \pi(dx)$.

Multiple stochastic integrals (m.s.i.) with respect to Gaussian and/or Poisson random measure were discussed in the classical works of K. Itô [5, 6]. The definition of m.s.i. can easily be extended to more general random measures with finite variance [10]. In Section 5 (the Appendix) we recall the definition and some basic properties of m.s.i.

3. SOME COROLLARIES AND SPECIAL CASES

CASE 1.

$$\xi_j^{(n)} = \xi_j/\sqrt{n} \ , \quad \xi_j \ \text{i.i.d.}, \quad E\xi_j^2 < \infty \ , \quad E\xi_j = 0 \ ,$$

$$F(dx) \equiv P(\xi_j \in dx) ,$$

$$h_k^{(n)}(x_1,\ldots,x_k) = n^{-k/2} h_k(\sqrt{n}\,x_1, \ldots, \sqrt{n}\,x_k) .$$

In this case conditions (17) - (20) hold with $\gamma = \pi = 0$, $\mu(dx) = x^2F(dx)$. If the distribution F $(= \nu = \nu^{(n)})$ is continuous (i.e., $F(\{x\}) = 0 \quad \forall\, x \in \mathbb{R}$), then condition (A) of the Theorem is easily verified with $\bar{h}_k = h_k$ on $\mathbb{R}'^k$ and equals zero on $\bar{\mathbb{R}}^k \setminus \mathbb{R}'^k$. Assuming also (2) we have

COROLLARY 1 (cf. [1], Theorem 1). Under the conditions above,

$$(D(t))_{t\ge 0} \overset{d}{\to} \left(\sum \frac{1}{k!} \int_{((0,t]\times\mathbb{R})^k} h_k(x_1, \ldots, x_k) \times W(ds_1,dx_1) \cdots W(ds_k,dx_k)\right)_{t\ge 0} ,$$

where $W(ds,dx)$ is Gaussian random measure on $\mathbb{R}_+ \times \mathbb{R}$ with independent values on nonintersecting sets and variance $E(W(ds,dx))^2 = ds\, F(dx)$.

CASE 2. Let i.i.d. random variables $\xi_j^{(n)}$, $j \ge 1$ belong to the domain of attraction of an infinitely divisible law without a Gaussian component. In other words, we assume that conditions (17) - (19) hold with $b = 0$. (This implies also (20) with $\mu = 0$.) We have

COROLLARY 2. Let $h^{(n)} \equiv h_k^{(n)} : \mathbb{R}^k \to \mathbb{R}$, $n \ge 1$ satisfy conditions (1), (2). If, moreover,

(A') for each $\delta > 0$ there exist $\varepsilon = \varepsilon(\delta) > 0$, $n_o = n_o(\varepsilon,\delta) \ge 1$, and a simple*/ function $g: \mathbb{R}^k \to \mathbb{R}$; $g = 0$ on $\{(x_1, \ldots, x_k) \in \mathbb{R}^k: |x_i| \le \varepsilon$ for some $i = 1, \ldots, k\}$, such that for all $n \ge n_o$

$$\int_{\mathbb{R}^k} (h^{(n)} - g)^2 \, d^k \pi^{(n)} < \delta$$

and

$$\int_{\mathbb{R}^k} (h - g)^2 \, d^k \pi < \delta \quad ,$$

where h $(\in L^2(\mathbb{R}^k; \pi^k))$ is independent of n, ε, δ, then

$$\left(D_k^{(n)}(t)\right)_{t \ge 0} \overset{d}{\to} \left(\int_{((0,t] \times \mathbb{R}_o)^k} h(x_1, \ldots, x_k) \times q(ds_1, dx_1) \cdots q(ds_k, dx_k) \right)_{t \ge 0} ,$$

where $q(ds, dx)$ is the (centered) Poisson random measure on $\mathbb{R}_+ \times \mathbb{R}_o$ with variance $ds \, \pi(dx)$.

Conditions (A), (A') can be replaced by more transparent ones in some cases (see below).

<u>PROPOSITION 2</u>. Let $\pi = \pi^{(\infty)}$ be a continuous measure and let $h: \mathbb{R}^k \to \mathbb{R}$ be a continuous symmetric function such that $|h(x_1, \ldots, x_k)| \le h_1(x_1) \, h_2(x_2, \ldots, x_k)$, where

$$h_1(x) \le |x| \qquad |x| \le 1$$

$$\sup\left\{ \int_{|x|>k} h_1^2 \, d\pi^{(n)} : 1 \le n \le \infty \right\} \to 0 \qquad k \to \infty \; ;$$

$$\sup \left\{ \int h_2^2 \, d^{k-1}\pi^{(n)} : 1 \le n \le \infty \right\} < \infty \; .$$

*/ I.e., the function $\bar{g}: \bar{\mathbb{R}}^k \to \mathbb{R}$ given by $\bar{g}(x_1, \ldots, x_k) = g(x_1, \ldots, x_k)$ if $x_1, \ldots, x_k \in \mathbb{R}_o''$, $= 0$ if otherwise, is simple according to Definition 4.

Then $h^{(n)} \equiv h$ satisfy the condition (A').

We omit the proof of this Proposition, which is easy. Of course, some of its conditions, in particular the continuity of h, can be weakened.

CASE 3. $h_k^{(n)}(x_1,\ldots,x_k) = \tilde{h}^{(n)}(x_1), \ldots, \tilde{h}^{(n)}(x_k)$ seems to be rather particular because the convergence of $D_k^{(n)}(t;\ h_k^{(n)})$ can be proved without condition (20) (although the limit takes a somewhat different form). By turning to the new variables $\tilde{\xi}_j^{(n)} = \tilde{h}^{(n)}(\xi_j^{(n)})$ we reduce the problem to the case $h^{(n)}(x_1, \ldots, x_k) = x_1, \ldots, x_k$.

PROPOSITION 3. Assume that $\xi_j^{(n)}$, $n,j \geq 1$ satisfy conditions (17) - (19); $E\xi_j^{(n)} = 0$, $E(\xi_j^{(n)})^2 = 1/n$, $h^{(n)}(x_1, \ldots, x_k) = x_1, \ldots, x_k$. Then

$$\left(D_k^{(n)}(t;\ h^{(n)})\right)_{t \geq 0} \overset{d}{\to} \left(\frac{1}{k!}\int_{(0,t]^k} d^k X\right)_{t \geq 0},$$

where $X = (X_t)_{t \geq 0}$ is the same as in Proposition 1.

Proposition 3 can be proved in the same way as the Theorem, with the difference that Proposition 4 below should be replaced by Proposition 1.

4. PROOF OF THE THEOREM

For simplicity of notation we shall consider only the case $h_k^{(n)} = 0$ $(k \neq 2)$ and the convergence of one-dimensional distributions of $D_2^{(n)}(t)$ at $t = 1$. Set $h_2^{(n)} \equiv h^{(n)}$,

$$p^{(n)}(dx) = p^{(n)}((0,1] \times dx) ,$$

$$q^{(n)}(dx) = q^{(n)}((0,1] \times dx) \quad ,$$

$$V^{(n)}(dx) = V^{(n)}((0,1] \times dx) \quad ,$$

$$\bar{W}(dx) = \bar{W}((0,1] \times dx) = \begin{cases} W(dx) , & x \in \mathbb{R}' , \\ q(dx) , & x \in \mathbb{R}''_0 , \end{cases}$$

$$\bar{V}(dx) = \begin{cases} V(dx) , & x \in \mathbb{R}' , \\ q(dx) , & x \in \mathbb{R}''_0 , \end{cases}$$

$$V(dx) = W(dx) - W(\mathbb{R}')\, \nu(dx) \quad . \tag{31}$$

Here, $W(dx)$, $q(dx) = p(dx) - \pi(dx)$ are independent Gaussian and (centered) Poisson random measures with independent values on nonintersecting sets, with variances $E(W(dx))^2 = \nu(dx)$, $E(q(dx))^2 = \pi(dx)$, respectively. The random Gaussian measure $V(dx)$ (31) is not uncorrelated, in fact,

$$E\, V(A)V(B) = \nu(A \cap B) - \nu(A)\nu(B) , \qquad A,\, B \subset \mathbb{R} . \tag{32}$$

The process $(V(-\infty,t))_{t\in\mathbb{R}}$ can be transformed to the Brownian bridge on $(0,1)$ by the time change $t = \nu^{-1}s$; $\nu(-\infty,\nu^{-1}s) \equiv s$ $(0 < s < 1)$.

PROPOSITION 4. Let (17) - (20) hold. Let $\Delta'_1, \dots, \Delta'_m$ $(\subset \mathbb{R}')$, $\Delta''_1, \dots, \Delta''_m$ $(\subset \mathbb{R}''_0)$ be disjoint subsets of $\bar{\mathbb{R}}$ belonging to some partition $(\Delta)_N$ such that Δ'_i have no intersection with some neighborhood of $0 \in \mathbb{R}''$. Then

$$(V^{(n)}(\Delta'_i),\ q^{(n)}(\Delta''_i),\ i=1,\dots,m) \overset{d}{\to} (V(\Delta'_i,\ q(\Delta''_i),\ i=1,\dots,m) .$$

Proof. It suffices to show that

$$\sum_{\Delta} \bar{g}^{\Delta} \bar{V}^{(n)}(\Delta) \xrightarrow{d} \sum_{\Delta} \bar{g}^{\Delta} \bar{V}(\Delta) \tag{33}$$

for any simple function $\bar{g}; \bar{\mathbb{R}} \to \bar{\mathbb{R}}$, $\bar{g} \in \bigcup_{\varepsilon>0,n\geq 1} {}_{\varepsilon,n}\bar{G}_1$, where

$$\bar{V}^{(n)}(dx) = \begin{cases} V^{(n)}(dx), & x \in \mathbb{R}', \\ q^{(n)}(dx), & x \in \mathbb{R}''_o. \end{cases} \tag{34}$$

The left-hand side of (33) is the sum $\sum_{j\leq n} \eta_{n,j}$ of i.i.d. random variables $\eta_{n,j} = \eta'_{n,j} + \eta''_{n,j}$, where

$$\eta'_{n,j} = n^{-\frac{1}{2}} \sum_{\Delta'\subset\mathbb{R}'} \bar{g}^{\Delta'} \left[1\left(\xi_j^{(n)} \in \Delta'/\sqrt{n}\right) - F^{(n)}(\Delta'/\sqrt{n})\right],$$

$$\eta''_{n,j} = \sum_{\Delta''\subset\mathbb{R}''_o} \bar{g}^{\Delta''} \left[1\left(\xi_j^{(n)} \in \Delta''\right) - F^{(n)}(\Delta'')\right].$$

To prove the convergence (33), one verifies the conditions of Proposition 1, which follow easily from the assumptions of this Proposition. ///

Let δ, ε, n_o, $\bar{g} \equiv \bar{g}_2$, $\bar{h} \equiv \bar{h}_2$ satisfy condition (A) of Theorem. We'll denote by Δ' (Δ'') the elements of partition $(\Delta)_N$ which belong to $\mathbb{R}'$ (to $\mathbb{R}''_o$, respectively), and omit the subscript n in $\xi_i^{(n)}$ when this is obvious from the context. Set

$$g^{(n)}(\xi_i,\xi_j) = \begin{cases} n^{-1}\bar{g}^{\Delta'_1\Delta'_2} & \text{if } \xi_i \in \Delta'_1/\sqrt{n},\ \xi_j \in \Delta'_2/\sqrt{n},\ |\xi_i|\leq\varepsilon,\ |\xi_j|\leq\varepsilon, \\ n^{-\frac{1}{2}}\bar{g}^{\Delta''_1\Delta'_2} & \text{if } \xi_i \in \Delta''_1,\ \xi_j \in \Delta'_2/\sqrt{n},\ |\xi_i|>\varepsilon,\ |\xi_j|\leq\varepsilon, \\ n^{-\frac{1}{2}}\bar{g}^{\Delta'_1\Delta''_2} & \text{if } \xi_i \in \Delta'_1/\sqrt{n},\ \xi_j \in \Delta''_2,\ |\xi_i|\leq\varepsilon,\ |\xi_j|>\varepsilon, \\ \bar{g}^{\Delta''_1\Delta''_2} & \text{if } \xi_i \in \Delta''_1,\ \xi_j \in \Delta''_2,\ |\xi_i|>\varepsilon,\ |\xi_j|>\varepsilon, \end{cases}$$

$i,j \geq 1$. According to the definition, if $i \neq j$, then

$$E\left[g^{(n)}(\xi_i,\xi_j) \mid \xi_j\right] \tag{35}$$

$$= \begin{cases} n^{-1} \sum_{\Delta_1'} \bar{g}^{\Delta_1'\Delta_2'} F^{(n)}(\Delta_1'/\sqrt{n}) + n^{-\frac{1}{2}} \sum_{\Delta_1''} \bar{g}^{\Delta_1''\Delta_2'} F^{(n)}(\Delta_1'') & \text{if } \xi_j \in \Delta_2'/\sqrt{n}, \\ n^{-\frac{1}{2}} \sum_{\Delta_1'} \bar{g}^{\Delta_1'\Delta_2''} F^{(n)}(\Delta_1'/\sqrt{n}) + \sum_{\Delta_1''} \bar{g}^{\Delta_1''\Delta_2''} F^{(n)}(\Delta_1'') & \text{if } \xi_j \in \Delta_2''. \end{cases}$$

for any $i,j = 1,\ldots,n$, define

$$\tilde{g}^{(n)}(\xi_i,\xi_j) = g^{(n)}(\xi_i,\xi_j) - E\left[g^{(n)}(\xi_{n+1},\xi_j) \mid \xi_j\right] - E\left[g^{(n)}(\xi_i,\xi_{n+1}) \mid \xi_i\right] + Eg^{(n)}(\xi_i,\xi_j). \tag{36}$$

Then $E(\tilde{g}^{(n)}(\xi_i,\xi_j) \mid \xi_j) = E(\tilde{g}^{(n)}(\xi_i,\xi_j) \mid \xi_i) = 0$ for $i \neq j$, $i,j \leq n$ and therefore $(D^{(n)}(\cdot) \equiv D_2^{(n)}(1,\cdot))$

$$E(D^{(n)}(h^{(n)} - \tilde{g}^{(n)}))^2 = \tfrac{1}{2}n(n-1)\, E\left[h^{(n)}(\xi_1,\xi_2) - \tilde{g}(\xi_1,\xi_2)\right]^2$$

$$\equiv \tfrac{1}{2}n(n-1)\,\chi(n).$$

Here

$$\chi(n) \leq 2E\left[h^{(n)}(\xi_1,\xi_2) - g^{(n)}(\xi_1,\xi_2)\right]^2 + 2E\left[g^{(n)}(\xi_1,\xi_2) - \tilde{g}^{(n)}(\xi_1,\xi_2)\right]^2$$

$$\equiv 2\chi_1 + 2\chi_2.$$

By (36),

$$\chi_2 \leq 4E\left[E\left[g^{(n)}(\xi_1,\xi_2) \mid \xi_2\right]\right]^2 + 2\left[Eg^{(n)}(\xi_1,\xi_2)\right]^2. \tag{37}$$

As

$$\left(E\left[g^{(n)}(\xi_1,\xi_2)\mid\xi_2\right]\right)^2 = \left(E\left[h^{(n)}(\xi_1,\xi_2)-g^{(n)}(\xi_1,\xi_2)\mid\xi_2\right]\right)^2 \le E\left(\left[h^{(n)}(\xi_1,\xi_2)-g^{(n)}(\xi_1,\xi_2)\right]^2\mid\xi_2\right), \tag{38}$$

the first term on the right-hand side of (37) does not exceed $4\chi_1 \equiv 4E\{h^{(n)}(\xi_1,\xi_2)-g^{(n)}(\xi_1,\xi_2)\}^2$. Similarly, the second term is also less than $2\chi_1$. Therefore $\chi_2 \le 6\chi_1$. But

$$n^2\chi_1 = \int_{\bar{\mathbb{R}}^2}\left\{\varepsilon,{}_n\bar{h}^{(n)} - \bar{g}\right\}^2 d^2\,\bar{\nu}^{(n)} < \delta \tag{39}$$

by (28), (27) and (22). This implies

$$E(D^{(n)}(h^{(n)} - \tilde{g}^{(n)}))^2 \le 7\delta\ . \tag{40}$$

Next, $D^{(n)}(\tilde{g}^{(n)}) = \frac{1}{2}\sum_{i,j=1}^{n}\cdots - \frac{1}{2}\sum_{i=j=1}^{n}\cdots \equiv \sum_{1,n} + \sum_{2,n}$, where

$$\sum_{1,n} = \tfrac{1}{2}\sum_{\Delta_1,\Delta_2}\bar{g}^{\Delta_1\Delta_2}\,\bar{V}^{(n)}(\Delta_1)\,\bar{V}^{(n)}(\Delta_2)$$

$$\xrightarrow{d}\ \tfrac{1}{2}\sum_{\Delta_1,\Delta_2}\bar{g}^{\Delta_1\Delta_2}\,\bar{V}(\Delta_1)\,\bar{V}(\Delta_2) \equiv \tfrac{1}{2}\sum_1 \tag{41}$$

according to Proposition 4. Now, Σ_1 can be rewritten as

$$\Sigma_1 = \sum_{\Delta_1\Delta_2}\bar{g}^{\Delta_1\Delta_2}\,\bar{W}(\Delta_1)\,\bar{W}(\Delta_2) + i_1 + i_2 + i_3\ ,$$

where

$$i_1 = -2\sum_{\Delta_1',\Delta_2'}\bar{g}^{\Delta_1'\Delta_2'}\,\nu(\Delta_1')\,W(\Delta_2')\,W(\mathbb{R})\ ,$$

$$i_2 = \sum_{\Delta_1',\Delta_2'}\bar{g}^{\Delta_1'\Delta_2'}\,\nu(\Delta_1')\,\nu(\Delta_2')\,W^2(\mathbb{R})\ ,$$

$$i_3 = -2 \sum_{\Delta_1', \Delta_2''} \bar{g}^{\Delta_1' \Delta_2''} \nu(\Delta_1') \, q(\Delta_2'') \, W(\mathbb{R}) \quad .$$

Let us show that i_1, i_2, i_3 are small in probability. In fact, by the Cauchy inequality,

$$(E|i_1|)^2 \le 4\nu(\mathbb{R}) \sum_{\Delta_2'} \left(\sum_{\Delta_1'} \bar{g}^{\Delta_1' \Delta_2'} \nu(\Delta_1') \right)^2 \nu(\Delta_2') \quad ,$$

where

$$\sum_{\Delta_1'} \bar{g}^{\Delta_1' \Delta_2'} \nu(\Delta_1') = \lim \left(\sum_{\Delta_1'} \bar{g}^{\Delta_1' \Delta_2'} F^{(n)}(\Delta_1'/\sqrt{n}) + \sqrt{n} \sum_{\Delta_1''} \bar{g}^{\Delta_1'' \Delta_2'} F^{(n)}(\Delta_1'') \right) \quad .$$

By (35), (38) - (39)

$$\begin{aligned} \sum_{\Delta_2'} \left\{ \sum_{\Delta_1'} \bar{g}^{\Delta_1' \Delta_2'} \nu(\Delta_1') \right\}^2 \nu(\Delta_2') &= \lim n^2 E \left\{ \left[E\left(g^{(n)}(\xi_1, \xi_2) \mid \xi_2 \right) \right]^2 ; \ |\xi_2| \le \varepsilon \right\} \\ &\le \lim n^2 \chi_1 < \delta \quad . \end{aligned}$$

Similarly,

$$(E|i_k|)^2 \le \text{const } \delta \ , \qquad k = 1,2,3 \quad . \tag{42}$$

We turn now to the sum $\sum_{2,n}$. For $g^{(n)}(\xi_i, \xi_j) = 0$ (g vanishes on diagonals)

$$\sum_{2,n} = -2 \sum_{i=1}^{n} E\left(g^{(n)}(\xi_{n+1}, \xi_i) \mid \xi_i \right) + nEg^{(n)}(\xi_i, \xi_i) \quad . \tag{43}$$

Again, taking the mean square of the first term on the right-hand side of (43), we see that it does not exceed

$$4n^2 E\left[E\left[g^{(n)}(\xi_1,\xi_2) \mid \xi_2\right]\right]^2 \le 4\delta ,$$

and, similarly, the second term does not exceed $\delta^{\frac{1}{2}}$.

Finally, by (29),

$$E\left(\sum_{\Delta_1,\Delta_2} \bar{g}^{\Delta_1\Delta_2} \bar{W}(\Delta_1)\bar{W}(\Delta_2) - \int_{\mathbb{R}^2} \bar{h}(x_1,x_2)\, \bar{W}(dx_1)\bar{W}(dx_2)\right)^2$$

$$\le 2\int_{\mathbb{R}^2} (\bar{g} - \bar{h})^2 \, d^2 \bar{\nu} < 2\delta . \quad (44)$$

Consequently, by (40) - (44),

$$\overline{\lim} \left|E \exp\{iaD^{(n)}(h^{(n)})\} - E \exp\{ia \textstyle\int \bar{h}\, d^2 W\}\right|$$

$$\le \text{const } \delta^{\frac{1}{2}} |a| \quad \text{for each } a \in \mathbb{R}.$$

Since $\delta > 0$ is arbitrarily small, this completes the proof. ///

5. Appendix: MULTIPLE STOCHASTIC INTEGRALS

Let $T \subset \mathbb{R}^m$ $(m \ge 1)$ be an open set, $B(T)$ be its Borel subsets and $B_c(T)$ be its relatively compact subsets. Let μ be a σ-finite measure such that $\mu(A) < +\infty$ $\forall A \in B_c(T)$ and $\mu(\{t\}) = 0$ $\forall t \in T$. Finally, let there be given a real random measure Z in T with independent values on nonintersecting sets. with zero mean and variance μ. In other words, $Z = (Z(A), A \in B_c(T))$ is a family of random variables such that for any $n \ge 1$ and any disjoint $A_1, \dots, A_n \in B_c(T)$, $Z(A_1), \dots, Z(A_n)$ are independent, $Z(A_1) + \cdots + Z(A_n) = Z\left(\bigcup_{i=1}^{n} A_i\right)$ and $\forall A \in B_c(T)$, $EZ(A) = 0$, $EZ^2(A) = \mu(A)$. Any such random measure Z is infinitely divisible and its characteristic function can be written in the Lévy-Hinčin form:

$$E \exp\{iaZ(A)\} = \exp\left\{\int_A \left(-\tfrac{1}{2} a^2 \sigma(t) + \int_{\mathbb{R}_0} (e^{iau} - 1 - iau)\pi(t,du)\right)\right\}, \tag{45}$$

where $\sigma(t) \geq 0$ and $\pi(t,du)$ is a kernel on $T \times \mathbb{R}_0$ such that

$$\sigma(t) + \int_{\mathbb{R}_0} u^2 \pi(t,du) = 1 \qquad \mu\text{-a.e.} \tag{46}$$

Let $L^2(T^k) = L^2(T^k;\mu^k)$ be the Hilbert space of all (measurable) symmetric functions $f\colon T^k \to \mathbb{R}$ with finite norm
$(f,f)_k^{\frac{1}{2}} = \left(\int_{T^k} f^2 \, d^k\mu\right)^{\frac{1}{2}}$.

<u>PROPOSITION 5</u>. For any $k \geq 1$ and $f \in L^2(T^k)$ there exists a real random variable $I^{(k)}(f) = \int f \, d^k Z$, called the k-tuple stochastic integral of f with respect to Z, with the following properties:

$$EI^{(k)}(f) = 0 \tag{47}$$

$$EI^{(k)}(f)\, I^{(k)}(g) = \delta_{kn} k!(f,g)_k \tag{48}$$

for any $n \geq 1$ and $g \in L^2(T^n)$, where δ_{kn} is Kronecker's symbol.

<u>Proof</u>. Let $(\Delta)_N$, $N \geq 1$ be a monotone sequence of partitions of T by sets $\Delta \in B_c(T)$ such that $\sup(\operatorname{diam} \Delta \mid \Delta \in (\Delta)_N) \to 0$ $(N \to \infty)$. A function $g\colon T^k \to \mathbb{R}$ is said to be simple if it satisfies conditions (1) and (2) of Definition 4. For simple $g \in L^2(T^k)$ set

$$I^{(n)}(g) = \sum_{\Delta_1,\ldots,\Delta_k \in (\Delta)_N} g^{\Delta_1 \cdots \Delta_k} Z(\Delta_1) \cdots Z(\Delta_k) .$$

The sum on the right-hand side is well defined for sufficiently large N and does not depend on N. It can easily be verified that it satisfies (47) and (48). For arbitrary $f \in L^2(T^k)$ set $I^{(k)}(f) = \lim I^{(k)}(g_j)$ $(j \to \infty)$, where $(g_j)_{j \geq 1}$ is a sequence of simple functions convergent to f in $L^2(T^k)$. Due to (48), such a limit exists and satisfies (47), (48) as well. ///

REFERENCES

[1] Dynkin, E.B., and Mandelbaum, A. "Symmetric Statistics, Poisson Point Processes, and Multiple Wiener Integrals." *Ann. Stat.*, vol.11 (1983): 733-745.

[2] Fillipova, A.A. "Mises' Theorem on the Asymptotic Behavior of Functionals of Empirical Distribution Functions and Its Statistical Applications." *Theory Probab. Applications*, vol.7 (1961): 24-27.

[3] Gnedenko, B.V., and Kolmogorov, A.N. *Limit Distributions for Sums of Independent Random Variables*. Reading, Mass.: Addison-Wesley, 1954.

[4] Hoefding, W. "A Class of Statistics With Asymptotically Normal Distribution." *Ann. Math. Stat.*, vol.19 (1948): 239-325.

[5] Itô, K. "Multiple Wiener Integral." *J. Math. Soc. Japan*, vol.3 (1951): 157-164.

[6] Itô, K. "Spectral Type of Shift Transformations of Differential Process With Stationary Increments." *Trans. Americ. Math. Soc.*, vol.81 (1956): 253-263.

[7] Major, P. Multiple Wiener-Itô Integrals. *Lecture Notes in Math.*, vol.849. Berlin Heidelberg New York: Springer-Verlag Inc., 1981.

[8] Rubin, H., and Vitale, R.A. "Asymptotic Distribution of Symmetric Statistics." *Ann. Stat.*, 8 (1980): 165-170.

[9] Serfling, R.J. *Approximation Theorems: Theorems of Mathematical Statistics*. New York: John Wiley, 1980.

[10] Surgailis, D. "On L^2 and Non-L^2 Multiple Stochastic Integration." *Lecture Notes in Control and Information Sci.*, pp. 211-226, vol.36. Berlin Heidelberg New York: Springer-Verlag Inc., 1981.

B. GRIGELIONIS

MULTIPLE RANDOM TIME CHANGES OF SEMIMARTINGALES

INTRODUCTION

Let $(\Omega,\mathcal{F},\mathbb{F},P)$ be a filtered probability space (for the standard terminology and notations, see e.g. [1]). Let us recall the following two well-known remarkable results.

THEOREM A. (F.B. Knight [2], [3]) Let $X(t) = (X_1(t),\ldots,X_m(t))$, $t \geq 0$, $X(0) = 0$, be a continuous m-dimensional $(P,\mathbb{F})$-local martingale such that $\langle X_j,X_k\rangle_t \equiv 0$ as $j \neq k$ and for each j, $j = 1,\ldots,m$, $\langle X_j\rangle_t \uparrow \infty$, as $t \to \infty$. Let $\gamma_j(t) = \inf\{s: \langle X_j\rangle_s > t\}$, $t \geq 0$, $j = 1,\ldots,m$. Then $Y(t) = (X_1(\gamma_1(t)), \ldots, X_m(\gamma_m(t)))$, $t \geq 0$ is the standard m-dimensional Brownian motion.

THEOREM B. (P.A. Meyer [4]). Let $X(t) = (X_1(t),\ldots,X_m(t))$, $t \geq 0$, $X(0) = 0$ be a quasi-left continuous m-dimensional point process such that $[X_j,X_k]_t \equiv 0$ for $j \neq k$ and for each j, $j = 1,\ldots,m$, $A_j(t) \uparrow \infty$, as $t \to \infty$, where A_j is the $(P,\mathbb{F})$-compensator of the process X_j, $j = 1,\ldots,m$. Denote $\gamma_j(t) = \inf\{s: A_j(s) > t\}$, $t \geq 0$, $j = 1,\ldots,m$. Then $Y(t) = (X_1(\gamma_1(t)), \ldots, X_m(\gamma_m(t)))$, $t \geq 0$ is a Poisson process with $EY_j(t) = t$, $t \geq 0$, $j = 1,\ldots,m$.

Suppose we are given an m-dimensional $(P,\mathbb{F})$-semimartingale $X(t) = (X_1(t), \ldots, X_m(t))$, $t \geq 0$, having the canonical decomposition

$$X(t) = X(0) + \alpha(t) + X^c(t) + \int_0^t \int_{|x|\leq 1} x q(ds,dx)$$

$$+ \int_0^t \int_{|x|>1} x p(ds,dx) , \qquad t \geq 0 ,$$

with the triple (α, β, Π) of the $(P,\mathbb{F})$-predictable characteristics, where $p(dt,dx)$ is the jump measure of the process X, $q(dt,dx) = p(dt,dx) - \Pi(dt,dx)$, $\Pi(dt,dx)$ is the $(P,\mathbb{F})$-compensator of the measure $p(dt,dx)$, $B(t) = \| \beta_{jk}(t) \|_1^m$, $\beta_{jk}(t) = \langle X_j^c, X_k^c \rangle_t$, $j,k = 1,\ldots,m$, $t \geq 0$.

We shall make the following assumptions.

•I. Assume that $\Pi(\{t\} \times R^m \setminus \{0\}) \equiv 0$, $\beta_{jk}(t) \equiv 0$, for $j \neq k$ and $\operatorname{supp} \Pi \subseteq \bigcup_{j=1}^m (R_+ \times R_j)$, where

$$R_j = \{x \in R^m : x = (0,\ldots,0,x_j,0,\ldots,0),\ x_j \in R \setminus \{0\}\} .$$

By $\Pi_j(dt,dx_j)$, $j = 1,\ldots,m$ we denote the one-dimensional marginals of Π on $\mathcal{B}(R_+) \otimes \mathcal{B}(R \setminus \{0\})$, completely defining the measure Π.

•II. Assume that there exist a continuous $\mathbb{F}$-adapted random function $\tau(t) = (\tau_1(t), \ldots, \tau_m(t))$, $t \geq 0$, $\tau(0) = 0$, $\tau_j(t) \uparrow \infty$, as $t \to \infty$ and triples $(\hat{\alpha}_j, \hat{\beta}_j, \hat{\Pi}_j)$, $j = 1,\ldots,m$, such that $\alpha_j(t) = \hat{\alpha}_j(\tau_j(t))$, $\beta_{jj}(t) = \hat{B}_j(\tau_j(t))$, $\Pi_j([0,t] \times \Gamma) = \hat{\Pi}_j([0,\tau_j(t)] \times \Gamma)$, $t \geq 0$, $\Gamma \in \mathcal{B}(R \setminus \{0\})$, $j = 1,\ldots,m$.

Let $\gamma_j(t) = \inf\{s: \tau_j(s) > t\}$, $t \geq 0$, $j = 1,\ldots,m$ and

$Y(t) = (X_1(\gamma_1(t)), \ldots, x_m(\gamma_m(t)), \quad t \geq 0.$

Using some ideas from the paper [6] under Assumptions I and II we shall find the additional conditions on the triples $(\hat{\alpha}_j, \hat{\beta}_j, \hat{\Pi}_j)$, $j = 1,\ldots,m$ for the process Y to have the independent components. We will consider multiple random time transformations of semimartingales to the processes, components of which have independent increments or are infinitely divisible processes.

1. MULTIPLE RANDOM TIME TRANSFORMATIONS TO THE PROCESSES WITH INDEPENDENT INCREMENTS

The following statement generalizing Theorems A and B (see also [5], [6]) holds.

THEOREM 1. If Assumptions I and II are satisfied, the random variables $(X_1(0), \ldots, X_m(0))$ are mutually independent and the triples $(\hat{\alpha}_j, \hat{\beta}_j, \hat{\Pi}_j)$, $j = 1,\ldots,m$ are non-random, then the process Y has mutually independent components which are processes with independent increments. Also we have

$$E[\exp\{iz(Y_j(t) - Y_j(s))\}]$$

$$= \exp\left\{iz(\hat{\alpha}_j(t) - \hat{\alpha}_j(s)) - \tfrac{1}{2}z^2(\hat{B}_j(t) - \hat{B}_j(s)) + \int_{R\setminus\{0\}} \left(e^{izx} - 1 - iz\chi_{\{|x|\leq 1\}}\right) \hat{\Pi}_j([s,t]x\, dx)\right\},$$

$$0 \leq s < t, \quad j = 1,\ldots,m, \quad z \in R.$$

Proof. We shall use several times the following two well-known statements. A quasi-left continuous $\mathbb{F}$-adapted process $X = (X_1,\ldots,X_m)$ is a $(P,\mathbb{F})$-semimartingale with the triple of predictable characteristics (α, β, π) iff for each $z \in R^m$ the process

$$L_t^z = \exp\left\{ i(X(t)-X(0),\ z) - i(\alpha(t),\ z) + \tfrac{1}{2}(z,\ B(t)z) - \int_{R^m\setminus\{0\}} \left\{ e^{i(x,z)} - 1 - i(x,z)\chi_{\{|x|\leq 1\}} \right\} \Pi([0,t]\times dx) \right\},$$

$$t \geq 0\ ,$$

is a $(P,\mathbb{F})$-local martingale (see [7]). A $(P,\mathbb{F})$-semimartingale X with the triple of the predictable characteristics (α, β, Π) has independent increments iff (α, β, Π) are non-random (see [8]).

For $u = (u_1, \ldots, u_m) \in R_+^m$, $z \in R^m$ and $I \subset \{1, \ldots, m\} = I_m$ let $\gamma^I(u) = \bigwedge_{j \in I} \gamma_j(u_j)$,

$$\mathcal{G}_u = \sigma(Y_j(s_j),\ s_j \leq u_j,\ j \in I_m) \vee \mathcal{F}_{\gamma^{I_m}(u)}\ ,$$

$$\Phi_z^I(u) = \prod_{j \in I} \exp\left\{ iz_j(Y_j(u_j) - Y_j(0)) - iz_j\hat{\alpha}_j(u_j) + \tfrac{1}{2}z_j^2\hat{B}_j(u_j) - \int_{R\setminus\{0\}} \left\{ e^{iz_jx} - 1 - iz_jx\chi_{\{|x|\leq 1\}} \right\} \hat{\Pi}_j([0,u_j]\times dx) \right\}.$$

The statement of Theorem 1 will obviously follow if we prove that for each $v \geq u$, $u, v \in R_+^m$, $z \in R^m$ and $I \in I_m$

$$E[\Phi_z^I(v) \mid \mathcal{G}_u] = \Phi_z^I(u)\ , \tag{1}$$

where $v \geq u$ means that $v_j \geq u_j$, $j \in I_m$.

For any $\mathbb{F}$-predictable bounded functions ϕ_j, $j \in I_m$ let

$$M_j(t) = \exp\left\{ i\int_0^t \phi_j(s)\, dx_j(s) - i\int_0^t \phi_j(s)\, d\alpha_j(s) + \tfrac{1}{2}\int_0^t \phi_j^2(s)\, d\beta_{jj}(s) - \int_0^t \int_{R\setminus\{0\}} \left(e^{i\phi_j(s)x} - 1 - i\phi_j(s)x\, \chi_{\{|x|\le 1\}}\right) \Pi_j(ds,dx) \right\}, \quad t \ge 0, \tag{2}$$

$$M^I(t) = \prod_{j\in I} M_j(t), \qquad t \ge 0, \quad I \subset I_m .$$

Using the exponential Doléans-Dade formula, we find that

$$M^I(t) = 1 + \int_0^t M^I(s-)\, dL^I(s), \quad t \ge 0,$$

where

$$L^I(t) = \sum_{j\in I} i\int_0^t \phi_j(s)\, dX_j^c(s) + \int_0^t \int_{R\setminus\{0\}} \left(e^{i\phi_j(s)x} - 1\right) q_j(ds,dx),$$

$$q_j(dt,dx) = p_j(dt,dx) - \Pi_j(dt,dx)$$

and $p_j(dt,dx)$ is the jump measure of the process X_j.

Therefore for each $I \subset I_m$, M^I is a $(P,\mathbb{F})$-local martingale. Letting

$$T_n = \inf\left\{ t : \sum_{j\in I_m} \left[\beta_{jj}(t) + \int_{R\setminus\{0\}} x^2 \wedge 1\,\Pi_j([0,t]x\, dx)\right] = n \right\} \wedge n, \qquad n \ge 1,$$

we have that $M^I(t \wedge T_n)$, $t \ge 0$ are bounded $(P,\mathbb{F})$-martingales. Applying Lemma 3.18 from [6], we obtain that for each $u \le v$, $u,v \in R_+^m$, $n \ge 1$, $I \subset I_m$

$$E\left[\prod_{j\in I} M_j(\gamma_j(v_j) \wedge T_n) \mid F_{\gamma^I(u)}\right] = \prod_{j\in I} M_j(j^I(u) \wedge T_n) \tag{3}$$

and

$$E\left[\prod_{j\in I} M_j(\gamma_j(u_j)\wedge T_n) \mid F_{\gamma^I(u)}\right] = \prod_{j\in I} M_j(\gamma^I(u)\wedge T_n) . \quad (4)$$

Since

$$\sup_{0\le t\le\gamma_j(v_j)} |M_j(v_j)|$$

$$\le \exp\left\{\sup_{t\ge 0} \phi_j^2(t)\left(\tfrac{1}{2}\hat{B}_j(v_j) + 2\int_{R\setminus\{0\}} x^2 \wedge 1\hat{\Pi}_j([0,v_j]\times dx)\right)\right\}$$

$$< \infty \quad , \qquad j \in I_m \quad ,$$

the equalities (3) and (4) imply that

$$E\left[\prod_{j\in I} M_j(\gamma_j(v_j)) \mid F_{\gamma^I(u)}\right] = E\left[\prod_{j\in I} M_j(\gamma_j(u_j)) \mid F_{\gamma^I(u)}\right] . \quad (5)$$

Taking $0 = t_0^j < t_1^j < \cdots < t^j_{n(j)} = u_j$ and choosing

$$\phi_j(s) = \phi_{jk} \quad \text{for} \quad \gamma_j(t_k^j) < s \le \gamma_j(t_{k+1}^j) , \quad 0 \le k < n^{(j)} ,$$

and equals z_j for $s > \gamma_j(u_j)$, where ϕ_{jk} are bounded and $F\gamma_j(t_k^j)$-measurable, we find that

$$\prod_{j\in I} M_j(\gamma_j(v_j)) = \frac{\Phi_z^I(v)}{\Phi_z^I(u)} \prod_{j\in I} M_j(\gamma_j(u_j)) , \quad (6)$$

$$M_j(\gamma_j(u_j))$$

$$= \exp\left\{\sum_{k=0}^{n^{(j)}-1} i\phi_{jk}\left(Y_j(t_{k+1}^j) - Y_j(t_k^j)\right) - i\phi_{jk}\left(\hat{\alpha}_j(t_{k+1}^j) - \hat{\alpha}(t_k^j)\right)\right.$$

$$+ \tfrac{1}{2}\phi_{jk}^2\left(\hat{B}_j(t_{k+1}^j) - \hat{B}_j(t_k^j)\right) \quad (7)$$

$$\left. - \int_{R\setminus\{0\}} \left(e^{i\phi_{jk}x} - 1 - i\phi_{jk}x\, \chi_{\{|x|\le 1\}}\right)\hat{\Pi}_j((t_k^j, t_{k+1}^j]\times dx)\right\} .$$

Assuming that ϕ_{jk}, $k = 0,1,\ldots,n^{(j)}-1$ are non-random, from the formulas (5) - (7) we have that for each $A \in F_{\gamma^I(u)}$

$$E\left[\frac{\Phi_z^I(v)}{\Phi_z^I(u)} \chi_A \exp\left\{i \sum_{k=0}^{n^{(j)}-1} \phi_{jk}\left(Y_j(t_{k+1}^j) - Y_j(t_k^j)\right)\right\}\right]$$
$$= E\left[\chi_A \exp\left\{i \sum_{k=0}^{n^{(j)}-1} \phi_{jk}\left(Y_j(t_{k+1}^j) - Y_j(t_k^j)\right)\right\}\right]$$

and easily conclude that $E\left(\Phi_z^I(v) \mid F_u\right) = \Phi_z^I(u)$, (cf. [6]). The proof of Theorem 1 is completed.

REMARK 1. Let each component X_j of the semimartingale X be d_j-dimensional, $d_j \geq 1$, $j \in I_m$, $d_1 + \cdots + d_m = d$. Let $X_j^d = \left(X_1^{(j)}, \ldots, X_{d_j}^{(j)}\right)$, let p_j be the jump measure of X_j, and let Π_j the $(P,\mathbb{F})$-compensator of the measure of p_j,

$$\alpha_j = \left(\alpha_1^{(j)}, \ldots, \alpha_{d_j}^{(j)}\right), \qquad B_j = \|\beta_{k\ell}^{(j)}\|_{k,\ell=1}^{d_j},$$

$$R^j = \{x \in R^d : x = (0,\ldots,0,x_j,0,\ldots,0),\ x_j \in R^{d_j} \setminus \{0\}\}.$$

We shall use later the following assumptions.

▲ I'. Assume that $\Pi(\{t\} \times R^d \setminus \{0\}) \equiv 0$, $\operatorname{supp} \Pi \subseteq \bigcup_{j=1}^m (R_+ \times R_j)$,

$$B = \begin{vmatrix} B_1 & 0 & \cdots & 0 \\ 0 & B_2 & \cdots & 0 \\ \cdots & \cdots & \cdots & \cdots \\ 0 & 0 & \cdots & B_m \end{vmatrix}.$$

▲ II'. Assume that there exists a continuous $\mathbb{F}$-adapted random function $\tau(t) = (\tau_1(t), \ldots, \tau_m(t))$, $t \geq 0$, $\tau(0) = 0$, $\tau_j(t) \uparrow \infty$ as $t \to \infty$ and triples $(\hat{\alpha}_j, \hat{\beta}_j, \hat{\Pi}_j)$, $j \in I_m$, such that

$$\alpha_j(t) = \hat{\alpha}_j(\tau_j(t)) , \qquad B_j(t) = \hat{B}_j(\tau_j(t)) ,$$

$$\Pi_j([0,t] \times \Gamma_j) = \hat{\Pi}_j([0,\tau_j(t)] \times \Gamma_j) , \quad t \geq 0 ,$$

$$\Gamma_j \in \mathcal{B}(R^{d_j} \setminus \{0\}) , \qquad j \in I_m .$$

Denote $\gamma_j(t) = \inf \{s: \tau_j(s) > t\}$, $Y_j(t) = X_j(\gamma_j(t))$, $j \in I_m$, $Y(t) = (Y_1(t), \ldots, Y_m(t))$, $t \geq 0$.

After trivial modifications in the proof of Theorem 1 we can derive the following statement.

THEOREM 1'. If Assumptions I' and II' are satisfied, the random vectors $(X_1(0), \ldots, X_m(0))$ are mutually independent and the triples $(\hat{\alpha}_j, \hat{B}_j, \hat{\Pi}_j)$, $j \in I_m$ are non-random, then the processes Y_j, $j \in I_m$, are mutually independent with the independent increments and the following formulas hold:

$$E[\exp\{i(z_j, Y_j(t) - Y_j(s))\}]$$
$$= \exp\Big\{ i(z_j, \hat{\alpha}_j(t) - \hat{\alpha}_j(s)) - \tfrac{1}{2}(z_j, (\hat{B}_j(t) - \hat{B}_j(s))z_j) \tag{8}$$
$$+ \int_{R^{d_j}\setminus\{0\}} \left[e^{i(z_j,x)} - 1 - i(z_j,x)\, \chi_{\{|x|\leq 1\}} \right] \hat{\Pi}_j((s,t] \times dx) \Big\},$$

$$0 \leq s < t , \quad z_j \in R^{d_j} , \quad j \in I_m .$$

REMARK 2. We denote by $\{X_j \underset{t\to\infty}{\to} \cdot\}$ the set of ω for which $X_j(t,\omega)$ converges as $t \to \infty$. Assuming that $\{X_j \underset{t\to\infty}{\to} \cdot\} \subseteq \{\tau_j(\infty) < \infty\}$ P – a.e., we can define

$$Y_j(t) = \begin{cases} X_j(\gamma_j(t)) & \text{for } \tau_j(\infty) > t, \\ X_j(\infty) + \tilde{Y}(t - \tau_j(\infty)) & \text{for } \tau_j(\infty) \le t, \ t \ge 0, \end{cases}$$

where $\tilde{Y}_j$ are mutually independent processes with independent increments and independent of the process X, $\tilde{Y}_j(0) = 0$, $j \in I_m$, having characteristic functions given by the formulas (8), respectively. If the space $(\Omega, F, \mathbb{F}, P)$ is rich enough, such processes $\tilde{Y}_j$ always exist. This modified version of Theorem 1' holds without the assumption that $P\{\tau_j(\infty) = \infty\} = 1$, $j \in I_m$ (see [2] - [5] for special cases).

2. MULTIPLE RANDOM TIME TRANSFORMATIONS TO THE LOCALLY INFINITETY DIVISIBLE PROCESSES

Under Assumptions I and II we suppose that for each $t > 0$ there exists a constant $K(t) \to \infty$ such that

$$\left[\mathrm{Tr}\hat{B}(t) + \sum_{j \in I_m} \int_{R \setminus \{0\}} x^2 \wedge 1 \hat{\Pi}_j([0,t] \times dx) \right] \le K(t) \quad . \tag{9}$$

Then obviously we shall have that the equality (5) holds without the assumption that the triples $(\hat{\alpha}_j, \hat{B}_j, \hat{\Pi}_j)$, $j \in I_m$ are non-random. Let $F_t^{Y_j} = \sigma(Y_j(s), s \le t)$, $t \ge 0$, $\mathbb{F}^{Y_j} = \{F_t^{Y_j}, t \ge 0\}$, $j \in I_m$.

•III. Assume further that the triple $(\hat{\alpha}_j, \hat{B}_j, \hat{\Pi}_j)$ is $\mathbb{F}^{Y_j}$ - predictable and the $\sigma(L^\infty, L^1)$ -linear span of the random variable

$$\exp\left\{\sum_{k=0}^{n-1} i\phi_k(Y_j(t_{k+1})-Y_j(t_k)) - i\phi_k(\hat{\alpha}_j(t_{k+1})-\hat{\alpha}_j(t_k)) + \tfrac{1}{2}\phi_k^2(\hat{B}_j(t_{k+1})-\hat{B}_j(t_k)) - \int_{R\setminus\{0\}}\left(e^{i\phi_k x} - 1 - i\phi_k x\,\chi_{\{|x|\le 1\}}\right)\hat{\Pi}_j((t_k,t_{k+1}]\times dx)\right.,$$

$$0 = t_0 < t_1 < \cdots < t_n = t\ ,$$

ϕ_k is bounded $F_{t_k}^{Y_j}$-measurable, $n \ge 1$, contains all bounded $\sigma(Y_j(s) - Y_j(0),\ s \le t)$-measurable random variables, $j \in I_m$, $t > 0$.

From Assumptions I - III, (9) and the equalities (5) - (7) we obtain that for each $u \le v$, $u,v \in R_+^m$, $z \in R^m$, $I \subset I_m$ the equality (1) holds, which implies that the processes Y_j are semimartingales having the triples $(\hat{\alpha}_j, \hat{B}_j, \hat{\Pi}_j)$, $j \in I_m$ of predictable characteristics with respect to the natural filtrations.

• IV. Assume that

$$\hat{\alpha}_j(t) = \int_0^t \hat{a}_j(s,Y_j)\,ds\ ,\qquad \hat{B}_j(t) = \int_0^t \hat{A}_j(s,Y_j)\,ds,$$

$$\hat{\Pi}_j([0,t]\times\Gamma) = \int_0^t \hat{\pi}_j(s,Y_j,\Gamma)\,ds\ ,$$

$t \ge 0$, $\Gamma \in B(R\setminus\{0\})$, $j \in I_m$, i.e., Y_j are locally infinitely divisible processes with the triples $(\hat{a}_j, \hat{A}_j, \hat{\pi}_j)$ of local characteristics, and the corresponding Itô stochastic equations defining the processes Y_j, $j \in I_m$ (see, e.g., [1], [9], [10]) have a unique strong solution.

THEOREM 2. If Assumptions I - IV and the inequality (9) are satisfied and the random variables $(X_1(0), \ldots, X_m(0))$ are mutually independent, then the process Y has mutually independent components Y_j that are locally infinitely divisible processes having the triples $(\hat{a}_j, \hat{A}_j, \hat{\pi}_j)$, $j \in I_m$ of local characteristics with respect to the natural filtrations.

Proof. According to the above Remarks it is enough to check that the processes Y_j, $j \in I_m$, are mutually independent. This property can be derived from the equality (1) and Assumption IV.

REMARK 3. It looks as if the assumptions of Theorem 2 can be relaxed and made more explicit.

REMARK 4. The analogue of Theorem 2 can be easily obtained in the case when the components X_j of the process X are d_j-dimensional, $d_j \geq 1$, $j \in I_m$.

REFERENCES

[1] Jacod, J. Calcul stochastique et problèmes des martingales. *Lecture Notes in Math.*, vol.714. Berlin Heidelberg New York: Springer-Verlag Inc., 1979.

[2] Knight, F.B. "A Reduction of Continuous Square Integrable Martingales to Brownian Motion." *Lecture Notes in Math.*, pp. 19-31, vol.190. Berlin Heidelberg New York: Springer-Verlag Inc., 1970.

[3] Knight, F.B. "An Infinitesimal Decomposition for a Class of Markov Processes." *Ann. Math. Stat.*, vol.41, no.5 (1970): 1510-1529.

[4] Meyer, P.A. "Démonstration simplifiée d'un théorème de Knight." *Lecture Notes in Math.*, pp. 191-195, vol.191. Berlin Heidelberg New York: Springer-Verlag Inc., 1971.

[5] Coccozza, C., and Yor, M. "Démonstration d'un théorème de F. Knight à l'aide de martingales exponentielles." *Lecture Notes in Math.*, pp. 496-499, vol.784. Berlin Heidelberg New York: Springer-Verlag Inc., 1980.

[6] Kurtz, T.G. "Representations of Markov Processes as Multiparameter Time Changes." *Ann. Probab.*, vol.8, no.4 (1980): 682-715.

[7] Grigelionis, B., and Mikulevicius. "Weak Convergence of Stochastic Point Processes." *Lithuanian Math. Journal*, vol.21, no.4 (1981): 297-300.

[8] Grigelionis, B. "Martingale Characterization of Random Processes by Independent Increments." *Lithuanian Math. Journal*, vol.17, no.1 (1977): 52-60.

[9] Grigelionis, B. "On the Representation of Integer-Valued Measures by Means of Stochastic Integrals with Respect to Poisson Measure." *Litovskij matem. sbornik*, vol.11, no.1 (1971): 93-108.

[10] Karoui, N.El., and Lepeltier, J.P. "Représentation des processus ponctuels multivariés à l'aide d'un processus de Poisson." *Z. Wahrscheinlichkeitstheorie und Verw. Gebiete*, B.39 (1977): 111-135.

M.L. KLEPTSINA and A.Yu. VERETENNIKOV

ON FILTERING AND PROPERTIES OF CONDITIONAL LAWS OF ITO-VOLTERRA PROCESSES

INTRODUCTION

Recently there have appeared intensive studies of Itô-Volterra processes (cf. [1 - 6], etc.). Such processes play an important role in control theory and various applications; they comprise a rather general class of stochastic processes with delay. The necessity of constructing a theory for handling such processes was emphasized, e.g., by A.V. Balakrishnan (cf., for instance, [7]). Our paper is devoted to two related problems for Itô-Volterra processes: the first problem consists in constructing the optimal filter in the conditionally Gaussian case, and the second problem consists in investigating smoothness properties of conditional laws. The filter of the Kalman-Bucy type for the Gaussian case has been constructed by one of the authors in [5]. The smoothness of conditional laws has been established with the aid of the Malliavin calculus version (cf. [9 - 13]).

1. CONDITIONALLY GAUSSIAN FILTER

Let (Ω,F,P) be a complete probability space with an increasing right-continuous family of σ-algebras (F_t), and let (w_t^1,F_t)

and (w_t^2, F_t) be independent Wiener processes. Let (θ_t, ξ_t) be a continuous solution of the system

$$\theta_t = \int_0^t b(t,s,\xi)\, dw_s^2 + \int_0^t (a_0(t,s,\xi) + a_1(t,s,\xi)\theta_s)\, ds ,$$

$$\xi_t = \int_0^t B(s,\xi)\, dw_s^1 + \int_0^t (A_0(s,\xi) + A_1(s,\xi)\theta_s)\, ds , \qquad (1)$$

$$0 \le t \le T .$$

The functionals $B(s,x)$, $b(t,s,x)$, $a_i(t,s,x)$, $A_i(s,x)$, $i = 0,1$, are bounded, F_s-non-anticipatory, continuous in t uniformly in (s,x),

$$\inf_{x \in C[0,T]} B(t,x) \geq c > 0 , \quad 0 \le t \le T , \qquad (2)$$

and the inequality

$$|B(t,x) - B(t,y)|^2 \leq L_1 \int_0^t |x_s - y_s|^2\, dK_s + L_2 |x_t - y_t|^2 \qquad (3)$$

holds for every $x,y \in C[0,T]$, where K_s is an increasing continuous function, $0 \le K_s \le 1$. Let $L(\eta_0^t \mid F_t^\xi)$ be the regular conditional distribution of a process η_s, $0 \le s \le t$, with respect to σ-algebra F_t^ξ.

THEOREM 1. The conditional law $L(\theta_0^t \mid F_t^\xi)$ is Gaussian.

Proof. Let $(\tilde\theta_t, \tilde\xi_t)$ satisfy the following system:

$$\tilde\theta_t = \int_0^t b(t,s,\tilde\xi)\, dw_s^2 + \int_0^t (a_0(t,s,\tilde\xi) + a_1(t,s,\tilde\xi)\tilde\theta_s)\, ds ,$$

$$\tilde\xi_t = \int_0^t B(s,\tilde\xi)\, dw_s^1 . \qquad (4)$$

By virtue of (2) and (3) system (4) has a unique solution $\tilde{\xi}_t = \tilde{\xi}_t[w^1]$, $\tilde{\theta}_t = \tilde{\theta}_t[\tilde{\xi},w^2]$, and $\theta_t = \theta_t[\xi,w^2] = \tilde{\theta}_t[\xi,w^2]$.

LEMMA 1. The conditional law $L((\tilde{\theta},w^2)_0^t \mid F_t^{\tilde{\xi}})$ is Gaussian.

Proof. Consider the following processes: $\theta_t^0 = 0$,

$$\theta_t^{n+1} = \int_0^t b(t,s,\tilde{\xi})dw_s^2 + \int_0^t (a_0(t,s,\tilde{\xi}) + a_1(t,s,\tilde{\xi})\theta_s^n)ds .$$

Then $\lim_{n\to\infty} E \sup_{0\le t\le T} |\theta_t^n - \tilde{\theta}_t|^2 = 0$, $E \sup_{n\ge 1, t\le T} |\theta_t^n|^2 \le C < \infty$, and the conditional law $L((\theta^0,w^2)_0^t \mid F_t^{\tilde{\xi}})$ is Gaussian.

Let the conditional law $L((\theta^n,w^2)_0^t \mid F_t^{\tilde{\xi}})$ be Gaussian. Then

$$E\left(\exp\left(i\sum_{j=1}^m z_j w_{t_j}^2 + i\sum_{\ell=1}^k u_\ell \theta_{t_\ell}^{n+1}\right) \Bigm| F_t^{\tilde{\xi}}\right)$$

$$= E\left(\exp\left(i\sum_{j=1}^m z_j w_{t_j}^2 + i\sum_{\ell=1}^k u_\ell\left(\int_0^{t_\ell} b(t_\ell,s,\tilde{\xi})dw_s^2 + \int_0^{t_\ell}(a_0(t_\ell,s,\tilde{\xi}) + a_1(t_\ell,s,\tilde{\xi})\theta_s^n)ds\right)\right) \Bigm| F_t^{\tilde{\xi}}\right) . \tag{5}$$

As $L((\theta^n,w^2)_0^t \mid F_t^{\tilde{\xi}})$ is Gaussian, then by virtue of (5) $L((\theta^{n+1},w^2) \mid F_t^{\tilde{\xi}})$ is also Gaussian and

$$E\left(\exp\left(i\sum_{j=1}^m z_j w_{t_j}^2 + i\sum_{\ell=1}^k u_\ell \theta_{t_\ell}^n\right) \Bigm| F_t^{\tilde{\xi}}\right)$$

$$= \exp\left(i\sum_{j=1}^m z_j m_{t_j}^w(\tilde{\xi}) + i\sum_{\ell=1}^k u_\ell m_{t_\ell}^{\theta^n}(\tilde{\xi}) - \tfrac{1}{2}\gamma_n(\tilde{\xi})\right) ,$$

where $\gamma_n(\tilde{\xi})$ is a positive definite quadratic form and $m(\tilde{\xi})$ is some functional. Thus, the conditional law $L((\tilde{\theta},w)^2 \mid F_t^{\tilde{\xi}})$ is Gaussian.

Let us continue the proof of Theorem 1. By virtue of Lemma 11.5 of [8] we have

$$E\left(\exp\left[i\sum_{j=1}^{m} z_j\theta_{t_j}[\xi,w^2]\right] \mid F_t^{\xi}\right)$$

$$= \int_{C[0,T]} \exp\left(i\sum_{j=1}^{m} z_j\theta_{t_j}[\xi,c]\right) \rho_T(c,\xi)\, d\mu_w(c) \equiv I(\xi) ,$$

where

$$\rho_T(c,\xi) = \exp\left\{\int_0^T \left(A_1(s,\xi)\frac{\theta_s[\xi,c]-m_s(\xi)}{G(s,\xi)}\right) d\bar{w}_s - \tfrac{1}{2}\int_0^T \left(A_1^2(s,\xi)\frac{(\theta_s[\xi,c]-m_s(\xi))^2}{B^2(s,\xi)}\right) ds\right\} ,$$

$$m_s(\xi) = E(\theta_s \mid F_s^{\xi}) ,$$

$$d\bar{w}_s = \frac{d\xi_s - (A_0(s,\xi) + A_1(s,\xi)m_s(\xi))\, ds}{B(s,\xi)} .$$

Let us calculate $I(\tilde{\xi})$. We have

$$I(\tilde{\xi}) = E\left\{\exp\left(i\sum_{j=1}^{m} z_j\theta_{t_j}[\tilde{\xi},w^2] + \left[\int_0^T \frac{A_1(s,\tilde{\xi})}{B^2(s,\xi)}(\theta_s[\tilde{\xi},w^2]-m_s(\tilde{\xi}))d\tilde{\xi}_s - \tfrac{1}{2}\int_0^T \left(\frac{A_1(s,\tilde{\xi})}{B(s,\xi)}\right)^2 (\theta_s[\tilde{\xi},w^2]-m_s(\tilde{\xi}))^2\, ds - \frac{A_1(s,\tilde{\xi})(A_0(s,\tilde{\xi})+A_1(s,\tilde{\xi})m_s(\tilde{\xi}))}{B^2(s,\tilde{\xi})}(\theta_j[\tilde{\xi},w^2]-m_s(\tilde{\xi}))\, ds\right]\right\} \mid F_t^{\tilde{\xi}}\right).$$

Since $\theta_s[\tilde{\xi},w^2] = \tilde{\theta}_s[\tilde{\xi},w^2]$, then by virtue of Lemma 11.6 of [8] and Lemma 1 we obtain

$$I(\tilde{\xi}) = \exp\left(-\int_0^T \frac{A_1(s,\tilde{\xi})}{B^2(s,\tilde{\xi})}\left(m_s(\tilde{\xi})\,d\xi_s + A_1(s,\tilde{\xi})A_0(s,\tilde{\xi})m_s(\tilde{\xi})\,ds\right)\right)$$

$$\times \exp\left(i\sum_{j=1}^m z_j m_{t_j}^{\tilde{\theta}}(\tilde{\xi}) + \Phi(\tilde{\xi}) - \tfrac{1}{2}\sum_{j,q=1}^m z_j z_q \gamma_{jq}(\tilde{\xi})\right.$$

$$\left. + i\sum_{q=1}^m z_q\delta_q(\tilde{\xi}) + \tfrac{1}{2}\psi(\tilde{\xi})\right) ,$$

where explicit expressions for the functionals $\psi(\tilde{\xi})$, $\delta(\tilde{\xi})$ and $\Phi(\tilde{\xi})$ may be found by the formula (11.48) of [8]. Since $I(\xi) = I(\tilde{\xi})|_{\tilde{\xi}=\xi}$, then

$$I(\xi) = \exp\left(i\sum_{q=1}^m z_q\delta_z(\xi) - \tfrac{1}{2}\sum_{q,j=1}^m \gamma_{qj}(\xi)z_q z_j\right) . \quad ///$$

COROLLARY. Let

$$v_s^t = \int_0^s b(t,r,\xi)dw_r^2 + \int_0^s (a_0(t,r,\xi) + a_1(t,r,\xi)\theta_r)dr .$$

Then the conditional law $L((v_{s_1}^{t_1}, \ldots, v_{s_n}^{t_n}) \mid F_t^\xi)$, $s_1 \le s_2 \le \cdots \le s_n \le t$ is Gaussian. Let

$$m_t = E(\theta_t \mid F_t^\xi) ,$$

$$\gamma_t = E((\theta_t - m_t)^2 \mid F_t^\xi) ,$$

$$K(t,s) = E((\theta_t - m_t)(\theta_s - m_s) \mid F_t^\xi) , \qquad s \le t ,$$

$$m_s^t = E(v_s^t \mid F_s^\xi) ,$$

$$\delta_s = \theta_s - m_s ,$$

$$\delta_s^t = v_s^t - m_s^t ,$$

$$f(t,\xi,s) = E(\delta_s\delta_s^t \mid F_s^\xi) .$$

THEOREM 2. The following equations hold:

$$m_t = \int_0^t (a_0(t,s,\xi) + a_1(t,s,\xi)m_s)ds + \int_0^t \frac{A_1(s,\xi)}{B(s,\xi)} f(t,\xi,s)\, d\bar{w}_s , \tag{6}$$

$$\gamma_t = \int_0^t \left(2a_1(t,s,\xi)f(t,\xi,s) + b^2(t,s,\xi) - \frac{A_1^2(s,\xi)}{B^2(s,\xi)} f^2(t,\xi,s)\right) ds , \tag{7}$$

$$K(t,s) = f(t,\xi,s) + \int_s^t \left(a_1(t,r,\xi) - \frac{A_1^2(s,\xi)}{B^2(s,\xi)} f(t,\xi,r)K(r,s)\right) dr , \tag{8}$$

$$\begin{aligned} f(t,\xi,s) = & \int_0^s b(t,r,\xi)b(s,r,\xi)\, dr \\ & + \int_0^s (a_1(t,r,\xi)f(s,\xi,r) + a_1(s,r,\xi)f(t,\xi,r))\, dr \\ & - \int_0^s \frac{A_1^2(r,\xi)}{B^2(r,\xi)} f(t,\xi,r)f(s,\xi,r)\, dr . \end{aligned} \tag{9}$$

This system has a unique solution (m,γ,K,f) with $\gamma_t \geq 0$, continuous in t.

Proof. By virtue of Theorem 8.6 of [8] we obtain

$$\begin{aligned} m_s^t = & \int_0^s (a_0(t,r,\xi) + a_1(t,r,\xi)m_r)\, dr \\ & + \int_0^s \frac{A_1(r,\xi)}{B(r,\xi)} E((v_r^t\theta_r - m_r^t m_r) \mid \mathcal{F}_r^\xi)\, d\bar{w}_r . \end{aligned}$$

Assuming $s = t$, we obtain (6):

$$m_t = \int_0^t (a_0(t,r,\xi) + a_1(t,r,\xi)m_r)\, dr + \int_0^t \frac{A_1(r,\xi)}{B(r,\xi)} f(t,\xi,r)\, d\bar{w}_r .$$

Let us derive (9). By definition, $f(t,\xi,s) = E(\delta_s \delta_s^t \mid \mathcal{F}_s^\xi)$. By Itô's formula

$$d_s \delta_s^u = \left\{a_1(u,s,\xi) - \frac{A_1^2(s,\xi) f(u,\xi,s)}{B^2(s,\xi)}\right\} \delta_s \, ds + b(u,s,\xi)\, dw_s^2$$

$$+ \frac{A_1(s,\xi)}{B(s,\xi)} f(u,\xi,s)\, dw_s^1 \quad ,$$

$$d_s \delta_s^u \delta_s^t = \delta_s^t \left\{a_1(u,s,\xi) - \frac{A_1^2(s,\xi) f(u,\xi,s)}{B^2(s,\xi)}\right\} \delta_s \, ds$$

$$+ \delta_s^u \left\{a_1(t,s,\xi) - \frac{A_1^2(s,\xi) f(t,\xi,s)}{B^2(s,\xi)}\right\} \delta_s \, ds$$

$$+ b(u,s,\xi) b(t,s,\xi) + \frac{A_1^2(s,\xi)}{B^2(s,\xi)} f(u,\xi,s)\, f(t,\xi,s)\, ds$$

$$- \delta_s^t \left\{\frac{A_1(s,\xi)}{B(s,\xi)} f(u,\xi,s)\, dw_s^1 - b(u,s,\xi)\, dw_s^2\right\}$$

$$- \delta_s^u \left\{\frac{A_1(s,\xi)}{B(s,\xi)} f(t,\xi,s)\, dw_s^1 - b(t,s,\xi)\, dw_s^2\right\} .$$

By virtue of Theorem 8.6 of [8] we have

$$E(\delta_s^u \delta_s^t \mid \mathcal{F}_s^\xi)\Big|_{u=s} = \int_0^s \Big[a_1(s,r,\xi) E(\delta_r^t \delta_r \mid \mathcal{F}_r^\xi) + a_1(t,r,\xi) E(\delta_r^s \delta_r \mid \mathcal{F}_r^\xi)$$

$$- \frac{A_1^2(r,\xi)}{B^2(r,\xi)} f(t,\xi,r)\, f(s,\xi,r)\Big]\, dr$$

$$+ \int_0^s \frac{A_1(r,\xi)}{B(r,\xi)} E(\delta_r^s \delta_r^t \delta_r \mid \mathcal{F}_r^\xi)\, d\bar{w}_r \quad .$$

By virtue of Theorem 1, $E(\delta^s_r \delta^t_r \delta_r \mid F^\xi_r) = 0$. Since $E(\delta^t_r \delta_r \mid F^\xi_r) = f(t,\xi,r)$, then by (9) we obtain (7).

Now let us prove (8). Let

$$\sigma^t_s = \int_0^t I(r\le s)\left[a_1(s,r,\xi) - \frac{A_1^2(r,\xi)}{B^2(r,\xi)} f(s,\xi,r)\right]\delta_r\, dr + \int_0^t I(r\le s)\, b(s,r,\xi)\, dw^2_r - \int_0^t I(r\le s)\frac{A_1(r,\xi)}{B(r,\xi)} f(s,\xi,r)\, dw^1_r .$$

We have

$$K(t,s) = E(\delta_s \delta_t \mid F^\xi_t) = E(\delta^s_s \delta^t_s \mid F^\xi_t) ,$$

$$K(t,s) = E(\sigma^s_t \delta^u_t \mid F^\xi_t)\Big|_{u=t} .$$

By virtue of Itô's formula we have

$$\begin{aligned} d_r \sigma^s_r \delta^u_r = {} & \delta^u_r I(r\le s)\left[a_1(s,r,\xi) - \frac{A_1^2(r,\xi)}{B^2(r,\xi)} f(s,\xi,r)\right]\delta_r\, dr \\ & + \sigma^s_r\left[a_1(u,r,\xi) - \frac{A_1^2(r,\xi)}{B^2(r,\xi)} f(u,\xi,r)\right]\delta_r\, dr \\ & + I(r\le s)\, b(s,r,\xi)\, b(u,r,\xi)\, dr \\ & + I(r\le s)\frac{A_1^2(r,\xi)}{B^2(r,\xi)} f(s,\xi,r)\, f(u,\xi,r)\, dr \\ & + I(r\le s)\left[b(s,r,\xi)\, dw^2_r - \frac{A_1(r,\xi)}{B(r,\xi)} f(s,\xi,r)\, dw^1_r\right]\delta^u_r \\ & + \sigma^s_r\left[b(u,r,\xi)\, dw^2_r - \frac{A_1(r,\xi)}{B(r,\xi)} f(u,\xi,r)\, dw^1_r\right] . \end{aligned}$$

Applying the general filtering equations again with $u = t \geq s$, we obtain

$$K(t,s) = \int_0^s \left\{ a_1(s,r,\xi) f(t,\xi,r) - \frac{A_1^2(r,\xi)}{B^2(r,\xi)} f(s,\xi,r)\, f(t,\xi,r) \right\} dr$$

$$+ \int_0^s \left\{ b(s,r,\xi)\, b(t,r,\xi) + \frac{A_1^2(r,\xi)}{B^2(r,\xi)} f(t,\xi,r)\, f(s,\xi,r) \right\} dr$$

$$+ \int_0^t \left(\left(a_1(t,r,\xi) - \frac{A_1^2(r,\xi)}{B^2(r,\xi)} \right) E(\sigma_r^s \delta_r \mid F_r^\xi) \right) dr$$

$$+ \int_0^t \frac{A_1(r,\xi)}{B(r,\xi)} \left| E(\delta_r \delta_r^t \sigma_r^s \mid F_r^\xi) - E(\sigma_r^s \mid F_r^\xi) f(t,\xi,r) \right| d\bar{w}_r \; .$$

Since $\sigma_r^s = \delta_r^s, \quad r \leq s, \quad \sigma_r^s = \delta_s, \quad r \geq s,$ we obtain

$$\begin{aligned} K(t,s) = & \int_0^s \Big\{ a_1(s,r,\xi)\, f(t,\xi,r) + a_1(t,r,\xi) f(s,\xi,r) \\ & \quad - \frac{A_1^2(r,\xi)}{B^2(r,\xi)} f(t,\xi,r)\, f(s,\xi,r) + b(t,r,\xi)\, b(s,r,\xi) \Big\} dr \\ & + \int_s^t \left\{ a_1(t,r,\xi) - \frac{A_1^2(r,\xi)}{B^2(r,\xi)} f(t,\xi,r) \right\} K(r,s)\, dr \\ & + \int_s^t \frac{A_1(r,\xi)}{B(r,\xi)} \left| E(\delta_r \delta_r^t \delta_s \mid F_r^\xi) - E(\delta_s \mid F_r^\xi) f(t,\xi,r) \right| d\bar{w}_r \; . \end{aligned}$$

By virtue of Theorem 1

$$\begin{aligned} E(\delta_s \delta_r \delta_r^t \mid F_t^\xi) &= E(\delta_s \mid F_r^\xi)\, E(\delta_r \delta_r^t \mid F_r^\xi) \\ &= E(\delta_s \mid F_r^\xi)\, f(t,\xi,r) \; . \end{aligned}$$

Taking (9) into consideration, we obtain

$$K(t,s) = f(t,\xi,s) + \int_s^t \left\{ a_1(t,r,\xi) - \frac{A_1^2(r,\xi)}{B^2(r,\xi)} f(t,\xi,r) \right\} K(r,s)\, dr .$$

Uniqueness can be obtained in a standard way. ///

2. SMOOTHNESS OF CONDITIONAL DISTRIBUTIONS

Consider the Itô-Volterra equation

$$x_t = x_0 + \int_0^t \Big\{ X_0(x_s,z_s)ds + \sum_{i=1}^m X_i(x_s,z_s) \circ dw_s^i + \sum_{j=1}^n \tilde{X}_j(x_s,z_s)(\circ\, d\tilde{w}_s^j + \ell^j(t,x_s,z_s)ds) \Big\} ,$$

$$z_t = z_0 + \int_0^t \Big\{ Z_0(t,z_s) + \sum_{j=1}^n Z_j(z_s)(\circ\, d\tilde{w}_s^j + \ell^j(t,x_s,z_s)\, ds \Big\} , \quad t \ge 0 \tag{10}$$

(we have a new notation), where w^i, $\tilde{w}^j$ are independent Wiener processes on a complete probability space $(\Omega,F,(F_t),P)$, $d \ge 1$, $\int \circ\, dw$ and $\int \circ\, dw$ are stochastic Stratonovitch integrals; later we also use the Itô integrals $\int dw$ and $\int d\tilde{w}$. Every vector field X_0, X_i, $\tilde{X}_j$, Z_0, Z_j, ℓ^j of dimension d is supposed to belong to C_b^∞ in all variables. The matrix XX^* (here $X = (X_i)_1^m$) is assumed to be nondegenerate.

Let $T \ge 0$, $t > 0$. By virtue of the results in [14] the regular conditional laws

$$P(x_t \in A \mid z_0^t) , \qquad A \in B^d ,$$

are well defined (here B^d is a Borel σ-algebra in E^d).

THEOREM 3. For almost every trajectory $z_\bullet$ with respect to P for arbitrary $t > 0$ and $T \geq 0$ the conditional law $P(x \in A \mid z_0^T)$ has a density $p(x) = p^{z_\bullet}(t,T,x)$ of class $C_b^{0,0,\infty}([0,K] \times [\delta \times K] \times E^d)$ for arbitrary $0 < \delta \leq K$.

In the Markovian case an analogous result is proved in [13]. Instead of nondegeneracy of XX^* in [13], the weaker condition of the Hörmander type is assumed: the fields $(X_i,0)^*$ and all Lie brackets of the fields $(X_i,0)^*$, generate in (x_0,z_0) the linear space of a full dimension.

Our theorem under such a condition (instead of the nondegeneracy) remains true. Some modifications are necessary in Proof IV only.

In this paper, Bismut's approach is used as in [13].

Proof. Let us divide the proof into steps. Assume $d = 1$ for simplicity (cf. [13]).

1.

Under the conditions of the Theorem, equations (10) may be expressed in the form

$$x_t = x_0 + \int_0^t \Bigg(X_0(x_s,z_s)\,ds + \sum_{i=1}^m X_i(x_s,z_s) \circ dw_s^i + \sum_{j=1}^n \tilde{X}_j(x_s,z_s) \Bigg\{ \circ\, d\tilde{w}_s^j + \ell^j(s,x_s,z_s)\,ds + \left[\int_0^s \ell_s^j(s,x_r,z_r)\,dr \right] ds \Bigg\} \Bigg) ,$$

$$z_t = z_0 + \int_0^t \left[Z_0(s,z_s)ds + \left(\int_0^s Z_{0s}(s,z_r)dr \right) ds + \sum_{j=1}^n Z_j(z_s) \left[\circ\, d\tilde{w}_s^j + \ell^j(s,x_s,z_s)ds + \left(\int_0^s \ell_s^j(s,x_r,z_r)dr \right) ds \right] \right] . \tag{11}$$

Here Z_{0s}, ℓ_s^j are derivatives of the fields Z_0 and ℓ^j in s. Let $K > 0$, $0 \le T \le K$, $0 < \tau \le K$, $\varepsilon \ge 0$ be real numbers, and let $g \in C_b^\infty(E^1)$, $f \in B(C[0,K];E^1)$, u_t^i $(1 \le i \le m)$ be a random process integrable in (t,ω) with every degree,

$$\hat{w}_t^j = \tilde{w}_t^j + \int_0^t \left(\ell^j(s,x_s,z_s) + \int_0^s \ell_s^j(s,x_r,z_r)dr \right) ds .$$

Consider a perturbed stochastic equation

$$x_t^\varepsilon = x_0 + \int_0^t X_0(x_s^\varepsilon,z_s)ds + \sum_{i=1}^m X_i(x_s^\varepsilon,z_s) \circ dw_s^i + \varepsilon u_s^i ds + \sum_{j=1}^n \tilde{X}_j(x_s^\varepsilon,z_s) \circ d\hat{w}_s^j . \tag{12}$$

2.

Let
$$y_t^\varepsilon = \frac{dx^\varepsilon}{d\varepsilon} , \qquad y_t = y_t^0 ,$$

$$\rho_\varepsilon = \exp\left(-\int_0^K \sum_{i=1}^m \varepsilon u_s^i\, dw_s^i - \frac{\varepsilon^2}{2} \int_0^K \sum_{i=1}^m |u_s^i|^2\, ds \right) ,$$

$$\phi = \phi[x_\cdot,z_\cdot] = \exp\left(-\int_0^K \sum_{j=1}^n \ell^j(s,x_s,z_s) + \left(\int_0^s \ell_s^j(s,x_r,z_r dr \right) d\tilde{w}_s^j - \tfrac{1}{2} \int_0^K \sum_{j=1}^n \left| \ell^j(s,x_s,z_s) + \int_0^s \ell_s^j(s,x_r,z_r)\, dr \right|^2 \right) ,$$

$$dP^\phi = \phi dP , \ E^\phi$$

be an expectation with respect to the measure P^ϕ. By virtue of Girsanov's theorem ([8]) the processes $\hat{w}_t^j$ and w_t^i $(1 \le i \le m,$ $1 \le j \le n,$ $0 \le t \le K)$ are independent Wiener processes with respect to the measure P^ϕ, and the following equality holds:

$$E^\phi f[z_\bullet]\, \rho_\varepsilon\, \phi^{-1}[x_\bullet^\varepsilon, z_\bullet] g(x_\tau^\varepsilon) \;=\; E^\phi f[z_\bullet]\, \phi^{-1}[x_\bullet, z_\bullet] g(x_\tau) \,. \tag{13}$$

By virtue of the theorem on an integral derivative, equality (13) may be differentiated in ε. When $\varepsilon = 0$ we obtain

$$E^\phi f[z_\bullet]\phi^{-1}[x_\bullet, z_\bullet] g'(x_\tau) y_\tau + E^\phi f[z_\bullet] g(x_\tau) \left.\frac{d\phi^{-1}[x_\bullet^\varepsilon, z_\bullet]}{d\varepsilon}\right|_{\varepsilon=0} \tag{14}$$
$$+ E^\phi f[z_\bullet] g(x_\tau)\phi^{-1}[x_\bullet, z_\bullet] \left.\frac{d\rho_\varepsilon}{d\varepsilon}\right|_{\varepsilon=0} = 0 \,.$$

It is not hard to write explicit formulas for the expressions $\frac{d\phi^{-1}}{d\varepsilon}$ and $\frac{d\rho_\varepsilon}{d\varepsilon}$ here, but only integrability of these values in every degree $p > 0$ is important. By virtue of this integrability we have

$$|Ef[z_\bullet] g'(x_\tau) y_\tau| \;\le\; C\|g\|_C \,. \tag{15}$$

3.

To verify the first part of the Theorem, it suffices to prove the inequalities

$$|Ef[z_\bullet] g^{(k)}(x_\tau)| \;\ge\; C_k\|g\|_C \,, \qquad k = 1,2,\ldots \tag{16}$$

(cf. [13]). When $k = 1$ inequality (16) is similar to (15). Let us eliminate the factor y_τ in (14) and (15). Let ψ be in $C_b^\infty(E^1)$, let $\psi(y) = y^{-1}$ if $y \ne 0$ and $|y^{-1}| \le 1$, and let $\psi(y) = 0$ if $|y^{-1}| \ge 2$. Let $\psi_N(y) = N\psi(Ny)$. Then $\psi_N(y) \to y^{-1}$ $(N \to \infty)$ if $y \ne 0$.

Let us take the derivatives on both sides of the following equality instead of (13):

$$E^{\phi}\rho_{\varepsilon}\phi^{-1}[x^{\varepsilon}_{\cdot},z_{\cdot}]f[z_{\cdot}]g(x^{\varepsilon}_{\tau})\psi_N(y^{\varepsilon}_{\tau}) = E^{\phi}\phi^{-1}[x_{\cdot},z_{\cdot}]f[z_{\cdot}]g(x_{\tau})\psi_N(y_{\tau}).$$

We have (denote $v_t = \left.\frac{dy^{\varepsilon}_t}{d\varepsilon}\right|_{\varepsilon=0}$)

$$\begin{aligned} &E^{\phi}f[z_{\cdot}]\phi^{-1}[x_{\cdot},z_{\cdot}]g'(x_{\tau})y_{\tau}\psi_N(y_{\tau}) \\ &\quad + E^{\phi}f[z_{\cdot}]g(x_{\tau})\psi_N(y_{\tau})\left.\frac{d\phi^{-1}[x^{\varepsilon}_{\cdot},z_{\cdot}]}{d\varepsilon}\right|_{\varepsilon=0} \\ &\quad + E^{\phi}f[z_{\cdot}]g(x_{\tau})\phi^{-1}[x_{\cdot},z_{\cdot}]\psi'_N(y_{\tau})v_{\tau} \\ &\quad + E^{\phi}f[z_{\cdot}]g(x_{\tau})\phi^{-1}[x_{\cdot},z_{\cdot}]\psi_N(y_{\tau})\left.\frac{d\rho_{\varepsilon}}{d\varepsilon}\right|_{\varepsilon=0} = 0 . \end{aligned} \tag{17}$$

4.

By (12) the following equation for the process y_t holds:

$$\begin{aligned} y_t = \int_0^t \Bigg(& X_{0x}(x_s.z_s)y_s\,ds \\ & + \sum_{i=1}^m \left[X_{ix}(x_s,z_s)y_s \circ dw^i_s + X_i(x_s,z_s)u^i_s\,ds\right] \\ & + \sum_{j=1}^n \tilde{X}_{jx}(x_s,z_s)y_s \circ d\hat{w}^j_s \Bigg) . \end{aligned}$$

Let $x^{x'}_t$ be a solution of the equation

$$\begin{aligned} x^{x'}_t = x' + \int_0^t \Big[& X_0(x^{x'}_s,z_s)\,ds + \sum_{i=1}^m X_i(x^{x'}_s,z_s)\circ dw^i_s \\ & + \sum_{j=1}^n \tilde{X}_j(x^{x'}_s,z_s)\circ d\hat{w}^j_s \Big] , \end{aligned}$$

where z_t is already found from (10). Let $\xi_t = \left.\frac{dx_t^{x'}}{dx'}\right|_{x'=x_0}$. The following equation is satisfied:

$$\xi_t = 1 + \int_0^t \left(X_{0x}(x_s,z_s)\xi_s\, ds + \sum_{i=1}^m X_{ix}(x_s,z_s)\xi_s \circ dw_s^i + \sum_{j=1}^n \tilde{X}_j(x_s,z_s)\xi_s \circ d\hat{w}_s^j \right) .$$

Hence ξ_t is nondegenerate and ξ_t^{-1} satisfies the equation

$$\xi_t^{-1} = 1 - \int_0^t \left(X_{0x}(x_s,z_s)\xi_s^{-1}\, ds - \sum_{i=1}^m X_{ix}(x_s,z_s)\xi_s^{-1} \circ dw_s^i - \sum_{j=1}^n \tilde{X}_j(x_s,z_s)\xi_s^{-1} \circ d\hat{w}_s^j \right) .$$

Therefore, the process y_t can be expressed in the form

$$y_t = \xi_t \int_0^t \sum_{i=1}^m X_i(x_s,z_s)\xi_s^{-1} u_s^i\, ds .$$

Let

$$u_t^i = \xi_t X_i(x_t,z_t) , \qquad 1 \le i \le m .$$

Then

$$y_t = \xi_t \int_0^t \sum_{i=1}^m X_i^2(x_s,z_s)\, ds .$$

Since $\xi_t^{-1} \in L_p(\Omega)$ for every $p > 0$ with respect to the measure P^ϕ and hence with respect to the measure P, then y_t is nondegenerate if $t > 0$ (almost surely) and $\sup_{\delta \le t \le K} E^P \|y_t^{-1}\|^p < \infty$ for arbitrary $0 < \delta \le K$, $p > 0$. By virtue of (17) and the Lebesgue theorem we have

$$\lim_{N\to\infty} \Big\{ E^{\phi} f[z_\cdot]\, \bar{\phi}^{-1}[x_\cdot, z_\cdot] g'(x_\tau) y_\tau \psi_N(y_\tau)$$

$$+ \quad E^{\phi} f[z_\cdot] g(x_\tau) \psi_N(y_\tau) \left.\frac{d\phi^{-1}[x_\cdot, z_\cdot]}{d\varepsilon}\right|_{\varepsilon=0}$$

$$+ \quad E^{\phi} f[z_\cdot] g(x_\tau) \psi_N(y_\tau)\, \bar{\phi}^{-1}[x_\cdot, z_\cdot] \left.\frac{d\rho_\varepsilon}{d\varepsilon}\right|_{\varepsilon=0} \Big\}$$

$$= \quad E^{\phi} f[z_\cdot]\, \bar{\phi}^{-1}[x_\cdot, z_\cdot] g'(x_\tau)$$

$$+ \quad E^{\phi} f[z_\cdot] g(x_\tau) \left.\frac{d\phi^{-1}[x_\cdot^{\varepsilon}, z_\cdot]}{d\varepsilon}\right|_{\varepsilon=0} y^{-1}$$

$$+ \quad E^{\phi} f[z_\cdot] g(x_\tau)\, \bar{\phi}^{-1}[x_\cdot, z_\cdot] \left.\frac{d\rho_\varepsilon}{d\varepsilon}\right|_{\varepsilon=0} y^{-1} .$$

Let us prove that the third term in (17) tends to zero as $N \to \infty$. We have

$$|\psi_N'(y)| \quad = \quad N^2 |\psi'(Ny)| \quad \le \quad C N^2\, I(|y^{-1}| \ge 2N) \quad .$$

Hence

$$E^{\phi} |\psi_N'(y_\tau)|^p \quad \le \quad C N^{2p} P^{\phi}(|y_\tau^{-1}| \ge 2N) \quad \le \quad C N^{2p} (2N)^{-a} E^{\phi} |y_\tau^{-1}|^a \quad .$$

Let $a > 2p$. Then $E^{\phi} |\psi_N'(y_\tau)|^p \to 0$, $N \to \infty$. The latter assertion holds for an arbitrary $p > 0$. By virtue of Hölder's inequality the third term in (17) does tend to zero as $N \to \infty$. Thus we have (16) when $k = 1$. In analogous way by induction (cf. [15]) inequalities (16) can be obtained.

The first part of the theorem is proved. The second part can be proved as was done in [13].

REFERENCES

[1] Itô, K. "On the Existence and Uniqueness of Solutions of Stochastic Integral Equations of the Volterra Type." *Kodai Math. J.*, vol.2 (1979): 158-170.

[2] Kravets, T.N. "On the Existence and Uniqueness of Solutions of Stochastic Integral Equations of the Volterra Type." *Teoriya sluchajnykh protsessov*, vol.7 (1979): 57-62.

[3] Kravets, T.N. "On a Weak Continuity in a Parameter of Solutions of Stochastic Integral Equations." *Teoriya Sluchajnykh Protsessov*, vol.11 (1983): 57-62.

[4] Fedorenko, I.V. "On the Existence of Solutions of Stochastic Integral Itô-Volterra Equations." In *Uravneniya na mnogo-obraziyakh*, pp. 129-133. Voronezh, 1982.

[5] Kleptsina, M.L. "On an Optimal Linear Nonstationary Filtering of Itô-Volterra Processes." In *Upravlenie nelinejnykh sistem*, pp. 41-42. Moskva: Nauka, 1984.

[6] Kleptsina, M.L., and Veretennikov, A.Yu. "On Strong Solutions of Stochastic Itô-Volterra Equations." *Theory Probab. Applications*, vol.29, no.1 (1984): 154-158.

[7] Balakrishnan, A.V. "On Stochastic Bang-Bang Control." *Lecture Notes in Control and Information Sci.*, pp. 221-238, vol. 25. Berlin Heidelberg New York: Springer-Verlag Inc., 1980.

[8[Liptser, R.S., and Shiryayev, A.N. *Statistics of Random Processes*, I. Berlin Heidelberg New York: Springer-Verlag Inc., 1977.

[9] Malliavin, P. "Stochastic Calculus of Variations and Hypoelliptic Operators." *Proc. Intern. Conf. Stochastic Differential Equations* (Kyoto, 1976), pp. 195-263. Tokyo: Kinokuniya and New York: Wiley, 1978.

[10] Malliavin, P. "C^k-Hypoellipticity with Degeneracy." In *Stochastic Analysis. Proc. Intern. Conf. Stochastic Analysis.* Edited by Avner Friedman and Mark Pinsky. New York: Academic Press, 1978.

[11] Bismut, J. "Martingales, the Malliavin Calculus, and Hypoellipticity Under General Hörmander's Conditions." *Z. Wahrscheinlichkeitstheorie und Verw. Gebiete*, 56 (1981): 469-505.

[12] Ikeda, N. and Watanabe, Sh. *Stochastic Differential Equations and Diffusion Processes*. Amsterdam, New York: North-Holland Publ. Co., 1981.

[13] Bismut, J. and Mishel, D. "Diffusions conditionnelles. I. Hypoellipticité partielle." *J. Funct. Analysis*, vol.44, no. 2 (1981): 174-211. II. "Générateur Conditionnel. Application au filtrage." Ibid., vol.45, no.2 (1982): 274-292.

[14] Krylov, N.V. "On a Regularity of Conditional Probabilities for Random Processes." *Theory Probab..Applications*, vol.28 no.1 (1983): 151-155.

[15] Veretennikov, A.Yu. "A Probabilistic Approach to Hypoellipticity." *Russian Math. Surveys*, vol.38, no.3 (1983): 127-140.

N.V. KRICHAGINA, R.Sh. LIPTSER, and E.Ya. RUBINOVICH

KALMAN FILTER FOR MARKOV PROCESSES

1. INTRODUCTION

This paper deals with a filtering problem for a Markov process $\theta = (\theta_t)$ with a finite state space $J = (a_1, \dots, a_k)$ and given transition probabilities, an observable process Y being the sum of the process θ and Gaussian white noise.

This is a feasible mathematical model for real-valued processes with step trajectories.

The optimal nonlinear filter (optimality will be treated in the mean square sense) is known for this case [1, 2]. But for applications, it is reasonable to consider an optimal linear estimate because of its simplicity in computations and lower sensitivity with respect to data inaccuracy.

An optimal linear filter for a non-Gaussian process is known to be the optimal linear filter for Gaussian analogs of original observable and unobservable processes, i.e., for Gaussian processes with the same mathematical expectation and correlation (see, for example, [1]].

The main difficulty in constructing a Gaussian analog of a Markov process with a large number of states consists in the explicit computation of its correlation (in the case of two states this is a well-known and easy task [1]).

In this paper, a method for constructing Gaussian Markov analogs of the processes (θ, Y) is suggested. This allows us to avoid the explicit computation of the correlation of θ. This method is used to construct Kalman filters. In the general case, the dimension of the Kalman filter is $k-1$, where k is the dimension of J. In Section 4, for a homogeneous Markov process we consider the case where the Kalman filter has the dimension less than $k-1$.

2. DISCRETE-TIME CASE

Let us introduce the following notation:

•1) $\Lambda(t)$ is a matrix of transition probabilities of the Markov process $\theta = (\theta_t)$ with elements

$$\Lambda_{ij}(t) = P(\theta_{t+1} = a_j \mid \theta_t = a_i), \quad i,j = 1,\dots,k,$$

•2) $p_i(t) = P(\theta_t = a_i)$.

Put $X(t) = (X_1(t), \dots, X_k(t))$ with

$$X_i(t) = I(\theta_t = a_i), \qquad i = 1,\dots,k.$$

The following result is needed for constructing the Gaussian analog of the processes $(X_1(t))$, $i = 1, \dots, k$.

<u>LEMMA 1</u>. The sequences $\varepsilon_j = (\varepsilon_j(t))$, $j = 1,\dots,k$, with

$$\varepsilon_j(t+1) = X_j(t+1) - \sum_{i \le k} \Lambda_{ij}(t) X_i(t),$$

are the martingale-differences with respect to the family of σ-algebras $\mathbb{F} = (F_t)$, $F_t = \sigma\{X(0), \ldots, X(t)\}$, i.e.,

$$E(\varepsilon_j(t+1) \mid F_t) = 0 \qquad (P \text{ a.s.}).$$

In addition,

$$q_{jj}(t) = E\varepsilon_j^2(t+1) = p_j(t+1) - \sum_{i \le k} \Lambda_{ij}^2(t)\, p_i(t) \quad , \tag{1}$$

$$q_{ij}(t) = E\varepsilon_j^2(t+1)\varepsilon_j(t+1) = -\sum_{\ell \le k} \Lambda_{\ell i}(t)\Lambda_{\ell j}(t) p_\ell(t) \tag{2}$$

when $i \neq j$.

Proof. Since θ is a Markov process, we have

$$\begin{aligned} E(X_j(t+1) \mid F_t) &= E(X_j(t+1) \mid \theta_t) \\ &= E(X_j(t+1) \mid X(t)) \\ &= \sum_{i \le k} \Lambda_{ij}(t)\, X_i(t) \quad . \qquad (P \text{ a.s.}). \end{aligned}$$

The last equality implies $E(\varepsilon_j(t+1) \mid F_t) = 0$ (P a.s.) and

$$\begin{aligned} E\varepsilon_j^2(t+1) &= EX_j^2(t+1) - E\Big[\sum_{i \le k} \Lambda_{ij}(t) X_i(t)\Big]^2 \\ &= EX_j(t+1) - \sum_{i \le k} \Lambda_{ij}(t)\, EX_i(t) \quad . \end{aligned}$$

Hence the representation (1) follows immediately. The representation (2) is proved in a similar way. Indeed, for $i \neq j$

$$E\varepsilon_i(t+1)\varepsilon_j(t+1) = E\Big[X_j(t+1) - \sum_{\ell\le k}\Lambda_{\ell j}(t)X_\ell(t)\Big] \times \Big[X_i(t+1) - \sum_{\ell\le k}\Lambda_{\ell i}(t)X_\ell(t)\Big]$$

$$= -E\Big[-\sum_{\ell\le k}\Big\{X_j(t+1)\Lambda_{\ell i}(t)X_i(t) + X_i(t+1)\Lambda_{\ell j}(t)X_j(t) - \Lambda_{\ell i}(t)\Lambda_{\ell j}(t)X_\ell(t)\Big\}\Big]$$

$$= -E\sum_{\ell\le k}\Lambda_{\ell i}(t)\Lambda_{\ell j}(t)X_\ell(t) \quad .$$

Denote*/ $\varepsilon(t) = (\varepsilon_1(t), \ldots, \varepsilon_k(t))$. From the definition of $\varepsilon_j(t)$ it follows that

$$X(t+1) = \Lambda^T(t)X(t) + \varepsilon(t+1) \quad . \tag{3}$$

Evidently, the Gaussian analog of the process $X = (X(t))$ is the vector process $\tilde{X} = (\tilde{X}(t))$ defined by

$$\tilde{X}(t+1) = \Lambda^T(t)\tilde{X}(t) + \tilde{\varepsilon}(t+1) \quad , \tag{4}$$

where $\tilde{\varepsilon}(t) = (\tilde{\varepsilon}_1(t), \ldots, \tilde{\varepsilon}_k(t))$, $t = 1,2,\ldots$ is the sequence of independent Gaussian vectors with $E\tilde{\varepsilon}(t) = 0$ and correlation matrix $Q(t) = ||q_{ij}(t)||$ and $\tilde{X}(0)$ is a Gaussian vector independent of $(\tilde{\varepsilon}(t))$ with

$$E\tilde{X}_j(0) = p_j(0) \quad ,$$

$$\operatorname{cov}(\tilde{X}_i(0), \tilde{X}_j(0)) = \begin{cases} p_i(0)(1-p_i(0)), & i = j \ , \\ -p_i(0)p_j(0) \ , & i \ne j \ . \end{cases} \tag{5}$$

---

*/ Hereinafter all vectors are the column vectors, and T means transposition.

Since $X_k(t) = 1 - \sum_{i \le k-1} X_i(t)$, we have

$$\tilde{X}_k(t) = 1 - \sum_{i \le k-1} \tilde{X}_i(t) \qquad (P \text{ a.s.}).$$

Therefore, the representation (4) implies the following recurrence equation for $\bar{X}(t) = (\tilde{X}_1(t), \ldots, \tilde{X}_{k-1}(t))$:

$$\bar{X}(t+1) = \bar{\Lambda}^T(t)\bar{X}(t) + \lambda(t) + \bar{\varepsilon}(t+1) \quad ,$$

where $\bar{\Lambda}(t)$ is a $(k-1) \times (k-1)$-dimensional matrix with elements

$$\bar{\Lambda}_{ij}(t) = \Lambda_{ij}(t) - \Lambda_{kj}(t) \quad ,$$

$\lambda(t)$ and $\bar{\varepsilon}(t)$ are $(k-1)$-dimensional vectors:

$$\lambda(t) = (\Lambda_{k,1}(t), \ldots, \Lambda_{k,k-1}(t)) \quad ,$$

$$\bar{\varepsilon}(t) = (\tilde{\varepsilon}_1(t), \ldots, \tilde{\varepsilon}_{k-1}(t)) \quad .$$

From the definition of $X_i(t)$, $i = 1, \ldots, k$ it follows that

$$\theta_t = \sum_{i \le k} a_i X_i(t) = \sum_{i \le k-1} (a_i - a_k) X_i(t) + a_k$$

and therefore the Gaussian analog for θ is the process $\tilde{\theta} = (\tilde{\theta}_t)$ defined by

$$\tilde{\theta}_t = \bar{J}^T \bar{X}(t) + a_k \quad , \tag{6}$$

where $\bar{J} = (a_1 - a_k, \ldots, a_{k-1} - a_k)$.

Now we consider the filtering problem with an observable process (Y_t), $t = 0, 1, \ldots$

$$Y_0 = 0 \quad ,$$

$$Y_{t+1} = \theta_t + \delta_{t+1} \quad ,$$

(δ_t) being the sequence of independent Gaussian random variables, $E\delta_t \equiv 0$, $E\delta_t^2 = b_{t-1}^2$ independent of θ.

Our aim is to give an optimal linear estimate of θ_t on the basis of observations $(Y_0, \ldots, Y_t)$, $t = 0,1,\ldots$.

To solve this problem it is sufficient to use a Kalman filter for the process $\bar{X}$ by observations $(\bar{Y}_t)$, $t = 0,1,\ldots$ with $\bar{Y}_0 = 0$,

$$\bar{Y}_{t+1} = \bar{J}^T\bar{X}(t) + \bar{\delta}_{t+1} + a_k , \tag{7}$$

($(\bar{\delta}_t)$ being independent of $\bar{X}(0)$-sequence of random variables coinciding in the distribution sense with (δ_t)).

According to the Kalman filtering theory [1] the optimal linear estimate $m(t)$ for the vector $\bar{X}(t)$ by observations $(Y_0, \ldots, Y_t)$ is given by the following recurrence equations $(b_t^2 > 0)$:

$$\begin{aligned} m(t+1) = {} & \bar{\Lambda}^T(t)m(t) + \lambda(t) \\ & + (b_t^2 + \bar{J}^T\Gamma_t\bar{J})^{-1}\bar{\Lambda}^T(t)\Gamma_t\bar{J}(Y_{t+1} - \bar{J}^Tm(t) - a_k) , \end{aligned}$$

$$\begin{aligned} \Gamma_{t+1} = {} & \bar{\Lambda}^T(t)\Gamma_t\bar{\Lambda}(t) + \bar{Q}(t) \\ & - (b_t^2 + \bar{J}^T\Gamma_t\bar{J})^{-1}\bar{\Lambda}^T(t)\Gamma_t\bar{J}\bar{J}^T\Gamma_t\bar{\Lambda}(t) , \end{aligned}$$

with initial conditions $m(0)$ and Γ_0 calculated by the formulas (5). The elements of the matrix $\bar{Q}(t) = \|q_{ij}(t)\|$, $i,j = 1,\ldots,k-1$ are defined by (1) and (2), using Kolmogorov's equations for $p_j(t)$ ([3]):

$$p_j(t+1) = \sum_{i\le k} \Lambda_{ij}(t)\, p_i(t) .$$

In addition, the optimal linear estimate $\hat{\theta}_t$ for θ_t has the form

$$\hat{\theta}_t = \bar{J}^T m(t) + a_k \ . \tag{8}$$

3. CONTINUOUS-TIME CASE

Suppose that the Markov process $\theta = (\theta_t)$ with vector state J has right continuous trajectories and the intensity transition matrix $\Lambda(t) = \|\Lambda_{ij}(t)\|$, $\Lambda_{ij}(t)$ being time-continuous functions. According to Lemma 9.2 of [1] the vector-indicator process $X(t) = (X_1(t), \ldots, X_k(t))$ $(X_i(t) = I(\theta_t = a_i))$ is a semimartingale and has the following representation:

$$X(t) = X(0) + \int_0^t \Lambda^T(s)\, ds + M(t) \ , \tag{9}$$

with a square integrable purely discontinuous martingale $M(t) = (M_1(t), \ldots, M_k(t))$ with respect to the family $\mathbb{F} = (\mathcal{F}_t)_{t \geq 0}$ of σ-algebras $\mathcal{F}_t = \sigma(X(s), s \leq t)$ and the quadratic characteristic $\langle M \rangle_t = \|\langle M_i, M_j \rangle_t\|$,

$$\langle M_j, M_j \rangle_t = \int_0^t \sum_{i \leq k} |\Lambda_{ij}(s)| X_i(s)\, ds \ , \tag{10}$$

$$\langle M_i, M_j \rangle_t = -\int_0^t [\Lambda_{ij}(s) X_i(x) + \Lambda_{ji}(s) X_j(s)]\, ds \ , \quad i \neq j$$

(see [4]). From (9) and (10) it follows that the Gaussian analog $\tilde{X}(t) = (\tilde{X}_1(t), \ldots, \tilde{X}_k(t))$ for the process $X(t)$ is described by the linear differential Itô equation

$$\tilde{X}(t) = \tilde{X}(0) + \int_0^t \Lambda^T(s)\tilde{X}(s)\,ds + \int_0^t B^{\frac{1}{2}}(s)\,d\tilde{W}(s),$$

with a Wiener process $\tilde{W} = (\tilde{W}(t)) = (\tilde{W}_1(t), \ldots, \tilde{W}_k(t))$ and the matrix-valued

$$B(t) = \frac{d}{dt} E\langle M\rangle_t .$$

From (10) it follows also that the matrix $B(t)$ has the elements

$$B_{jj}(t) = \sum_{i\le k} |\Lambda_{ij}(t)|\, p_i(t), \qquad i = 1,\ldots,k$$

$$B_{ij}(t) = -[\Lambda_{ij}(t)p_i(t) + \Lambda_{ji}(t)p_j(t)], \qquad i \ne j \tag{11}$$

and $p_j(t) = P(\theta_t = a_j)$ can be found from Kolmogorov's equation [3]:

$$\dot{p}_j(t) = \sum_{i\le k} \Lambda_{ij}(t)\, p_i(t) .$$

As in the discrete-time case, one can introduce the vectors $\bar{X}(t)$ $\lambda(t)$ and the matrix $\bar{\Lambda}(t)$. Then by virtue of (9) $\bar{X}(t)$ is the random process with Itô's differential

$$d\bar{X}(t) = [\bar{\Lambda}^T(t)\bar{X}(t) + \lambda(t)]\,dt + \bar{B}^{\frac{1}{2}}(t)\,d\bar{W}(t)$$

with respect to a Wiener process $\bar{W}(t) = (\tilde{W}_1(t), \ldots, \tilde{W}_{k-1}(t))$, where $\bar{B}(t)$ is a $(k-1)\times(k-1)$-dimensional matrix with elements

$$\bar{B}_{ij}(t) = B_{ij}(t), \qquad i,j = 1,\ldots,k-1 .$$

As to the observable process $Y = (Y_t)$, it will be assumed that it is an Itô process

$$Y_t = \int_0^t \theta_s\,ds + W_t \tag{12}$$

with a θ-independent Wiener process $W = (W_t)$. Then, as in the discrete-time case, it is sufficient to consider an auxiliary filtering problem for the process $\bar{X}(t)$ by using the observation

$$\bar{Y}_t = \int_0^t [\bar{J}^T\bar{X}(s) + a_k]ds + \bar{W}_t \tag{13}$$

with $\bar{J} = (a_1 - a_k, \ldots, a_{k-1})$ and with a Wiener process $(\bar{W}_t)$ independent of $(\bar{X}(t))$.

According to the Kalman filtering theory, for the continuous-time case [1] the optimal linear estimate for the vector $(X_1(t), \ldots, X_{k-1}(t)$ on the basis of observations $(Y_s, s \leq t)$ is given by the following equations:

$$\begin{aligned} m(t) &= m(0) + \int_0^t [\bar{\Lambda}^T(s)m(s) + \lambda(s)]ds \\ &\qquad + \int_0^t \Gamma_s \bar{J}[dY_s - (\bar{J}^T m(s) + a_k)ds] , \\ \Gamma_t &= \Gamma_0 + \int_0^t [\bar{\Lambda}^T(s)\Gamma_s + \Gamma_s\bar{\Lambda}(s) + \bar{B}(s) - \Gamma_s \bar{J}\bar{J}^T\Gamma_s] ds , \end{aligned} \tag{14}$$

with the initial conditions $m(0)$ and Γ_0 calculated by the formulas (5).

In addition, the optimal linear estimate $\hat{\theta}_t$ for θ_t has the form

$$\hat{\theta}_t = \bar{J}^T m(t) + a_k .$$

4. THE CONSTRUCTION OF REDUCED-ORDER KALMAN FILTER

In Sections 2 and 3 we used the $(k-1)$-dimensional Kalman filter for estimation of $X(t) = (X_1(t), \ldots, X_k(t))$. According to (6)

$$\theta_t = \sum_{i \le k} a_i X_i(t) = J^T X(t) ,$$

hence it is necessary to estimate the scalar process $J^T X(t)$ only. Therefore, we want to consider the cases when a construction of a filter with dimension lower than $k - 1$ is possible.

From now on we assume the matrix $\Lambda(t)$ to be stationary, i.e., $\Lambda(t) \equiv \Lambda$.

First we consider a special case where the lowest dimension of the filter is 1.

THEOREM 1. Let $J = (a_1, \ldots, a_k)$ satisfy the relationship

$$\Lambda J = c_1 J + c_2 h ,$$

where c_1 and c_2 are constants, h is a vector with unit components.

Then the Kalman filter for the estimate of the process θ has a unit dimension.

Proof. *DISCRETE-TIME CASE.* Multiplying both sides of (3) by J^T and using the equality $h^T X(t) \equiv 1$, we have the recurrence equation

$$\theta_{t+1} = c_1 \theta_t + c_2 + J^T \varepsilon(t+1)$$

with the orthogonal sequence of random variables $(J^T \varepsilon(t))$ such that $E J^T \varepsilon(t) \equiv 0$, $E(J^T \varepsilon(t+1))^2 = J^T Q(t) J$ (the matrix $Q(t)$ is defined by (1), (2)).

Evidently, one-dimensional Kalman filter corresponds to this case.

CONTINUOUS-TIME CASE. As in the discrete-time case, multiplication of both sides of (9) by J^T and $\Lambda h = 0$ gives the following representation:

$$\theta_t = \theta_0 + \int_0^t c_1 \theta_s \, ds + J^T M(t)$$

with the square integrable martingale $(J^T M(t))$. The square characteristic of $(J^T M(t))$ is defined by $J^T \langle M \rangle_t J$, and $\langle M \rangle_t$ is the square characteristic of M (see (10)). In addition,

$$E J^T \langle M \rangle_t J = \int_0^t J^T B(s) J \, ds$$

and the matrix $B(t)$ is taken from (11).

Now we consider the general case. Let J, ΛJ, $\Lambda^2 J$, ... be the sequence of vectors from R^k. Obviously there exists a positive integer $r \le k$ such that the vectors

$$J, \quad \Lambda J, \quad \ldots, \quad \Lambda^{r-1} J \tag{15}$$

are linear independent, but vectors

$$J, \quad \Lambda J, \quad \ldots, \quad \Lambda^r J$$

are linear dependent. It means that

$$\Lambda^r J = c_1 J + c_2 \Lambda J + \cdots + c_r \Lambda^{r-1} J = Ac \tag{16}$$

with a nonzero vector $c = (c_1, \ldots, c_r)$ and a block $k \times r$-dimensional matrix $A = (J \mid \Lambda J \mid \cdots \mid \Lambda^{r-1} J)$ of the rank r.

Denote by L_r the linear subspace generated by the vectors (15), and by h the k-dimensional vector with unit components. The following theorem is true for the cases of both continuous and discrete time.

THEOREM 2. •1. Let $h \notin L_r$. Then in the estimation problem for the process θ one can construct the r-dimensional Kalman filter.

•2. If $h \in L_r$, the dimension of the Kalman filter is $r-1$.

Proof. •1. We shall start from discrete-time case. Multiplying both sides of (3) by matrix A^T, we obtain

$$A^T X(t+1) = A^T \Lambda^T X(t) + A^T \varepsilon(t+1) . \tag{17}$$

Note that by virtue of (16)

$$\Lambda A = [\Lambda J \mid \Lambda^2 J \mid \cdots \mid \Lambda^{r-1} J \mid \Lambda^r J] = AD \tag{18}$$

with the block $r\times r$-dimensional matrix

$$D = \left[\begin{array}{c|c} 0 & \\ \hline I & c \end{array}\right] \tag{19}$$

and $(r-1)\times(r-1)$-dimensional identity matrix I and the vector c, defined in (16). Thus $A^T \Lambda^T = D^T A^T$.

Introduce the r-dimensional vector $Z(t) = (Z_1(t), \ldots, Z_r(t))$, putting $Z(t) = A^T X(t)$.

Then it is possible to write the recurrence relation in the form

$$Z(t+1) = D^T Z(t) + A^T \varepsilon(t+1) .$$

In addition, the observation $Y = (Y_t)$ has the representation

$$Y_{t+1} = Z_1(t) + \delta_{t+1} \ .$$

Since $\theta_t = Z_1(t)$, the dimension of the Kalman filter is $r \leq k-1$.

Under the same notation this result remains true in the continuous-time case. So instead of (9) and (12) we have

$$Z(t) = Z(0) + \int_0^t D^T Z(s)ds + A^T M(t) \ ,$$

$$Y_t = \int_0^t Z_1(s)ds + W_t \ .$$

•2. Since $J, \Lambda J, \ldots, \Lambda^{r-1}J$ is the basis in L_r and $h \in L_r$, we have the following decomposition in this basis:

$$h = v_1 J + v_2 \Lambda J + \cdots + v_r \Lambda^{r-1} J = Av \tag{20}$$

with a non-zero vector $v = (v_1, \ldots, v_r)$ ($r \geq 3$, for the case $r \leq 2$ see Theorem 1).

In the discrete-time case $\Lambda h = h$ and by virtue of (20) $\Lambda Av = Av$ or $A(v-Dv) = 0$ (see (18)). From linear independence of columns of the matrix A it follows that $v - Dv = 0$, i.e., $v_1 - v_r c_1 = 0$, $v_i - v_{i-1} - v_r c_i = 0$, $2 \leq i \leq r$. It implies that $v_r \neq 0$. This and (20) mean that $J, \Lambda J, \ldots, \Lambda^{r-2}J$, h is also the basis in L_r.

Because of (20)

$$\Lambda^{r-1}J = \tilde{c}_1 J + \cdots + \tilde{c}_{r-1}\Lambda^{r-2}J + \tilde{c}_r h \ , \tag{21}$$

where $\tilde{c}_r = v_r^{-1}$ and $\tilde{c}_i = -v_i v_r^{-1}$, $i = 1,\ldots,r-1$.

Denote the block matrix $\tilde{A} = [J \mid \Lambda J \mid \cdots \mid \Lambda^{r-2}J]$. According to (21) we have

$$\Lambda\tilde{A} = \tilde{A}\tilde{D} + \tilde{c}_r H \tag{22}$$

with the block $(r-1)\times(r-1)$-dimensional matrix

$$\tilde{D} = \left[\begin{array}{c|c} 0 & \\ \hline \tilde{I} & \tilde{c} \end{array}\right],$$

where $\tilde{I}$ is a $(r-2)\times(r-2)$-dimensional identity matrix, the vector $\tilde{c}$ is $\tilde{c} = (\tilde{c}_1,\dots,\tilde{c}_{r-1})$ and a block $k\times(r-1)$-dimensional matrix H is

$$H = [0 \,|\, 0 \,|\, \cdots \,|\, 0 \,|\, h] \; .$$

The multiplication of both sides of (3) by the matrix $\tilde{A}^T$ gives for $\tilde{Z}(t) = \tilde{A}^T X(t)$ (keeping in mind $h^T X(t) \equiv 1$) the recurrence equation

$$\tilde{Z}(t+1) = \tilde{D}^T \tilde{Z}(t) + \tilde{c}_r \tilde{h} + \tilde{A}^T \varepsilon(t+1)$$

with $\tilde{h} = (0,\dots,0,1) \in R^{r-1}$. In addition, the observation $Y = (Y_t)_{t=1,2,\dots}$ has the form

$$Y_{t+1} = \tilde{Z}_1(t) + \delta_{t+1} \; .$$

Consequently, in this case the dimension of Kalman filter is $r-1$.

In the continuous-time case $\Lambda h = 0$ and thus from (20) and (18) we obtain $\Lambda A v = 0$ or $ADv = 0$.

From the linear independence of columns of the matrix A it follows that $Dv = 0$, i.e., $v_r c_1 = 0$, $v_{i=1} + v_r c_i = 0$, $i = 2,\dots,r$. Hence $v_r \neq 0$ and as the basis in L_r one can take the vectors J, ΛJ, ..., $\Lambda^{r-2}J$, h. Retaining the notation introduced in the discrete-time case, we have (see (9), (12))

$$\tilde{Z}(t) = \tilde{Z}(0) + \int_0^t [\tilde{D}^T\tilde{Z}(s) + \tilde{c}_r\tilde{h}]ds + \tilde{A}M(t) ,$$

$$Y_t = \int_0^t \tilde{Z}_1(s)\, ds + W_t , \qquad (\theta_t = \tilde{Z}_1(t)) ,$$

i.e., the dimension of the Kalman filter is then $r - 1$. ///

In the previous discussion we used the change of variables of the form $F^TX(t) = Z(t)$ satisfying the relations

$$F^T\Lambda^TX(t) = G^TZ(t) + d(t) , \tag{23}$$

$$J^TX(t) = \ell^TZ(t) + a , \tag{24}$$

where $Z(t)$ had the dimension r or $r - 1$. For example, in Theorem 2 we had $F = A$ or $\tilde{A}$, $G = D$ or $\tilde{D}$, $d(t) = 0$ or $\tilde{c}_r\tilde{h}$, $a = 0$.

Now we wish to prove that there is no $k\times m$-dimensional matrix F with $m < r$ (or $m < r-1$) satisfying the relations (23) and (24).

To prove this statement we need the following lemma.

<u>LEMMA 2</u>. L_r is the minimal-order invariant subspace with respect to Λ ($u \in L_r$ implies $\Lambda u \in L_r$) containing the vector J.

<u>Proof</u>. Let $u \in L_r$, i.e., $u = Av$ where $v \in R^r$ and $v \neq 0$. Then according to (18)

$$\Lambda u = \Lambda Av = ADv$$

and hence $\Lambda u \in L_r$. So L_r is a subspace invariant with respect to Λ.

Suppose there exists a linear subspace L_m invariant with respect to Λ with $m \leq r-1$, containing J. All the vectors (15) belong to L_m because L_m is an invariant subspace. The set of (15) contains r linear independent vectors. The dimension of L_m cannot be less than r. This contradicts our assumption $m \leq r - 1$. ///

Now suppose that there exists a $k \times m$-dimensional matrix F (rank $F = m < k$) which does not coincide with A and $\tilde{A}$ and satisfies the relations (23), (24). Since rank $F = m$, the columns $f_1, \ldots, f_m$ of the matrix F are linear independent and generate a subspace L_m. Let us consider two cases: $h \notin L_m$ and $h \in L_m$, where h is the k-dimensional vector defined above with unit components.

If $h \notin L_m$, then the linear subspace L_{m+1} generated by the vectors $f_1, \ldots, f_m, h$ is invariant with respect to Λ. Actually, h is an eigenvector of the matrix Λ, and therefore it suffices to prove that $\Lambda f_i \in L_{m+1}$, $i = 1, \ldots, m$.

We have $\Lambda F = [\Lambda f_1 | \cdots | \Lambda f_m]$. From this and (23) it follows that

$$\Lambda f_i = F g_i + d_i(t) h \quad , \qquad i = 1, \ldots, m \; , \tag{25}$$

where $d_i(t)$ is the ith component of the vector $d(t)$ and g_i is the ith column of the matrix G. The right side of (25) is the decomposition of the vector Λf_i by the basis $f_1, \ldots, f_m, h$, i.e., $\Lambda f_i \in L_{m+1}$ and L_{m+1} is an invariant subspace.

The relation (24) implies

$$J = F\ell + ah \quad , \tag{26}$$

i.e., J belongs to L_{m+1}. Then by Lemma 2 the only possible minimal m is $m = r - 1$.

If $h \in L_m$, then the decomposition h by the vectors $f_1, \ldots, f_m$ and (25) imply that $\Lambda f_i \in L_m$, i.e., L_m is a subspace invariant with respect to Λ. From (26) it follows that $J \in L_m$ and hence by Lemma 2 the only possible minimal m is $m = r$.

REFERENCES

[1] Liptser, R.S., and Shiryayev, A.N. *Statistics of Random Processes*. I,II. Berlin Heidelberg New York: Springer-Verlag Inc., 1977.

[2] Kapylov, A.K. "Nonlinear Filtering of Stochastic Processes From Discrete-time Data." *Problems of Control and Information Theory*, vol.8, no.1 (1979): 39-54. (In Russian.)

[3] Gihman, I.I., and Skorohod, A.V. *Introduction to the Theory of Random Processes*. Philadelphia, Pa.: Saunders, 1969.

[4] Liptser, R.S. "On a Functional Limit Theorem for Markov Processes with Finite State Space." *In this volume*, pp.305-316.

N.V. KRYLOV

ONCE MORE ABOUT THE CONNECTION BETWEEN ELLIPTIC OPERATORS AND ITÔ'S STOCHASTIC EQUATIONS

1. INTRODUCTION

Let $E_d = \{x = (x^1,\dots,x^d)\}$ be a Euclidean space and for all $x \in E_d$ we are given $a(x) = (a^{ij}(x))$ symmetric nonnegative $d\times d$-matrix, $b(x) = (b^i(x))$ d-vector, real $c(x) \geq 0$. Throughout the paper, a, b, c are supposed to be Borel bounded functions. Define

$$L = \tfrac{1}{2} a^{ij}(x) \frac{\partial^2}{\partial x^i \partial x^j} + b^i(x) \frac{\partial}{\partial x^i} - c(x) . \tag{1}$$

On some probability space $(\Omega, \mathcal{F}, P)$ supporting a d-dimensional Wiener process $(w_t, \mathcal{F}_t)$, we consider also the Itô stochastic equation

$$dx_t = \sqrt{a(x_t)}\, dw_t + b(x_t)dt , \quad t \geq 0, \quad x_0 = x , \tag{2}$$

where x is nonrandom. Of course, only $\mathcal{F}_t$-adapted solutions x_t of this equation are considered.

It has been known for a long time [1] that if $a(x)$ is uniformly nondegenerate, then (2) can be solved on an appropriate probability space, and if $(L - \lambda)C_0^\infty(E_d)$ is dense in $L_d(E_d)$ for every $\lambda > 0$, then all solutions of (2) coincide in finite-dimensional distributions. Unfortunately, this density property is known from PDE theory only for continuous a, b or for $d = 2$. In Section 6 of this paper, another property of L is given as a condition for weak uniqueness of a solution of (2). At present, it is not clear whether the new criterion is effective or not. Nevertheless the method used in the proof is undoubtedly of interest.

In a sense, this method sometimes makes it possible to divide the problem into two parts: one part to be solved by purely probabilistic methods and the other part by purely PDE methods. For instance, in [2] the mixed methods are used to prove the L_p-estimate of distributions of x_t. It is the same for the proof given by Stroock and Varadhan [3], of weak uniqueness of a solution of (2) with a continuous nondegenerate a. Now we have the possibility to make reference to PDE theory after we have performed some probabilistic transformations (cf. Section 3 and Section 6).

Our method is introduced in Section 2 and the possibilities of the method are studied in Sections 3 and 4, where two more problems are considered. In Section 4, it is proved that if $c = 0$, μ is a probability measure on E_d and $L^*\mu = 0$ in the distribution sense, then μ is an invariant measure for some solution of (2) with a random x_0. S.R.S. Varadhan discussed a similar result in his report at the Vilnius Conference in June 1981. Our proof seems to be simpler than the proof mentioned by Dr. Varadhan. The material of Section 3 has been spurred by a study of nonuniqueness property of solutions to (2). It appears that under broad assumptions, for each solution of (2) there exists a Markov solution of

(2) with the same one-dimensional distribution. Furthermore, if $dy_t = \sqrt{\tilde{a}_t} dw_t + \tilde{b}_t dt$, $y_0 = 0$, where $\tilde{a}$, $\tilde{b}_t$ are random, then there exists a solution of

$$dx_t = \sqrt{a(t, x_t)} + b(t, x_t)\, dt \, , \quad x_0 = 0$$

such that $P(x_t \in \Gamma) = P(y_t \in \Gamma)$ for every $t \geq 0$, Borel $\Gamma \subseteq E_d$ and

$$a(t,x) = E(\tilde{a}_t \mid y_t = x) \, , \quad b(t,x) = E(\tilde{b}_t \mid y_t = x) \; .$$

In fact, this result is not formulated in Section 3, but it can easily be derived from Section 3 if one takes $\tilde{c} \equiv \delta$ and (t,y_t) instead of y_t. In this connection, note that we consider only the time-homogeneous equation (2), though our method can also be applied to a time-dependent case, for example, by formally adding one more coordinate t to the state coordinates x. Note also that the basic idea of our method can be found in [4, Proof of Lemma 3.3].

2. AUXILIARY RESULT

Let $\zeta(x) = \zeta(|x|)$ be a nonnegative infinitely differentiable function with $\int \zeta \, dx = 1$. Define $\zeta_\varepsilon(x) = \varepsilon^{-d}\zeta(\varepsilon^{-1}x)$. For a function $f(x)$ and a measure μ on E_d denote by $f \circ \mu$ the indefinite integral of f with respect to μ ($f \circ \mu$ is a measure, $f \circ \mu(\Gamma) = \int_\Gamma f\mu(dx)$), $f^{(\varepsilon)} = f * \zeta_\varepsilon$, $\mu^{(\varepsilon)} = \zeta_\varepsilon * \mu$.

<u>LEMMA 1</u>. Let $\nu \geq 0$, $\mu \geq 0$ be two finite measures on E_d such that $\mu^{(\varepsilon)} > 0$ on E_d for every $\varepsilon > 0$ and

$$\int u\nu(dx) \geq -\int L u \, \mu(dx) \qquad (3)$$

for every $u \in C_0^\infty(E_d)$, $u \geq 0$. Suppose $c \geq \delta = \text{const} > 0$. For $\varepsilon > 0$ define

$$(a_\varepsilon, b_\varepsilon, c_\varepsilon) = (\mu^{(\varepsilon)})^{-1}((a \circ \mu)^{(\varepsilon)}, (b \circ \mu)^{(\varepsilon)}, (c \circ \mu)^{(\varepsilon)}) . \tag{4}$$

Let $x_t^{\varepsilon,x}$ be a solution of the stochastic equation

$$dx_t = \sqrt{a_\varepsilon(x_t)}\, dw_t + b_\varepsilon(x_t)dt , \quad t \geq 0 , \quad x_0 = x , \tag{5}$$

$$R_\lambda^\varepsilon f(x) = E\int_0^\infty f(x_t^{\varepsilon,x}) \exp\left[-\int_0^t c_\varepsilon(x_r^{\varepsilon,x})dr - \lambda t\right] dt . \tag{6}$$

Then

$$\underline{\lim}_{\varepsilon \downarrow 0} \int (R_0^\varepsilon f)^{(\varepsilon)} \nu(dx) \geq \int f\mu(dx) \tag{6}$$

for all $f \in C_0^\infty(E_d)$, $f \geq 0$. If in (3) the inverse inequality holds (for every $u \in C_0^\infty(E_d)$, $u \geq 0$), then

$$\overline{\lim}_{\varepsilon \downarrow 0} \int (R_0^\varepsilon f)^{(\varepsilon)} \nu(dx) \leq \int f\mu(dx) \tag{7}$$

for all $f \in C_0^\infty(E_d)$, $f \geq 0$.

Proof. First of all note that a_ε, b_ε, c_ε are infinitely differentiable with bounded derivatives ($\varepsilon > 0$ fixed). As is known from [5], $\sqrt{a_\varepsilon}$ satisfies a Lipschitz condition in this situation. Therefore, $x_t^{\varepsilon,x}$ exists; it is unique, continuous (in probability) in x and $x_t^{\varepsilon,x}$ is a Markov process. Let

$$L_\varepsilon = a_\varepsilon^{ij} \frac{\partial^2}{\partial x^i \partial x^j} + b_\varepsilon^i \frac{\partial}{\partial x^i} - c_\varepsilon \tag{8}$$

and let us prove the Lemma under the additional condition that

$$a^{ij}\xi^i\xi^j \geq \gamma|\xi|^2 , \qquad \gamma = \text{const} > 0 \tag{9}$$

for all $\xi \in E_d$. Obviously a_ε satisfies the same condition (9), and from PDE theory we obtain the existence of the solution $u_\varepsilon \in C_b^2(E_d)$, $u_\varepsilon \geq 0$ of the equation $L_\varepsilon u + f = 0$ for every $f \in C_0^\infty(E_d)$, $f \geq 0$. By the Itô formula $u_\varepsilon = R_0^\varepsilon f$.

Now replace u in (3) by $u^{(\varepsilon)}$. Then for all $u \in C_b^2(E_d)$, $u \geq 0$ we get

$$\int u\nu^{(\varepsilon)}\, dx \geq -\int L_\varepsilon\, u\, \mu^{(\varepsilon)}\, dx . \tag{10}$$

Hence for $u = u_\varepsilon$

$$\int (R_0^\varepsilon f)^{(\varepsilon)} \nu(dx) \geq \int f^{(\varepsilon)}\, \mu(dx) \tag{11}$$

and (7) is proved. If (9) is not satisfied, additional considerations are needed.

Given $\varepsilon > 0$ one knows from [6] about the existence of $\lambda_\varepsilon > 0$ such that for $\lambda \geq \lambda_\varepsilon$, $f \in C_0^\infty(E_d)$, $f \geq 0$ the equation $\lambda u - L_\varepsilon u = f$ is solvable in $C_b^2(E_d)$. In the same way as above $u = R_\lambda^\varepsilon f$ and (10) leads to

$$\int R_\lambda^\varepsilon\, f\, \nu^{(\varepsilon)}\, dx \geq \int (-\lambda R_\lambda^\varepsilon f + f)\mu^{(\varepsilon)}\, dx \tag{12}$$

for $f \in C_0^\infty(E_d)$, $f \geq 0$. Standard arguments prove (12) for all Borel bounded $f \geq 0$.

Now recall that $x_t^{\varepsilon,x}$ is a Markov process which implies $R_\kappa^\varepsilon = R_\lambda^\varepsilon + (\lambda - \kappa)R_\lambda^\varepsilon R_\kappa^\varepsilon$,

$$R_\kappa^\varepsilon = \sum_{n=1}^{\infty} (\lambda - \kappa)^{n-1}(R_\lambda^\varepsilon)^n \tag{13}$$

if $|\lambda - \kappa| < \lambda + \delta$, where the series is convergent in the norm of operators from $C_b(E_d)$ to $C_b(E_d)$ (in this norm $|R_\lambda^\varepsilon| \le (\delta + \lambda)^{-1}$). Thus by (12) for bounded $f \ge 0$ we have

$$\int \nu^{(\varepsilon)} R_0^\varepsilon f \, dx = \sum_{n=1}^{\infty} \lambda^{n-1} \int \nu^{(\varepsilon)} (R_\lambda^\varepsilon)^n f \, dx$$

$$\geq \sum_{n=1}^{\infty} \lambda^{n-1} \int \mu^{(\varepsilon)} [-\lambda R_\lambda^\varepsilon (R_\lambda^\varepsilon)^{n-1} f + (R_\lambda^\varepsilon)^{n-1} f] \, dx$$

$$= \int f \, \mu^{(\varepsilon)} \, dx \quad .$$

It gives (11) again. The last statement of our Lemma could be proved in the same way.

3. THE GREEN MEASURES OF STOCHASTIC INTEGRALS

Fix $\delta > 0$ and let A be a bounded closed convex set of triplets (a,b,c), where $a = (a^{ij}) = a^*$ is a $d\times d$-matrix, $a \geq 0$, $b - (b^i)$ is a d-vector, real-valued $c \geqslant \delta$. Let $\tilde{a}(s)$, $\tilde{b}(s)$, $\tilde{c}(s)$ be some processes measurable with respect to (s,ω), $\mathcal{F}_s$-measurable for each $s \geqslant 0$ and such that $(\tilde{a}(s), \tilde{b}(s), \tilde{c}(s)) \in A$ for all (s,ω).

Let

$$y_t = \int_0^t \sqrt{\tilde{a}(s)} \, dw_s + \int_0^t \tilde{b}(s) \, ds \quad ,$$

$$\phi_t = \int_0^t \tilde{c}(s) \, ds \quad ,$$

$$\mu(\Gamma) = E \int_0^\infty e^{-\phi_t} \chi_\Gamma(y_t) \, dt \quad ,$$

where Borel $\Gamma \subseteq E_d$. The measure μ is called the Green $\tilde{c}$-measure of y_t. Also, define

$$(\mu_a(\Gamma),\ \mu_b(\Gamma),\ \mu_c(\Gamma)) = E\int_0^\infty e^{-\phi_t}\chi_\Gamma(y_t)(\tilde{a}(t),\ \tilde{b}(t),\ \tilde{c}(t))\,dt ,$$

$$(a(x),\ b(x),\ c(x)) = \frac{(\mu_a(dx),\ \mu_b(dx),\ \mu_c(dx))}{\mu(dx)} .$$

Since A is a bounded closed convex set, one may choose a, b, c in such a way that $(a,b,c) \in A$ for all $x \in E_d$ and a, b, c are Borel.

LEMMA 2. Suppose that $\mu^{(\varepsilon)} > 0$ on E_d for every $\varepsilon > 0$ define a_ε, b_ε, c_ε by (4) and let x_t^ε be a solution of (5) with random initial condition x_0 independent of w and with density equal to ζ_ε. Then the Green $\tilde{c}$-measure of y_t is the weak limit, as $\varepsilon \downarrow 0$, of the Green c_ε-measure of x_t^ε, where $c_\varepsilon(s,\omega) = c_\varepsilon(x_s^\varepsilon(\omega))$.

Proof. Denote $\nu(\Gamma) = \chi_\Gamma(0)$, use (1) and the Itô formula. Then for all $u \in C_0^\infty(E_d)$ we get

$$\int u\nu(dx) = u(0)$$

$$= E\int_0^\infty e^{-\phi_t}[\tilde{c}(t)u(y_t) - \tfrac{1}{2}\tilde{a}^{ij}(t)u_{x^ix^j}(y_t) - \tilde{b}^i(t)u_{x^i}(y_t)]\,dt$$

$$= \int [u\,\mu_c(dx) - \tfrac{1}{2}u_{x^ix^j}\,\mu_a^{ij}(dx) - u_{x^i}\mu_b^i(dx)]$$

$$= -\int L\,u\,\mu(dx) .$$

Hence for $f \in C_0^\infty(E_d)$ by Lemma 1

$$\int f\ \mu(dx) = \lim_{\varepsilon \downarrow 0} (R_0^\varepsilon f)^{(\varepsilon)}(0) \ . \tag{14}$$

The Lemma is thus proved since $(R_0^\varepsilon f)^{(\varepsilon)}$ is evidently the integral of f with respect to the Green c_ε-measure of x_t^ε.

This lemma allows us to consider only solutions of stochastic equations in prooving certain estimates for distributions of stochastic integrals. For instance from [2], [7] we know that if $\tilde{a}^{ij}\xi^i\xi^j \geq \gamma|\xi|^2$ for all $\xi \in E_d$, s, ω, where $\gamma = \text{const} > 0$, then

$$E \int_0^\infty e^{-\delta t} f(y_t)\, dt \leq N\|f\|_d \tag{15}$$

for every Borel $f \geq 0$, where $\|f\|_d$ is the $L_d(E_d)$-norm of f and N depends only on d, γ, δ and on the upper bound of $|\tilde{a}|$, $|\tilde{b}|$. Though the complete proof of (15) given in [2], [7] is sufficiently brief, it is based on the quite unexpectedly used convex polyhedrons.

Now (15) can be obtained in a much more natural way. First note that it is enough to prove (15) for $f \in C_0^\infty(E_d)$ and by Lemma 2 (with $\tilde{c} \equiv \delta$) it is enough to estimate $R_0^\varepsilon f(x)$ by the right-hand side of (15). Of course it must be pointed out that the condition $\mu^{(\varepsilon)} > 0$ of Lemma 2 can be satisfied if one takes $\zeta > 0$ on E_d. Since a_ε, b_ε $(c_\varepsilon \equiv \delta)$ are smooth bounded and $a^{ij}\xi^i\xi^j \geq \gamma|\xi|^2$, PDE theory gives a smooth bounded solution u^ε of $L_\varepsilon u + f = 0$. By Alexandrov's estimate [8] (cf. also [9]), u^ε is less than the right-hand side of (15), and to complete the proof of (15) it suffices to note that $u^\varepsilon = u_0^\varepsilon f$ by the Itô formula.

In this proof of (15) the question is not how Alexandrov's estimate has been obtained. It is important only that it has been proved entirely by PDE methods.

In another place we shall prove by PDE methods that if

• (1) the coefficients a_ε, b_ε, c_ε of L_ε from (8) are smooth uniformly bounded for $\varepsilon > 0$ (a_ε, b_ε, c_ε need not be connected by (4) with some a, b, c);

• (2) $c_\varepsilon \geq \delta$, $a_\varepsilon^{ij}\xi^i\xi^j \geq \gamma|\xi|^2$ for all $\xi, x \in E_d$, $\varepsilon > 0$ and some constants $\delta, \gamma > 0$;

• (3) $f_\varepsilon \in C_0^\infty(E_d)$, $f_\varepsilon \geq 0$, $f_\varepsilon = 0$ in the exterior of a fixed ball;

• (4) u_ε is a smooth bounded solution of $L_\varepsilon u + f_\varepsilon = 0$ and $u_\varepsilon(x(\varepsilon)) \to 0$ for $\varepsilon \to 0$ where $\sup\limits_\varepsilon |x(\varepsilon)| < \infty$ then $f_\varepsilon \to 0$ in measure. By virtue of Lemma 2, this immediately proves

COROLLARY 1. If $\tilde{a}^{ij}\xi^i\xi^j \geq \gamma|\xi|^2$ for all $\xi \in E_d$, s, ω, then the Lebesgue measure is absolutely continuous with respect to μ. On the other hand, $\mu(dx) << dx$ by (15) so that the Green $\tilde{c}$-measure of y_t is equivalent to the Lebesgue measure.

Recall next that by the Lebesgue theorem, for every finite measure μ we have $\mu^{(\varepsilon)} \to \mu(dx)/dx$ as $\varepsilon \downarrow 0$ (a.e. with respect to Lebesgue measure. Moreover, $\mu(dx)/dx > 0$ (a.e. if $dx << \mu(dx)$. Therefore, in Corollary 1 also a_ε, b_ε, $c_\varepsilon \to a,b,c$ as $\varepsilon \downarrow 0$ (a.e.). One can use this fact and (15) in order to prove, as in [1], using the Skorokhod embedding method, the existence of a sequence $\varepsilon_n \downarrow 0$ such that the distributions of $x_t^{\varepsilon_n}$ in $C_{[0,T]}$ converge in the weak sense for every T to the distribution of a solution of (2) with $x = 0$. Hence we obtain

THEOREM 1. Suppose $\tilde{a}^{ij}\xi^i\xi^j \geq \gamma|\xi|^2$ for all $\xi \in E_d$, s, ω, where $\gamma = \text{const} > 0$. Then the Green $\tilde{c}$-measure of y_t coincides with the Green c-measure of the process x_t, $c(s,\omega) = c(x_s(\omega))$ and x_t is a solution of (2) with $x = 0$ on an appropriate probability space.

REMARK 1. The results obtained by Krylov and Safonov [10] can easily be used to prove that x_t mentioned in Theorem 1 and constructed before this theorem is a strong Markov as well as strong Feller process.

4. INVARIANT MEASURES

Let $c \equiv 0$ and let μ be a probability measure on E such that

$$\int Lu\,\mu(dx) = 0 \tag{16}$$

for all $u \in C_0^\infty(E_d)$, where L is defined by (1).

THEOREM 2. Suppose one of the following conditions is fulfilled:

• (a) a, b are continuous, $\mu(y: |y-x| < r) > 0$ for all $x \in E_d$, $r > 0$ or

• (b) $a^{ij}\xi^i\xi^j \geq \gamma|\xi|^2$ for all $x, \xi \in E_d$, where $\gamma = \text{const} > 0$.

Then on some probability space there exists a solution x_t of equation (2) with random x_0 such that $P\{x_t \in \Gamma\} = \mu(\Gamma)$ for all $t \geq 0$ and Borel $\Gamma \subset E_d$.

Proof. Take $\lambda > 0$. It is clear that $\int \lambda u\mu(dx) = \int (\lambda - L)u\mu(dx)$ if $u \in C_0^\infty(E_d)$. Hence by Lemma 1

$$\lim_{\varepsilon \downarrow 0} \lambda \int (R_\lambda^\varepsilon f)^{(\varepsilon)}\,\mu(dx) = \int f\,\mu(dx) \tag{17}$$

for every $f \in C_0^\infty(E_d)$. It must be made clear that to satisfy the condition $\mu^{(\varepsilon)} > 0$ of Lemma 1 we take $\zeta > 0$ on E_d in the case (b) and simply $\zeta \in C_0^\infty(E_d)$ in the case (a). It is clear that

$$\int (R_\lambda^\varepsilon f)^{(\varepsilon)} \mu(dx) = \int R_\lambda^\varepsilon f \mu^{(\varepsilon)} dx = \int_0^\infty e^{-\lambda t} Ef(x_t^\varepsilon)\, dt \quad ,$$

where x_t^ε is a solution of (5) with random x_0 which is independent of w and the density of x_0 is $\mu^{(\varepsilon)}$.

In the case (a) it is easy to see that a_ε, $b_\varepsilon \to a, b$ uniformly on every compact subset of E_d. By the Skorokhod embedding method one obtains a sequence $\varepsilon_n \downarrow 0$ such that the distributions of $x_t^{\varepsilon_n}$ converge in the weak sense to the distribution of a solution x_t of (2) with random x_0. The process x_t is in no way connected with λ and by (17) we have that

$$\lambda \int_0^\infty e^{-\lambda t} Ef(x_t)\, dt = \int f\mu\, (dx) \quad .$$

Now $Ef(x_t)$ is continuous in t, λ is arbitrary positive and the uniqueness of the Laplace transform is known. Therefore, $Ef(x_t = \int f\mu(dx)$ for every $f \in C_0^\infty(E_d)$, and thus we complete the proof.

In the case (b) the same arguments hold true if we derive, as in Corollary 1, the equivalence of dx and $\mu(dx)$ and next use the procedure mentioned before Theorem 1.

<u>REMARK 2</u> (cf. Remark 1). In the case (b) one can choose x_t to be a strong Markov as well as strong Feller process.

<u>REMARK 3</u>. In the proof we have not used the condition $c \equiv 0$. Hence, if $c \geq 0$ and, in addition, in the case (a) c is contin-

uous, there exists a process x_t such that

$$Ef(x_t)\exp\left(-\int_0^t c(x_s)ds\right) = \int f\mu(dx) \qquad (18)$$

for every $t \geq 0$ and Borel $f \geq 0$. Take $f = c$ and integrate (18) with respect to t over $(0,\infty)$. Then the left-hand side will be less than one, and therefore the right-hand side will be finite and $c\mu(dx) = 0$. In the case (a) it obviously implies $c \equiv 0$. In the case (b), $\mu(dx) \sim dx$ and $c = 0$ almost everywhere with respect to Lebesgue measure.

5. SOME RESULTS FROM PARTIAL DIFFERENTIAL EQUATION THEORY

Define L by (1) and suppose that $c \geq \delta = \text{const} > 0$ and condition (b) of Theorem 2 is satisfied. For $n = 1,2,\ldots$ we assume also that uniformly bounded functions $a_n^{ij}(x) = a_n^{ji}(x)$, $b_n^i(x)$, $c_n(x)$ infinitely differentiable with respect to x, are defined. Assume that $c_n \geq \delta$ and a_n satisfies condition (b) of Theorem 2 with the same γ. Let $a_n, b_n, c_n \to a, b, c$ (a.e.) if $n \to \infty$. Using a_n, b_n, c_n instead of a, b, c, we introduce L_n by (1). It is well known that for every $f \in C_b(E_d) \cap C^\infty(E_d)$ in $C_b(E_d) \cap C^\infty(E_d)$ there exists a unique solution of $L_n u + f = 0$. Denote this solution by $R^n f$. The operators R^n thus defined can be extended on $B(E_d)$. For every $f \in B(E_d)$ the sequence $R^n f$ is equicontinuous on each compact subset of E_d (cf. [10]). Therefore, by Alexandrov's estimate there exists a subsequence $n_k \to \infty$ (independent of f) such that for every $f \in B(E_d)$, $x \in E_d$ the sequence $R^{n_k} f(x)$ is convergent. Denote by $Rf(x)$ its limit and by $\mathcal{R}(L)$ the set of all operators R which can be constructed in this way, starting with some a_n, b_n, c_n. If $R(L)$ contains only one element, we shall say that the R-weak uniqueness holds for L.

This is true in the case where a^{ij} is continuous. In fact, this follows from the solvability of $Lu + f = 0$ in the Sobolev spaces and from Alexandrov's estimate again. For the same reasons the R-weak uniqueness holds for L if $d = 2$. Note also that for $\lambda > -\delta$ formula (13) implies the equivalence of the R-weak uniqueness for L to the R-weak uniqueness for $L - \lambda$.

6. ON THE WEAK UNIQUENESS OF SOLUTIONS OF (2)

Recall that the solution of (2) is said to be unique in the weak sense for the given initial data x if for every probability space (Ω, F, P) carrying a Wiener process (w_t, F_t) every solution x_t of (2) (if it exists) has the same finite-dimensional distributions (which are thus independent of the choice of Ω, F, P, w_t, F_t, x_t).

THEOREM 3. Let condition (b) of Theorem 2 be satisfied and let $c \equiv \delta = \text{const} > 0$. Then the following statements are equivalent:

- (a) R-weak uniqueness holds for L;
- (b) the solution of (2) is unique in the weak sense for every x.

Proof. (b) => (a). Let u_n, $f \in C_b(E_d) \cap C^\infty(E_d)$, $L_n u_n + f = 0$. By the Itô formula, $u_n(x) = R_0^n f(x)$, where $R_0^n f$ are constructed in the obvious way by a formula similar to (6). The Skorokhod embedding method enables us to extract a subsequence of processes corresponding to L_n, which converges in the weak sense to some solution of (2). Since the distributions of (2) are uniquely defined, there is no need to extract a subsequence, and the entire sequence $u_n(x)$ is convergent. This proves (a), of course.

(a) => (b). For simplicity, take $x = 0$ and let μ be the Green δ-measure of x_t. Formula (14), in the context of Section 3, defines $f\mu(dx)$ as the limit of the solutions of the equation $L_\varepsilon u + f = 0$. In view of (a) this limit is independent of the special construction of a_ε, b_ε; it is important only that a_ε, $b_\varepsilon \to a, b$ (a.e.). Thus all the solutions of (2) have the same Green δ-measure. Moreover, the R-weak uniqueness holds also for $L - \lambda$ if $\lambda >$ and all the solutions of (2) have the same Green $(\delta+\lambda)$-measure. It follows that one-dimensional distributions of the solutions of (2) are uniquely defined.

Now use the recurrence. Suppose that for some n and all $f_1,\ldots,f_n \in B(E_d)$, $0 \le t_1,\ldots,t_n$ the expression $Ef_1(x_{t_1}) \cdots f_n(x_{t_n})$ does not depend on the choice of a solution of (2).

Fix $f_i > 0$, $0 \le t_1 \le \cdots \le t_n$ and define

$$\nu(\Gamma) = Ef_1(x_{t_1}) \times \cdots \times f_n(x_{t_n}) \chi_\Gamma(x_{t_n}) ,$$

$$\mu_n(\Gamma) = Ef_1(x_{t_1}) \times \cdots \times f_n(x_{t_n}) \int_{t_n}^{\infty} \chi_\Gamma(x_t) e^{-\delta(t-t_n)} dt .$$

By the Itô formula

$$\int u\nu(dx) = -\int Lu \, \mu_n(dx) .$$

Lemma 1, as above, implies that $\int f\mu_n(dx)$ for $f \in C_0^\infty(E_d)$ is defined only by ν and the limit of the solutions of

$L_\varepsilon u + f = 0$. Our assumption means that ν and this limit do not depend on the choice of a solution of (2). Therefore it is the same for $f\mu_n(dx)$ which equals

$$Ef_1(x_{t_1}) \times \cdots \times f_n(x_{t_n}) \int_{t_n}^{\infty} f(x_t)\, e^{-\delta(t-t_n)}\, dt \, . \tag{19}$$

We can consider $\lambda > 0$ instead of δ and prove that (19) with λ instead of δ does not depend on the choice of a solution. Thus $Ef_1(x_{t_1}) \times \cdots \times f_n(x_{t_n}) f(x_t)$ for $t > 0$ is uniquely defined and the Theorem is proved.

In the following remarks we assume that condition (b) of Theorem 2 is satisfied.

REMARK 4. The weak uniqueness means that the distributions of every solution of (2) coincide with those of the solution which can be constructed by the Skorokhod method on the basis of smooth a_n, b_n. The results of [10] show then that the limit process and all the solutions of (2) are strong Markov as well as strong Feller processes.

REMARK 5. We know the notion of the weak uniqueness for (2) with fixed initial data $x = 0$. One can introduce also a definition of R-weak uniqueness for $x = 0$, in which the convergence of $R^n f(x)$ (cf. Section 5) only at the point $x = 0$ is required (note that the R-weak uniqueness for L is equivalent to uniform convergence of $R^n f(x)$ on compact subsets for every approapriate a_n, b_n, c_n). As can be proved, the R-weak uniqueness for $x = 0$ is equivalent to the weak uniqueness of the solution of (2) with $x = 0$ and is equivalent to properties (a) and (b) of Theorem 3.

REFERENCES

[1] Krylov, N.V. "On Itô's Stochastic Integral Equations." *Theory Probab. Application.*,vol.14, no.2 (1969):330-336.

[2] Krylov, N.V. "An Inequality in the Theory of Stochastic Integrals." *Theory Probab. Application*, vol.16, no.3 (1971): 438-448.

[3] Stroock, D.W., and Varadhan, S.R.S. "Diffusion Processes with Continuous Coefficients, I, II." *Comm. on Pure and Appl. Math.*, *Math.*, 22, 3 (1969): 345-400; 479-530.

[4] Krylov, N.V. "On Passing to the Limit in Degenerate Bellman Equations, II." *Math. USSR Sbornik*, vol.35, no.3 (1979): 351-362.

[5] Freidlin, M.I. "On the Factorization of Non-negative Definite Matrices." *Theory Probab. Application*, vol.13, no.2 (1968): 354-356.

[6] Olejnik, O.A., and Radkevich, E.A. "Second Order Equations with Nonnegative Characteristic Form." *Itogi nauki, seriya matematiki, matem. analiz.* Moskva: VINITI, 1971.

[7] Krylov, N.V. "Control of a Solution of a Stochastic Integral Equation." *Theory Probab. Application*, vol.17, no.1 (1972): 114-131.

[8] Alexandrov, A.D. "Conditions for Uniqueness and Estimates of Solutions of the Dirichlet Problem." *Vestnik Leningradskogo Gosudarstvennogo Universiteta, seriya matematiki, mekhaniki, astronomii*, vyp.13, no.3 (1963): 5-19.

[9] Krylov, N.V. "On an Estimate of the Solution of an Elliptic Equation." *Vestnik Moskovskogo Gosudarstvennogo Universiteta, seriya matematiki i mekhaniki*, no.3 (1969): 11-13.

[10] Krylov, N.V., and Safonov, M.V. "A Certain Property of Solutions of Parabolic Equations with Measurable Coefficients." *Math. USSR Izvestiya, vol.16, no.1 (1981): 151-164.*

Yu.A. KUTOYANTS

ON NONPARAMETRIC ESTIMATION OF TREND COEFFICIENTS IN A DIFFUSION PROCESS

1. INTRODUCTION

Let the diffusion process $x_\varepsilon = \{x_\varepsilon(t),\ 0 \le t \le T\}$ be given with the stochastic differential [1]

$$dx_\varepsilon = S(x_\varepsilon(t))dt + \varepsilon dw(t)\ , \qquad x_\varepsilon(0) = x_0\ , \qquad 0 \le t \le T\ , \qquad (1)$$

where $w(t)$ is the Wiener process and the trend coefficient function $S(X)$ is to be estimated from realization X_ε. We are interested in the asymptotic properties of the estimate as $\varepsilon \to 0$. The case when the trend coefficient $S(x)$ is known up to the value of some parameter θ, i.e., $S(x) = S(\theta,x)$ and it is required to estimate θ and describe the asymptotic of estimates, was considered in [2], [3], where the consistency and asymptotic normality of the maximum likelihood and Bayes estimates were established.

Let us denote by $x_0(t)$ the solution of the ordinary differential equation

$$\frac{dx_0(t)}{dt} = S(x_0(t))\ , \qquad x_0(0) = x_0\ , \qquad 0 \le t \le T. \quad (2)$$

It is easy to see that the function $S(x)$, $x \in [a,b]$, is restored by $x_0(t)$, $0 \le t \le T$, where

$$a = \min_{0 \le t \le T} x_0(t)\ , \qquad b = \min_{0 \le t \le T} x_0(t)$$

and the estimation of $S(x)$, $x \in [a,b]$, is equivalent to that of $f_0(t) = S(x_0(t))$, $0 \le t \le T$, because the functions $\{f_0(t), x_0(t), 0 \le t \le T\}$ define $S(x)$, $x \in [a,b]$. Nevertheless, the estimation procedure for the function $S(x)$ differs from that for the function $f_0(t)$. In this paper, we suggest some kernel-type estimates $\hat{S}_\varepsilon(x)$ and $\hat{f}_\varepsilon(t)$ of the functions $S(x)$ and $f_0(t)$. The consistency of these estimates is proved and the optimal rate of convergence for these estimates is discussed.

The main result is the following: let the function $S(x)$ have k bounded derivatives on $x \in [a,b]$ and let the kth derivative satisfy a Hölder condition of α-order. Then the distributions of

$$[f_\varepsilon(t) - f(t)]\varepsilon^{-\frac{2\beta}{2\beta+1}}, \qquad [\hat{S}_\varepsilon(x) - S(x)]\varepsilon^{-\frac{2\beta}{2\beta+1}},$$

where $\beta = k + \alpha$, are asymptotically normal as $\varepsilon \to 0$.

We assume that the conditions

$$|S(x) - S(y)| \le L|x-y|, \qquad |S(x)| \le L(1 + |x|) \tag{3}$$

for the existence and uniqueness of the solution of the stochastic differential equation (1) are satisfied. We shall use condition (3) in proving the Theorems. In particular, condition (3) yields the following inequalities:

$$|x_\varepsilon(t) - x_0(t)| \le \varepsilon|w(t)| + \varepsilon L\int_0^t e^{L(t-s)}|w(s)|\,ds,$$

$$\sup_{0\le t\le T}|x_\varepsilon(t) - x_0(t)| \le \varepsilon e^{LT}\sup_{0\le t\le T}|w(t)| \tag{4}$$

(see, for example [2, p. 126]).

We construct kernel-type estimates for $S(\cdot)$ and $f(\cdot)$ by analogy with the Parzen-Rosenblatt [4] estimates of the density function $f(x)$. Let $x_1,\ldots,x_n$ be the realizations of the random variable ξ. The sample distribution function

$$\mathcal{F}_n(x) = \frac{1}{n}\sum_{j=1}^{n} \chi_{\{X_j<x_j\}} , \tag{5}$$

where $\chi_{\{B\}}$ is the indicator of B, can be expressed in the form

$$\mathcal{F}_n(x) = \mathcal{F}(0) + \int_0^x f(y)\,dy + \frac{1}{\sqrt{n}}\,\varepsilon_n(x) . \tag{6}$$

Here $\frac{1}{\sqrt{n}}\varepsilon_n(x)$ could be treated as small nondifferentiable noise. The kernel estimate

$$f_n(x) = \frac{1}{n\phi_n}\sum_{j=1}^{n} G\left(\frac{x-X_j}{\phi_n}\right) , \qquad \phi_n \to 0 , \quad n\phi_n \to \infty ,$$

where the kernel $G(\cdot)$ is some bounded function with

$$\int_{-\infty}^{\infty} G(u)\,du = 1 , \tag{7}$$

consistently estimates the density function. It can be treated as smoothed after formally differentiating $\mathcal{F}_n(x)$ on x.

The observations (1) written in integral form

$$x_\varepsilon(t) = x_0 + \int_0^t S(X_\varepsilon(\tau))\,d\tau + \varepsilon w(t) , \qquad 0 \le t \le T ,$$

bear obvious analogy to (6). Hence, we choose for example the $\hat{f}_\varepsilon(t)$-estimate of the function $f_0(t)$ in the form

$$\hat{f}_\varepsilon(t) = \frac{1}{\phi_\varepsilon}\int_0^T G\left(\frac{\tau-t}{\phi_\varepsilon}\right) dX_\varepsilon(\tau) . \tag{8}$$

The rate of convergence ϕ_ε to zero that we shall choose later depends on the smoothness of the function $S(x)$.

Some problems of nonparametric estimation of the trend $S(t)$, $0 \le t \le T$ by the observations

$$dX_\varepsilon(t) = S(t)dt + \varepsilon dw(t) , \quad X_\varepsilon(0) = 0 , \quad 0 \le t \le T ,$$

are considered in I.A. Ibragimov and R.Z. Has'minskii [4] and [5], where the consistency and asymptotic normality of the estimates in some metrics as well as the low bounds of the risk function have been established.

G. Banon [6] gives a preliminary consistent estimate of the one-dimensional density and its derivatives and also obtains the estimate of the trend coefficient from the Kolmogorov backward equation. For the same observations, Pham [7] suggests the estimate

$$\hat{S}_T(x) = \left\{\int_0^T H(t)\, G\left(\frac{X_t - x}{\phi_t}\right) dt\right\}^{-1} \int_0^T H(t)\, G\left(\frac{X_t - x}{\phi_t}\right) dX(t) ,$$

where $\phi_t \to 0$ and $\int_0^T H(t)\phi_t dt \to \infty$ and proves its consistency and asymptotic normality.

Other problems of nonparametrical estimation for diffusion observations are considered also in H.T. Nguen and T.D. Pham [7] (see Section 4).

2. ESTIMATION OF THE FUNCTION f(t)

Let us denote by $\theta(L)$ the set of functions on R, which are uniformly bounded and satisfy the Lipschitz condition (3). We suppose that (7) holds and the kernel $G(u)$ is bounded and $G(u) = 0$, $u \notin [A,B]$, where $A < 0$, $B > 0$. Also, we denote by $[c,d]$ the arbitrary interval $[c,d] \subset (0,T)$.

THEOREM 1. Let $S(\cdot) \in \theta(L)$, $\phi_\varepsilon \to 0$ and $\varepsilon^2\phi_\varepsilon \to 0$. Then the estimate $\hat{f}_\varepsilon(t)$ as $\varepsilon \to 0$ is uniformly consistent:

$$\lim_{\varepsilon\to 0}\ \sup_{S(\cdot)\in\theta(L)}\ \sup_{c\le t\le d} E|\hat{f}_\varepsilon(t) - f(t)|^2 = 0 .$$

Proof. We invoke the definition (8), equality (7) and the inequality $(a+b+c)^2 \le 3a^2 + 3b^2 + 3c^2$:

$$\begin{aligned}
E|\hat{f}_\varepsilon(t) - f_0(t)|^2 &= E\left\{\frac{1}{\phi_\varepsilon}\int_0^T G\left(\frac{\tau-t}{\phi_\varepsilon}\right)[S(X_\varepsilon(\tau)) - S(X_0(\tau))]\,d\tau\right.\\
&\quad + \frac{1}{\phi_\varepsilon}\int_0^T G\left(\frac{\tau-t}{\phi_\varepsilon}\right)S(X_0(\tau))\,d\tau - S(X_0(t))\\
&\quad \left. + \frac{\varepsilon}{\phi_\varepsilon}\int_0^T G\left(\frac{\tau-t}{\phi_\varepsilon}\right) dw(t)\right\}^2\\
&\le 3E\left\{\frac{1}{\phi_\varepsilon}\int_0^T G\left(\frac{\tau-t}{\phi_\varepsilon}\right)[S(X_\varepsilon(\tau)) - S(X_0(\tau))]\,d\tau\right\}^2\\
&\quad + 3\left\{\frac{1}{\phi_\varepsilon}\int_0^T G\left(\frac{\tau-t}{\phi_\varepsilon}\right)S(X_0(\tau))\,d\tau - S(X_0(\tau))\right\}^2\\
&\quad + 3\frac{\varepsilon^2}{\phi_\varepsilon^2}\int_0^T G\left(\frac{\tau-t}{\phi_\varepsilon}\right)^2 dt \,.
\end{aligned} \tag{9}$$

After a change of variable $u = (\tau-t)\phi_\varepsilon^{-1}$ and denoting $\varepsilon_1 = \min(\varepsilon', \varepsilon'')$, where

$$\varepsilon' = \sup\left\{\varepsilon : \phi_\varepsilon \le -\frac{C}{A}\right\}, \qquad \varepsilon'' = \sup\left\{\varepsilon : \phi_\varepsilon \le \frac{T-d}{B}\right\}, \tag{10}$$

for $\varepsilon < \varepsilon_1$ we obtain

$$\frac{\varepsilon^2}{\phi_\varepsilon^2}\int_0^T G\left(\frac{\tau-t}{\phi_\varepsilon}\right)^2 dt = \frac{\varepsilon^2}{\phi_\varepsilon}\int G(u)^2\,du = \frac{\varepsilon^2\sigma^2}{\phi_\varepsilon},$$

$$\begin{aligned}
&\left|\frac{1}{\phi_\varepsilon}\int_0^T G\left(\frac{\tau-t}{\phi_\varepsilon}\right)S(X_0(t))\,d\tau - S(X_0(t))\right|\\
&\qquad = \left|\int G(u)[S(X_0(t+u\phi_\varepsilon)) - S(X_0(t))]\,du\right|\\
&\qquad \le L\int |G(u)|\,|X_0(t+u\phi_\varepsilon) - X_0(t)|\,du\\
&\qquad \le L L_0\,\phi_\varepsilon \int |uG(u)|\,du\,,
\end{aligned} \tag{11}$$

where we omit the infinite limits of integration and use the condition

$$|S(x)| \leq L_0$$

in the inequality

$$|X_0(t+\delta) - X_0(t)| = \left|\int_0^\delta S(x_0(t+v))dv\right| \leq L_0\delta \quad .$$

Then, according to inequality (4) we have

$$E|X_\varepsilon(t) - X_0(t)|^2 \leq T\varepsilon^2 \exp\{LT^2\} \quad . \tag{12}$$

Hence

$$E\left\{\frac{1}{\phi_\varepsilon}\int_0^T G\left(\frac{\tau-t}{\phi_\varepsilon}\right)[S(X_\varepsilon(\tau)) - S(X_0(\tau))]\ d\tau\right\}^2$$

$$= E\{\int G(u)[S(X_\varepsilon(t+u\phi_\varepsilon)) - S(X_0(t+u\phi_\varepsilon))]\ du\}^2$$

$$\leq (B-A)\,L^2\int G(u)^2\,E|X_\varepsilon(t+u\phi_\varepsilon) - X_0(t+u\phi_\varepsilon)|^2\ du$$

$$\leq (B-A)\,L^2\sigma^2 T\,\varepsilon^2\exp\{LT^2\} \quad . \tag{13}$$

Now the uniform consistency of $\hat{f}_\varepsilon(t)$ follows from inequalities (9), (11), (13) and the conditions of Theorem 1.

REMARK. From this proof it follows that if instead of $[c,d]$ we consider the sequence of intervals $[c_\varepsilon, d_\varepsilon]$ such that

$$\frac{c_\varepsilon}{\phi_\varepsilon} \to \infty\ , \qquad \frac{T-d_\varepsilon}{\phi_\varepsilon} \to \infty\ ,$$

then the estimate $\hat{f}_\varepsilon(t)$ should be uniformly consistent for the function $f_0(t)$, $0 \leq t \leq T$.

Note also that

$$\bigcup_{S(\cdot)\in\Theta(L)} [a,b] = [x_0 - L_0T,\ x_0 + L_0T] \quad .$$

Let us denote by $[a,b] = [x_0-L_0T,\ x_0+L_0T]$ and Θ_β, $\beta = k+\alpha$, the set of k-times differentiable functions $Q(x)$, $x \in [a,b]$ such that k^{th} derivatives satisfy the Hölder condition

$$|q^{(k)}(x_2) - q^{(k)}(x_1)| \le L_\alpha |x_2 - x_1|^\alpha .$$

We also suppose that

$$\int u^j G(u)\, du = 0 , \qquad j = 1,\dots,k . \tag{14}$$

The following theorem establishes the optimal rate of convergence of ϕ_ε to 0 in the family of kernel estimates.

THEOREM 2. Let $S(\cdot) \in \Theta_\beta$. Then

$$\lim_{\varepsilon\to 0} \sup_{S(\cdot)\in\Theta_\beta} \sup_{c\le t\le d} E|\hat{f}_\varepsilon(t) - f(t)|^2 \, \varepsilon^{-\frac{4\beta}{2\beta+1}} < \infty . \tag{15}$$

Proof. From the condition of this Theorem it follows that the function $f(t)$ has k derivatives with respect to $t \in [0,T]$ and the kth derivative satisfies a Hölder condition of order α but with a different constant L'_α. Indeed, by the rules for differentiating a composite function we have, for example,

$$f^{(2)}(t) = S^{(2)}(X_0(t)) \, |S(X_0(t))^2 + S^{(1)}(X_0(t))^2 S(X_0(t))| .$$

By the Taylor formula

$$f(t+h) = f(t) + \sum_{j=1}^{k-1} \frac{f^{(j)}(t)}{j!} h^j + \frac{h^k}{k!}[f^{(k)}(t+\gamma h) - f^{(k)}(t)] ,$$

$$\gamma \in (0,1) ,$$

and by conditions (14) for $\varepsilon < \varepsilon_1$ we obtain

$$\left|\frac{1}{\phi_\varepsilon}\int_0^T G\left(\frac{\tau-t}{\phi_\varepsilon}\right) S(X_0(\tau))d\tau - S(X_0(t))\right| = \left|\int G(u)[f(t+u\phi_\varepsilon) - f(t)]\, du\right|$$

$$\le \frac{L'_\alpha}{k!}\phi_\varepsilon^{k+\alpha} \int |u|^k |G(u)|\, du . \tag{16}$$

Using (9), (11), (13) and (16), we obtain

$$\sup_{c\le t\le d} E|\hat{f}_\varepsilon(t) - f(t)|^2 \le c_1\varepsilon^2 + c_2\phi_\varepsilon^{2\beta} + \sigma^2 \frac{\varepsilon^2}{\phi_\varepsilon} .$$

Hence, if $\phi_\varepsilon = \varepsilon^{\frac{2}{2\beta+1}}$, then the above inequality proves (15).

For the $\hat{f}_\varepsilon(t)$ estimate the following representation holds $(\varepsilon < \varepsilon_1)$

$$\begin{aligned}\hat{f}_\varepsilon(t) &= \int S(X_0(t+u\phi_\varepsilon))G(u)\,du + \frac{\varepsilon}{\sqrt{\phi_\varepsilon}}\int G(u)\,d\tilde{w}(u) \\ &\quad + \int [S(X_\varepsilon(t+u\phi_\varepsilon)) - S(X_0(t+u\phi_\varepsilon))]G(u)\,du \\ &= \Phi(t,\phi_\varepsilon) + \varepsilon^{\frac{2\beta}{2\beta+1}}\xi + \varepsilon\xi\quad ,\end{aligned}$$

where $\tilde{w}(u) = \phi_\varepsilon^{-\frac{1}{2}}[w(t+u\phi_\varepsilon) - w(t)]$ is a Wiener process, ξ is a random variable bounded in probability and ξ is a Gaussian random variable with parameters $(0,\sigma^2)$. From (16) we have

$$\begin{aligned}&\varepsilon^{-\frac{2\beta}{2\beta+1}}[\Phi(t,\phi_\varepsilon) - f(t)] \\ &\qquad = (k!)^{-1}\int u^{k+\alpha}\,G(u)\,\frac{[f^{(k)}(t+\gamma u\phi_\varepsilon) - f^{(k)}(t)]\,du}{(u\phi_\varepsilon)^\alpha}\quad ,\end{aligned}$$

and by the condition $S \in \Theta\beta$ the integrand is bounded. Hence, denoting by $m(t)$ the limit of the right-hand expression as $\varepsilon \to 0$ (assuming that it exists), we can prove the asymptotic normality of $\hat{f}_\varepsilon(t)$:

$$L\left\{[\hat{f}_\varepsilon(t) - f(t)]\;\varepsilon^{-\frac{2\beta}{2\beta+1}}\right\} \;\Rightarrow\; N(m(t),\,\sigma^2)\quad .$$

Note that the simplest consistent estimate of the function $f(t)$ is the estimate obtained from the "asymptotic differentiation" of the observations:

$$\tilde{f}_\varepsilon(t) = \frac{X_\varepsilon(t+\phi_\varepsilon) - X_\varepsilon(t)}{\phi_\varepsilon}\quad ,$$

as $\phi_\varepsilon \to 0$ and $\varepsilon^2\phi_\varepsilon^{-1} \to 0$. This estimate is of the kernel type, with the kernel $G(u) = \chi_{\{-1<u<0\}}$, where $\chi_{\{\cdot\}}$ is the indicator.

There are no additional difficulties in estimating the function $f^*(t) = S(t, X_0(t))$, using the observation:

$$dX_\varepsilon(t) = S(t, X_\varepsilon(t))\,dt + \varepsilon\,dw(t)\,, \qquad X_\varepsilon(0) = X_0, \qquad 0 \le t \le T\,.$$

Let us denote by Θ_β the set of functions $f(t,x)$, which are k-times differentiable in t and x and

$$\sum_{r=0}^{k} \left| \left.\frac{\partial^k f(t_1, x_1)}{\partial t^r \partial x^{k-r}}\right|_{\substack{t=t_2\\x=x_2}} - \left.\frac{\partial^k f(t,x)}{\partial t^r \partial x^{k-r}}\right|_{\substack{t=t_1\\x=x_1}} \right| \le L_\alpha(|t_2 - t_1|^\alpha + |x_2 - x_1|^\alpha)\,.$$

<u>THEOREM 3</u>. Let $S(\cdot,\cdot) \in \Theta_\beta$. Then

$$\lim_{\varepsilon \to 0} \sup_{s \in \tilde{\Theta}_\beta} \sup_{c \le t \le d} E|\hat{f}_\varepsilon(t) - f^*(t)|^2 \,\varepsilon^{-\frac{4\beta}{2\beta+1}} < \infty\,.$$

The proof is similar to that of Theorem 2.

There are no additional difficulties in estimating the function $f^*(t)$, using the observations

$$dX_\varepsilon(t) = S(t, X_\varepsilon(t))\,dt + \varepsilon B(t, X_\varepsilon(t))\,dw(t)\,, \quad X_\varepsilon(0) = x_0\,, \quad 0 \le t \le T\,,$$

where $B(t,x)$ is a known diffusion coefficient.

It is also interesting to note that if the initial value $X_\varepsilon(0) = \eta$ is some random variable, $E\eta^2 < \infty$, which does not depend on the Wiener process, then $X_0(t)$ and $f(t) = S(X_0(t))$ are random processes. But all the above computations are valid if instead of mathematical expectations, we use conditional mathematical expectations, for example,

$$E\{[\hat{f}_\varepsilon(t) - f(t)]^2 \mid \eta\}\,.$$

3. ESTIMATION OF THE FUNCTION S(x)

For the same observations let

$$dX_\varepsilon(t) = S(X_\varepsilon(t))\,dt + \varepsilon dw(t) , \qquad X_\varepsilon(0) = x_0, \qquad 0 \le t \le T .$$

We have to construct the estimate of the function $S(x)$. This problem is more complicated than the previous problem. Indeed, it is not known for which x we can consistently estimate $S(x)$ uniformly on $S(\cdot)$, because the solution $\{X_0(t),\ 0 \le t \le T\}$ is not known to the observer and for different $S(\cdot)$ the regions $[a,b]$ can differ, for example $[c,x_0]$ and $[x_0,d]$ (with the only one generic point!). Therefore we make an additional assumption that $S(x)$ is positive. This condition then easily implies the monotonicity of the solution $X_0(t)$ and the equalities $a = x_0$, $b = X_0(T)$. But this condition is insufficient because the following relation holds:

$$\bigcap_{S\in\Theta(L)} [x_0, X_0(t)] = \{x_0\} .$$

To fix the domain of definition of the estimate $\hat{S}(x)$ common for all $S(\cdot) \in \Theta(L)$, we introduce the set

$$\Theta'(L) = \{S(\cdot) \in \Theta(L) : S(x) \ge \kappa\} ,$$

where the positive constant κ is the same for all $x \in [x_0, x_0 + L_0 T]$. Then it is easy to see that

$$\bigcup_{S\in\Theta'(L)} [x_0, X_0(T)] = [x_0, x_0 + L_0 T] ,$$

$$\bigcap_{S\in\Theta'(L)} [x_0, X_0(T)] = [x_0, x_0 + \kappa T] .$$

We denote the last interval by $[a,b]$ and estimate the function $S(X)$, $X \in [a,b]$. Also, we introduce the stopping time

$\tau_\varepsilon = \inf\{t: X_\varepsilon(t) = x\}$ and put $\tau_\varepsilon = T$ if the set $\{X_\varepsilon(t) = x,$ $t \in [0,T]\}$ is empty. Let $\tau = \inf\{t: X_0(t) = x\}$, $X \in [a,b]$ and let $[r,q]$ be an arbitrary interval $[r,q] \in (a,b)$.

THEOREM 4. Let $S(\cdot) \in \Theta'(L)$, $\phi_\varepsilon \to 0$ and $\varepsilon^2\phi_\varepsilon^{-1} \to 0$. Then the estimate

$$\hat{S}_\varepsilon(x) = \frac{1}{\phi_\varepsilon}\int_0^T G\left(\frac{t-\tau_\varepsilon}{\phi_\varepsilon}\right) dX_\varepsilon(t) \tag{17}$$

is uniformly consistent on $S(\cdot) \in \Theta'(L)$ and $x \in [r,q]$:

$$\lim_{\varepsilon\to 0} \sup_{S\in\Theta'(L)} \sup_{r\le x\le q} E|\hat{S}_\varepsilon(x) - S(x)|^2 = 0 .$$

Proof. First we note that by the equality

$$0 = X_\varepsilon(\tau_\varepsilon) - X_0(\tau) = X_\varepsilon(\tau_\varepsilon) - X_0(\tau_\varepsilon) + X_0(\tau_\varepsilon) - X_0(T)$$

and the mean-value theorem

$$X_\varepsilon(\tau_\varepsilon) - X_0(\tau_\varepsilon) = \int_{\tau_\varepsilon}^{\tau} S(X_0(t))\,dt = (\tau - \tau_\varepsilon)S(X_0(t^*))$$

we have the inequalities

$$|\tau_\varepsilon - \tau| \le \kappa^{-1}|X_\varepsilon(\tau_\varepsilon) - X_0(\tau_\varepsilon)| \le \kappa^{-1}\sup_{0\le t\le T}|X_\varepsilon(t) - X_0(t)|$$

$$\le \phi^{-1}e^{LT}\sup_{0\le t\le T}|w(t)|\,\varepsilon .$$

Let us introduce the set

$$F = \{w: \sup_{0\le t\le T}|w(t)| > \varepsilon^{-\gamma}\} ,$$

where $\gamma \in (0,1)$. The complement of this set has the probability

$$P\{F^c\} = P\{\sup_{0\le t\le T}|w(t)| > \varepsilon^{-\gamma}\} \le \exp\{-\frac{1}{2T}\varepsilon^{-2}\}$$

(see, for example, [2, p. 127]).

It is easy to see that

$$|\tau_\varepsilon - \tau|\chi_{\{F\}} \le \kappa^{-1} e^{LT} \varepsilon^{1-\gamma} = k\,\varepsilon^{1-\gamma}$$

Then we assume without loss of generality that ε' and ε'' determined by the relations

$$\varepsilon' = \sup\,\{\varepsilon: (-t_r^* + k\varepsilon^{1-\gamma})\phi_\varepsilon^{-1} < A\}$$

$$\varepsilon'' = \sup\,\{\varepsilon'': (T - t_q^* - k\varepsilon^{1-\gamma})\phi_\varepsilon^{-1} > B\}$$

are unique. Here

$$t_r^* = \inf\,\{t_r: X(t_r) = t,\ S(\cdot) \in \Theta'(L)\}$$

$$t_q^* = \sup\,\{t_q: X(t_q) = \dot{q},\ S(\cdot) \in \Theta'(L)\} \quad .$$

For the functions $S(\cdot) \in \Theta'(L)$

$$t_r^* = \frac{r-x_0}{L} \ , \qquad t_q^* = T - \frac{b-q}{L} \quad .$$

Let $\varepsilon_2 = \min(\varepsilon', \varepsilon'')$ and note that for $\varepsilon < \varepsilon_2$ and $w \in F$

$$\frac{1}{\phi_\varepsilon}\int_0^T G\left(\frac{t-\tau_\varepsilon}{\phi_\varepsilon}\right) S(X_\varepsilon(t))\,dt = \int G(u)\ S(X_\varepsilon(\tau_\varepsilon + u\phi_\varepsilon))\,du \qquad (18)$$

for all $x \in [r,q]$. Recall that $\tau_\varepsilon = \tau_\varepsilon(x)$ and $\tau = \tau(x)$.

In the sequel we substitute the observations (1) in (17), change the variables $u = \phi_\varepsilon^{-1}(t - \tau_\varepsilon)$ and use (18) for $\varepsilon < \varepsilon_2$ and $w \in F$:

$$\hat{S}_\varepsilon(x) - S(x)$$

$$= \frac{1}{\phi_\varepsilon}\int_0^T G\left(\frac{t-\tau_\varepsilon}{\phi_\varepsilon}\right) S(X_\varepsilon(t))dt - S(X_0(\tau)) + \frac{\varepsilon}{\phi_\varepsilon}\int_0^T G\left(\frac{t-\tau_\varepsilon}{\phi_\varepsilon}\right) dw(t)$$

$$= \int G(u)\,[S(X_\varepsilon(\tau_\varepsilon + u\phi_\varepsilon)) - S(X_0(\tau_\varepsilon + u\phi_\varepsilon))]\,du$$

$$+ \int G(u)[S(X_0(\tau_\varepsilon + u\phi_\varepsilon)) - S(X_0(\tau_\varepsilon))]\, du$$

$$+ \int G(u)[S(X_0(\tau_\varepsilon)) - S(X_0(\tau))]\, du + \frac{\varepsilon}{\sqrt{\phi_\varepsilon}} \int_0^T G(u) d\tilde{w}(u) .$$

Consequently

$$E|\hat{S}_\varepsilon(x) - S(x)|^2 = E|\hat{S}_\varepsilon(x) - S(x)|^2 I_{\{F^c\}} + E|\hat{S}_\varepsilon(x) - S(x)|^2 \chi_{\{F\}}$$

$$\leq 4L_0^2 \exp\left\{-\frac{1}{2T}\varepsilon^{-2\gamma}\right\} + 4E_F\{\int G(u)[S(X_0(\tau_\varepsilon+u\phi_\varepsilon)) - S(X_0(\tau_\varepsilon))]du\}^2$$

$$+ 4E_F\{\int G(u)[S(X_\varepsilon(\tau_\varepsilon+u\phi_\varepsilon) - S(X_0(\tau_\varepsilon+u\phi_\varepsilon))]\, du\}^2$$

$$+ 4E_F\{\int G(u)[S(X_0(\tau_\varepsilon)) - S(X_0(\tau))]\, du\}^2 + 4\frac{\varepsilon^2}{\phi_\varepsilon} \int G(u)^2 du .$$

Above we used the notation: $E_F\eta = E(\chi_{\{F\}}\eta)$ and $\tilde{w}(u) = \phi_\varepsilon^{-\frac{1}{2}}[w(\tau_\varepsilon+u\phi_\varepsilon) - w(\tau_\varepsilon)]$. It is known that $\tilde{w}(u)$ is also a a Wiener process.

Consider these summands separately:

$$E_F\{\int G(u)[S(X_\varepsilon(\tau_\varepsilon + u\phi_\varepsilon)) - S(X_0(\tau_\varepsilon + u\phi_\varepsilon))]\, du\}^2$$

$$\leq (B-A)L^2 \int G(u)^2\, E_F|X_\varepsilon(\tau_\varepsilon + u\phi_\varepsilon) - X_0(\tau_\varepsilon + u\phi_\varepsilon)|^2\, du$$

$$\leq (B-A)L^2\sigma^2 \sup_{0\leq t\leq T} E|X_\varepsilon(t) - X_0(t)|^2 \leq (B-A)L^2\sigma^2 T \exp(LT^2)\varepsilon^2.$$

Then

$$E_F\{ \int G(u)[S(X_0(\tau_\varepsilon + u\phi_\varepsilon)) - S(X_0(\tau_\varepsilon))]\, du\}^2$$

$$\leq L^2(B-A) \int G(u)^2 E_F|X_0(\tau_\varepsilon + u\phi_\varepsilon) - X_0(\tau_\varepsilon)|^2\, du$$

$$\leq L^2 L_0^2 (B-A)\, \phi_\varepsilon^2 \int [uG(u)]^2\, du \quad .$$

We thus obtain

$$E_F\{\int G(u)[S(X_0(\tau_\varepsilon)) - S(X_0(\tau))]\, du\}^2$$

$$= E_F[S(X_0(\tau_\varepsilon)) - S(X_0(\tau))]^2 \leq L^2 E_F |X_0(\tau_\varepsilon) - X_0(\tau)|^2$$

$$\leq L^2 L_0^2 E_F|\tau_\varepsilon - \tau|^2 \leq L^2 L_0^2 \varepsilon^{2(1-\gamma)} \quad .$$

Applying these inequalities, we obtain

$$E|\hat{S}_\varepsilon(x) - S(x)|^2$$

$$\leq 4L_0^2 \exp\{-\frac{1}{2T}\varepsilon^{-2\gamma}\} + c_1\varepsilon^2 + c_2\phi_\varepsilon^2 + c_3\varepsilon^{2-2\gamma} + 4\sigma^2 \frac{\varepsilon^2}{\phi_\varepsilon} \quad ,$$

where the constants c_i do not depend on $x \in [r,q]$ and $S(\cdot) \in \Theta'(L)$. The consistency of the estimate $\hat{S}_\varepsilon(x)$ now immediately follows from this inequality and the conditions of Theorem 4.

Assume that the kernel $G(\cdot)$ satisfies conditions (14). Let

$$\Theta'_\beta = \Theta'_\beta(L_\alpha) = \Theta_\beta(L_\alpha) \cap \Theta'(L) \quad .$$

Note that the intervals $[a,b]$ here and in Section 2 differ and in this definition of the set Θ'_β we use that of Section 2.

<u>THEOREM 5</u>. Let $S(\cdot) \in \Theta'_\beta(L_\alpha)$. Then

$$\lim_{\varepsilon\to 0} \sup_{S(\cdot)\in\Theta'_\beta} \sup_{r\le x\le q} E|\hat{S}_\varepsilon(x) - S(x)|^2 \varepsilon^{-\frac{4\beta}{2\beta+1}} < \infty .$$

The proof is similar to the previous one. The only difference is in the following estimate:

$$\begin{aligned}
E_F\{ \int G(u)[S(X_\varepsilon(\tau_\varepsilon + u\phi_\varepsilon)) - S(X_0(\tau))]\, du\}^2 \\
&\le 3(B-A) \int G(u)^2 E|S(X_\varepsilon(\tau_\varepsilon + u\phi_\varepsilon)) - S(X_0(\tau_\varepsilon + u\phi_\varepsilon))|^2\, du \\
&+ 3(B-A) \int G(u)^2 E|S(X_0(\tau_\varepsilon + u\phi_\varepsilon)) - S(X_0(\tau + u\phi_\varepsilon))|^2\, du \\
&+ 3\{ \int G(u)[S(X_0(\tau + u\phi_\varepsilon)) - S(X_0(\tau))]\, du\}^2 \\
&\le c_1\varepsilon^2 + c_2\varepsilon^{2-2\gamma} + c_3^2\phi_\varepsilon^{2\beta}
\end{aligned}$$

with some constants $c_i > 0$. The inequality

$$| \int G(u)[S(X_0(\tau + u\phi_\varepsilon)) - S(X_0(\tau))]\, du| \le c_3\phi_\varepsilon^\beta$$

follows by (16) if we set $t = \tau$.

Finally we get the inequality

$$\begin{aligned}
E|\hat{S}_\varepsilon(x) - S(x)|^2 \\
&\le 4L_0^2 \exp\{-\frac{1}{2T}\varepsilon^{-2\gamma}\} + c_1\varepsilon^2 + c_2\varepsilon^{2-2\gamma} + c_3^2\sigma_\varepsilon^{2\beta} + c_4\frac{\varepsilon^2}{\phi_\varepsilon}
\end{aligned}$$

from which the proposition of Theorem 5 follows if

$$\phi_\varepsilon = \varepsilon^{\frac{2}{2\beta+1}} \quad \text{and} \quad \gamma \le \frac{1}{2\beta+1} .$$

4. ESTIMATION OF THE FUNCTION $\theta(t)$

Consider the problem of estimating the function $\theta(t)$ on the basis of the observations $X_j = \{X_j(t),\ 0 \le t \le T\}$, $j = 1,\dots,n$ of the diffusion process

$$dX(t) = \theta(t)\,X(t)\,dt + dw(t)\ , \qquad X(0) = x_0\ , \quad 0 \le t \le T\ .$$

H.T. Nguyen and T.D. Pham [8] suggest the following procedure. The estimate $\theta_n^*(t)$ is determined by the method of sieves. Let V_n, $n = 1,2,\dots$ be the sequence of d_n-dimensional subspaces of the space $L_2[0,T]$, such that $\cup_{n\ge 1} V_n$ is dense in $L_2[0,T]$. The unknown function $\theta(t)$ is expanded in some series on the basis of $\phi_1(t), \phi_2(t),\dots$ and for each n the coefficients $\theta_1,\dots,\theta_{d_n}$ of this expansion are estimated, with

$$\theta_n^*(t) = \sum_{j=1}^{d_n} \hat{\theta}_j\ \phi_j(t)\ .$$

Under the assumptions $d_n^2\, n_{-1} \to 0$ and $d_n^3\, n^{-1} \to 0$ the consistency of

$$\lim_{n\to\infty} \int_0^T [\theta_n^*(t) - \theta_n(t)]^2\ EX(t)^2\ dt = 0$$

and asymptotic normality of the $\theta_n^*(t)$ have been proved [8].

We suggest a consistent estimate of $\theta(t)$ for all $t \in [0,T]$. Obviously it cannot be done without some assumptions on continuity of this function, because any change of this function on some set **A** with Lebesgue measure zero does not influence the observations

$$X(t) = x_0 + \int_0^T X(s)\,\theta(s)\,ds + w(t)\,, \qquad 0 \le t \le T\,.$$

Denote by Θ the set of functions $m(t)$ on $[0,T]$, which have k bounded derivatives on t and

$$|m^{(k)}(t_2) - m^{(k)}(t_1)| \le L_\alpha |t_2 - t_1|^\alpha\,.$$

Let $\theta(\cdot) \in \Theta$. It is impossible to determine the estimate

$$\tilde{\theta}_n(t) = \frac{1}{n\phi_n} \sum_{j=1}^{n} \int_0^T G\left(\frac{\tau - t}{\phi_n}\right) X_j(t)^{-1}\, dX_j(t)$$

because the corresponding stochastic integral does not exist. Thus we introduce the stochastic process

$$Y_n(t) = \begin{cases} X(t)^{-1} & \text{if } |X(t)| \ge x_n \\ x_n^{-1} & \text{if } |X(t)| < x_n \end{cases}$$

and the estimate

$$\hat{\theta}_n(t) = \frac{1}{n\phi_n} \sum_{j=1}^{n} \int_0^T G\left(\frac{\tau - t}{\phi_n}\right) Y_{n,j}(t)\, dX_j(t)\,.$$

This estimate has the following decomposition :

$$\hat{\theta}_n(t) - (t) = \frac{1}{\phi_n}\int_0^T G\left(\frac{\tau-t}{\phi_n}\right)\theta(\tau)\, d\tau - \theta(t)$$

$$+ \frac{1}{\phi_n}\int_0^T G\left(\frac{\tau-t}{\phi_n}\right)\left\{\frac{1}{n}\sum_{j=1}^n \chi_{\{|X_j(\tau)|\geq x_n\}} + \frac{1}{n}\sum_{j=1}^n X_j(\tau)x_n^{-1}\chi_{\{|X_j(\tau)|<x_n\}^{-1}}\right\}\theta(\tau)\, d\tau$$

$$+ \frac{1}{n\phi_n}\sum_{j=1}^n \int_0^T G\left(\frac{t-\tau}{\phi_n}\right) Y_{n,j}(\tau)\, dw_j(\tau) \quad .$$

Obviously

$$E\ Y_n(t)^2 \leq cx_n^{-1} \ .$$

Hence

$$E\left\{\frac{1}{n\phi_n}\sum_{j=1}^n \int_0^T G\left(\frac{\tau-t}{\phi_n}\right) Y_{n,j}(\tau) dw_j(\tau)\right\}^2 \leq \frac{c}{n\phi_n x_n}\int G(u)^2 du = \frac{c\sigma^2}{n\phi_n x_n} \ .$$

Then

$$\left|\frac{1}{\phi_n}\int_0^T G\left(\frac{\tau-t}{\phi_\varepsilon}\right)\theta(\tau)d\tau - \theta(t)\right| = \left|\int G(u)\left[\theta(t+u\phi_n) - \theta(t)\right] du\right|$$

$$\leq \frac{L_\alpha \phi^{k+\alpha}}{k!}\int |u^{k+\alpha}\ G(u)|\ du \quad .$$

Finally

$$E\left\{\frac{1}{n}\sum_{j=1}^{n} X_j(\tau)x_n^{-1}\chi_{\{|X_j(\tau)|<x_n\}}\right\}^2 \le E\chi_{\{|X_j(\tau)|<x_n\}}$$

$$= \frac{1}{\sqrt{2\pi h^2}} \int_{-x_n}^{x_n} \exp\left\{-\frac{x-m}{2n^2}\right\} dx \le \sqrt{\frac{2}{\pi}}\,\frac{x_n}{h} \le cx_n ,$$

where

$$m = x_0 \exp\left\{\int_0^T \theta(s)\, ds\right\} ,$$

$$h^2 = \int_0^T \exp\left\{2\int_S^t \theta(v)dv\right\} ds \ge \frac{1}{2L_0}\left[1 - \exp(-2L_0c)\right]$$

because $X(t)$ is a Gaussian random variable with parameters (m,h^2). This directly follows from the representation

$$X(t) = x_0 \exp\left\{\int_0^t \theta(s)ds\right\} + \int_0^t \exp\left\{\int_S^t \theta(v)dv\right\} dw(s) .$$

Applying the stated inequalities one can verify that

$$\sup_{c\le t\le d} E|\hat{\theta}_n(t) - \theta(t)|^2 \le c_1\phi_n^{2\beta} + c_2x_n + \frac{c_3}{n\phi_n x_n} .$$

Consequently we can set

$$x_n = n^{-\frac{1}{4\beta+1}} , \qquad x_n = n^{-\frac{2\beta}{4\beta+1}}$$

and prove the following.

THEOREM 6.

Let $\Theta(\cdot) \in \Theta$. Then for any $[c,d] \subset (0,T)$

$$\lim_{n\to\infty} \sup_{\Theta(\cdot)\in\Theta} \sup_{c\le t\le d} E|\hat{\theta}_n(t) - \theta(t)|^2 \, n^{\frac{2\beta}{4\beta+1}} < \infty .$$

REFERENCES

[1] Liptser, R.S., and Shiryayev, A.N. *Statistics of Random Processes*. I, II. Berlin Heidelberg New York: Springer-Verlag Inc., 1977.

[2] Kutouyants, Yu.A. *Otsenka parametrov sluchajnykh protsessov* (Estimation of Parameters of Random Processes). Erevan: 1980.

[3] Kutoyants, Yu.A. "Estimation of a Parameter of a Diffusion Type Process." *Theory Probab. Applications*, vol.23, no.3 (1978): 665-672.

Ibragimov, I.A., and Has'minskii, R.Z. *Statistical Estimation. Asymptotic Theory*. Berlin Heidelberg New York: Springer-Verlag Inc., 1981.

[5] Ibragimov, I.A., and Has'minskii, R.Z. "Asymptotic Properties of Some Nonparametric Estimates in a Gaussian White Noise." *Proceedings of the Summer School on Probability Theory and Mathematical Statistics*, pp. 29-64. Varna (Bulgaria), 1980.

[6] Banon, G. "Nonparametric Identification for Diffusion Processes." *SIAM J. Control and Optimization*, vol.16, no.3 (1978): 380-395.

[7] Pham, Tuan D. "Nonparametric Estimation of the Drift Coefficient in the Diffusion Equation." *Tech. Report Séminaire de Statistique*, pp. 103-124. Univers. Grenoble (France), 1978-79.

[8] Nguyen, Hung T., and Pham, Tuan D. "Identification of the Nonstationary Diffusion Model by the Method of Sieves." *SIAM J. Control and Optimization*, vol.20, no.5 (1982): 603-611.

N.L. LAZRIEVA

WEAK CONVERGENCE OF SEMIMARTINGALES

1. INTRODUCTION. FORMULATION OF THE MAIN RESULTS

Assume that a complete probability space (Ω, F, P) is given with filtrations $F_n = \{F_t^n, t \geq 0\}$, $n \geq 1$, satisfying the usual conditions ($F_s^n \subset F_t^n \subset F$, $s \leq t$, $F_t^n = F_{t+}^n = \bigcap_{s \geq t} F_s^n$, F_0^n is augmented by sets of zero measures from F). Let (D, D_∞) denote a measurable space of CORLOL functions $x = \{x(t), t \geq 0\}$ with Skorokhod's topology [1]. $D = \{D_t, t \geq 0\}$ denotes a family of σ-algebras

$$D_t = \bigcap_{\varepsilon > 0} \sigma\{x: x_s, 0 \leq s \leq t+\varepsilon\} .$$

Let $\xi_n = \{\xi_n(t), F_t^n, t \geq 0\}$ be the sequence of semimartingales given on (Ω, F, P) with trajectories from the space D allowing a canonical representation

$$\xi_n(t) = B_n(t) + \xi_n^c(t) + \int_0^t \int_{|x|>1} x\, \mu_n(ds, dx)$$
$$+ \int_0^t \int_{|x|>1} x(\mu_n - \nu_n)(ds, dx) ,$$

where $B_n = \{B_n(t), F_t^n, t \geq 0\}$ is a predictable process of locally bounded variation, $B_n(0) = 0$, $\xi_n^c = \{\xi_n^c(t), F_t^n, t \geq 0\}$ is a continuous local martingale with a square characteristic $\langle \xi_n^c \rangle$, $\xi_n^c(0) = 0$, μ_n is a measure of jumps of the process ξ_n, ν_n is the compensator of the measure μ_n.

Let P^{ξ_n} denote a probability measure induced by the process ξ_n in the space (D, D_x), i.e., a measure such that for any $\Gamma \in D_\infty$

$$P^{\xi_n}(\Gamma) = P\{\xi_n \in \Gamma\} \quad .$$

For any $t \geq 0$ and $\varepsilon \in (0,1]$ we put

$$A_\varepsilon^n(t) = \int_0^t \int_{|x|>\varepsilon} d\nu_n \quad ,$$

$$C_\varepsilon^n(t) = \langle \xi^{nc} \rangle_t + \int_0^t \int_{|x|\leqslant\varepsilon} x^2 \, d\nu_n - \sum_{0<s\leq t} \left(\int_{|x|\leqslant\varepsilon} x\nu_n(\{s\}, dx) \right)^2 .$$

A number of papers [1] - [7] (for detailed references see [6]) are devoted to the investigation of the conditions of weak convergence of measures P^{ξ_n}, $n \geq 1$, to the distribution P^{ξ_0} of the solution of the stochastic differential equation

$$d\xi_0(t) = a(t,\xi_0)dt + b(t,\xi_0)dw_t \,, \quad \xi_0(0) = 0 \,, \qquad (1)$$

where $a(t,x)$ and $b(t,x)$ are $B(R^+)\times D_\infty$ - measurable nonanticipatory (with respect to the family of σ-algebras D) functionals. The type of a condition depends on properties of the sequence $\{\xi_n\}$, $n \geqslant 1$, as well as on properties of the coefficients of the limit process. In particular, it has been proved in [6] that under the assumption of unique weak solvability of equation (1) the conditions

(A) $$A_\varepsilon^n(t) \overset{P}{\to} 0 ,$$

(sup B) $$\sup_{0\le s\le t} \left| B_s^n - \int_0^s a(u,\xi_n)du \right| \overset{P}{\to} 0 , \qquad (n \to \infty) ,$$

(C) $$C_\varepsilon^n(t) - \int_0^t b^2(u,\xi_n)du \overset{P}{\to} 0$$

(it is assumed to hold for every $t \ge 0$ and $\varepsilon \in (0,1]$) are sufficient as $P^{\xi_n} \overset{W}{\to} P^{\xi_0}$ $(n \to \infty)$ (this convergence is denoted by $\xi_n \overset{D}{\to} \xi_0$) under the following additional conditions imposed on the functionals $a(t,x)$ and $b(t,x)$:

(D) for every $t \ge 0$ they are continuous with respect to x in C (i.e.,

$$\rho(x_n,x) \to 0 \Rightarrow a(t,x_n) \to a(t,x) ,$$

$$b^2(t,x_n) \to b^2(t,x) , \qquad x_n \in D, \quad n \ge 1, \quad x \in C,$$

C is a space of continuous functions, ρ is a Skorokhod metric in D)

(E) for all $x \in D$

$$|a(t,x)| + b^2(t,x) \le k(t) ,$$

where $k = (k(t), t \ge 0)$ is a measurable function with

$$\int_0^t k(s)\, ds < \infty , \qquad t \ge 0 .$$

It is shown in [7] that condition (E) can be replaced by a less restrictive condition of linear growth:

(E') for any $T > 0$ the continuous increasing function $k_t = k_t(T) < \infty$, $k_0 = 0$, is such that for any $p,q \le t \le T$ with $|p-q| \le \Delta$

$$\left|\int_p^q a(s,x)dx\right| + \left|\int_p^q b^2(s,x)ds\right|^{\frac{1}{2}} \le k_\Delta(T)\left(1 + \sup_{0\le s\le t} |x_s|\right) .$$

But then the condition of the convergence of triplets is changed. In particular, condition (C) is replaced by condition (sup C), where

$$\text{(sup C)} \qquad \sup_{0\le s\le t}\left|C_\varepsilon^n(s) - \int_0^s b^2(u,\xi_n)ds\right| \xrightarrow{P} 0 , \qquad n \to \infty,$$

i.e., when conditions (D), (E'), (sup B) hold:

$$\text{(A), (sup C)} \iff \xi_n \xrightarrow{\mathcal{D}} \xi_0 , \qquad n \to \infty .$$

The objective of this paper is to investigate the possibility of weakening the additional conditions imposed on the functionals $a(t,x)$ and $b(t,x)$ in addition to the natural requirement of solvability of equation (1), as well as to determine the type of convergence conditions for a triplet of characteristics sufficient for $\xi_n \xrightarrow{\mathcal{D}} \xi_0$ $(n \to \infty)$ under conditions weaker than (E) and (E').

From the proof of Theorem 2 of the paper [6] it can be seen that condition (E) is used essentially to prove the relative compactness of the sequence ξ_n, $n \ge 1$ (i.e., relative compactness of the sequence of measures P^{ξ_n}, $n \ge 1$) and in particular to derive the relation

$$\lim_{N\to\infty} \overline{\lim_{n\to\infty}} P\left\{\sup_{0\le t\le T} |\xi_n(t)| \ge N\right\} = 0 \tag{2}$$

for any $0 < T < \infty$. If the sequence $\{\xi_n\}$, $n \geq 1$, however, is characterized by the latter property, then it can be seen from Theorem 1 given below, that condition (E) can be replaced by condition (E_L) that for any integer $N > 0$ a constant k_N is such that

$$|a(t,x)| + b^2(t,x) \leq k_N$$

for all $t \leq N$ and $||x||_N = \sup_{0 \leq s \leq N} |x(s)|| \leq N$ with the corresponding replacement of conditions (A), (sup B), (C) by local conditions (A_N), (sup B_N), (C_N) (see formula (12)).

In what follows we assume that a unique weak solution of equation (1) exists.

THEOREM 1. Let the sequence of semimartingales $\{\xi_n\}$, $n \geq 1$, have the property (2). Also let conditions (D) and (E_L) hold. Then

$$\text{(A), (sup B), (sup C)} \Rightarrow \xi_n \xrightarrow{\mathcal{D}} \xi_0 \qquad (n \to \infty) .$$

REMARK 1. From the proof of Theorem 1 it follows that as $\xi_n \xrightarrow{\mathcal{D}} \xi_0$ $(n \to \infty)$ it is sufficient, in fact, that conditions (A_N), (sup B_N), (C_N) hold for every $N > 0$.

Below we consider a case where $a(t,x) \equiv 0$. Applying the random-time change technique, we obtain the conditions for a weak convergence of the semimartingales $\{\xi_n\}$, $n \geq 1$ to a diffusion process being a solution of the equation

$$d\xi_t = b(t,\xi)dw_t , \qquad \xi_0 = 0 , \tag{3}$$

where the conditions imposed on the functional $b(t,x)$ are expressed in terms of solvability of the nonlinear equation

$$\Phi_t = \int_0^t [b(\Phi_s, T^\Phi x)]^{-2} ds \quad , \qquad t \geq 0, \qquad x \in D, \tag{4}$$

where $(T^\Phi x)(t) = x(\Phi_t^{-1})$, and Φ_t^{-1} is the inverse of Φ_t, i.e., $\Phi_{\Phi_t^{-1}} = t$.

The following theorem holds.

THEOREM 2. Let $\{\xi_n\}$, $n \geq 1$, be a sequence of semimartingales with a triplet of predictable characteristics $(B_n, \langle\xi_n^c\rangle, \nu_n)$, $n \geq 1$, and let the following conditions hold:

- (a) for all $t \geq 0$ and $x \in D$, $b^2(t,x) > 0$,

$$\int_0^t b^2(s,x)ds < \infty \quad \text{and} \quad \int_0^\infty b^2(s,x)ds = \infty ,$$

- (b) for any $x \in D$ the equation (4) has a unique solution $\Phi(x) = \{\Phi_t(x), \ t \geq 0\}$,
- (c) the mapping $\Phi^{-1}: x \in D \to \Phi^{-1}(x) = \{\Phi_t^{-1}(x), \ t \geq 0\} \in C$ is continuous with respect to x on C (i.e., $\rho(x_n, x) \Rightarrow \rho(\Phi^{-1}(x_n), \Phi^{-1}(x)) \to 0$, $n \to \infty$, $x \in D$, $x \in C$, ρ is the metric of the space C).

Then

$$(A_\tau), \ (\sup B_\tau), \ (C_\tau) \quad \Rightarrow \quad \xi_n \overset{\mathcal{D}}{\to} \xi_0, \qquad n \to \infty \quad ,$$

where

$$(A_\tau): \qquad A_\varepsilon^n(\tau_t(\xi_n)) \overset{P}{\to} 0 \ , \qquad t \geq 0, \qquad \varepsilon \in (0,1] \ ,$$

$$(\sup B_\tau): \qquad \sup_{0 \leq s \leq \tau_t(\xi_n)} |B_n(s)| \overset{P}{\to} 0 \quad , \qquad t \geq 0, \qquad \varepsilon \in (0,1] \ ,$$

$$(C_\tau): \qquad C_\varepsilon^n(\tau_t(\xi_n)) - \int_0^{\tau_t(\xi_n)} b^2(s, \xi_n)ds \overset{P}{\to} 0 \ ,$$

$$t \geq 0, \qquad \varepsilon \in (0,1] \ ,$$

$$\tau_t(x) = \inf\left\{s: \int_0^s b^2(u,x)du \geq t\right\}, \qquad x \in D .$$

The next theorem is a direct consequence of Theorem 2.

THEOREM 3. Let conditions (a) and (b) of Theorem 2 hold as well as conditions (D) and (E_L). Then

$$(A_\tau), (\sup B_\tau), (C_\tau) \Rightarrow \xi_n \xrightarrow{\mathcal{D}} \xi_0 , \qquad (n \to \infty) .$$

REMARK 2. If the functional $B(t,x) = \int_0^t b^2(s,x)ds$ is bounded below by a strictly increasing function, i.e., $B(t,x) > k_t > 0$, $k_t < \infty$, $t \geq 0$, $\lim_{t\to\infty} k(t) = \infty$, then for any $n > 0$

$$\tau_t(\xi_n) \leq \sigma_t < \infty ,$$

where $\sigma_t = \inf \{s: k_s \geq t\}$. This verifies the implication

$$(A), (\sup B), (\sup C) \Rightarrow (A_\tau), (\sup B_\tau), (C_\tau)$$

and hence under the conditions of Theorem 2

$$(A), (\sup B), (\sup C) \Rightarrow \xi_n \xrightarrow{\mathcal{D}} \xi_0 .$$

Finally, let us consider an example where no conditions are imposed on the functional $b(t,x)$ except for the condition of unique weak solvability of the equation (3), but conditions (a), (b) and (c) as well as condition (E_L) still hold.

EXAMPLE 1. Let $\beta(y)$, $y \in R^1$, be a continuous real-valued function.

Put $b(t,x) = \beta(x(t))$, $t \geq 0$, $x \in D$ and assume the function $\beta(y)$ to be such that the equation (3) has a unique weak solution. Then $b(t,x)$ satisfies conditions (a), (b) and (c) of Theorem 2.

The proof is given at the end of the paper.

2. PROOF OF THEOREMS

Proof of Theorem 1. For any $x \in D$ and $N > 0$ put

$$g_N(t,x) = \begin{cases} x(t) , & t < \tau_N(x) , \\ [x(\tau_N(x))]_{-N}^{N} & t \geq \tau_N(x) , \end{cases} \tag{5}$$

where $\tau_N(x) = \inf\{s: ||x||_s + f(s) \geq N\}$, $||x||_s = \sup_{0 \leq u \leq s} |x(u)|$, $f(s)$ is a strictly increasing continuous function with $f(0) = 0$ and $\lim_{s\to\infty} f(s) = 1$, $[x]_a^b = \min(b, \max(a,x))$. It is obvious that $\lim_{N\to\infty} \tau_N(x) = \infty$ for any $x \in D$.

Define now the nonanticipatory functionals $a_N(t,x)$ and $b_N(t,x)$ in the following manner:

$$a_N(t,x) = a(t, g_N(\cdot,x)) , \qquad b_N(t,x) = b(t, g_N(\cdot,x)) , \tag{6}$$

and show that for any $t \geq 0$ they are continuous with respect to x on C. It is sufficient to prove the continuity of the functional $\tau_N(x)$. Assume $\rho(x_n,x) \to 0$, $n \to \infty$, and $\lim_{n\to\infty} \tau_N(x_n) \neq \tau_N(x)$. Then a number $\varepsilon > 0$ and a subsequence $\{x_{n'}\}$ of the sequence $\{x_n\}$ can be found, such that either

$$\tau_N(x_{n'}) > \tau_N(x) + \varepsilon \tag{7}$$

or

$$\tau_N(x_{n'}) < \tau_N(x) - \varepsilon \quad . \tag{7'}$$

The inequality (7) ((7')) by the definition of the functional $\tau_N(x)$ implies

$$\|x_{n'}\|_{\tau_N(x)+\varepsilon} + f(\tau_N(x)+\varepsilon) < N$$

analogously, (7') implies

$$\left. \|x_{n'}\|_{\tau_N(x)-\varepsilon} + f(\tau_N(x)-\varepsilon) < N \right) .$$

Passing to the limit as $n' \to \infty$, from the last relation we have

$$\|x\|_{\tau_N(x)+\varepsilon} \leq N - f(\tau_N(x)+\varepsilon) < N - f(\tau_N(x))$$

(analogously,

$$\left. \|x\|_{\tau_N(x)-\varepsilon} > N - f(\tau_N(x)) \right) ,$$

while the inverse inequality holds and hence the assumption $\lim_{N\to\infty} \tau_N(x_n) \neq \tau_N(x)$ as $\rho(x_n,x) \to 0$ does not.

Since $\|g_N(\cdot,x)\|_t \leq N$ for any $t \geq 0$, then for the functionals $a_N(t,x)$ and $b_N(t,x)$ condition (E) (with a function $k(s)$ equal to k_N with $s \leq N$ and k_s with $s > N$) is satisfied.

Furthermore, let

$$\xi_0^N(t) = \begin{cases} \xi_0(t) , & t \leq \tau_N(\xi_0) , \\ \xi_0(\tau_N(\xi_0)) + \int_{\tau_N(\xi_0)}^{t} a_N(s,\xi_0)ds + \int_{\tau_N(\xi_0)}^{t} b_N(s,\xi_n)dw(s) , & \\ & t > \tau_N(\xi_0) , \end{cases}$$

$$\xi_n^N(t) = \begin{cases} \xi_n(t) , & t \le \tau_N(\xi_n) , \\ \xi_n(\tau_N(\xi_n)) + \int_{\tau_N(\xi_n)}^{t} a_N(s,\xi_n)ds + \int_{\tau_N(\xi_n)}^{t} b_N(s,\xi_n)dw(s) , & \\ & t > \tau_N(\xi_n) . \end{cases}$$

It can be easily seen that the process $\xi_0^N = \{\xi_0^N(t),\ t \ge 0\}$ is a solution of the equation

$$\xi_0^N(t) = \int_0^t a_N(s,\xi_0^N)ds + \int_0^t b_N(s,\xi_0^N)dw(s) , \qquad (8)$$

and the process $\xi_n^N = \{\xi_n^N(t),\ t \ge 0\}$ can be written as

$$\xi_n^N(t) = \begin{cases} \xi_n(t) , & t \le \tau_N(\xi_n) , \\ \xi_n(\tau_N(\xi_n)) + \int_{\tau_N(\xi_n)}^{t} b_N(s,\xi_n^N)dw(s) + \int_{\tau_N(\xi_n)}^{t} a_N(s,\xi_n^N)ds , & \\ & t > \tau_N(\xi_n) , \end{cases} \qquad (9)$$

and for any $T > 0$

$$P\left\{\lim_{N\to\infty} \rho_T(\xi_0,\xi_0^N) = 0\right\} = 1 , \qquad (10)$$

$$\lim_{N\to\infty} \overline{\lim_{N\to\infty}}\, P\{\rho_T(\xi_n^N,\xi_n) \ge \varepsilon\} = 0 , \qquad (11)$$

where $\rho_T(x,y)$ is Skorokhod's distance on $D[0,T]$. The relation (10) is obvious. Let us prove the relation (11).

It follows from (9) that

$$\begin{aligned} P\{\rho_T(\xi_n^N, \xi_n) \geq \varepsilon\} &\leq P\{\tau_N(\xi_n) < T\} \\ &= P\left\{ \sup_{0 \leq t \leq T} |\xi_n(t)| + f(T) \geq N \right\} \\ &\leq P\left\{ \sup_{0 \leq t \leq T} |\xi_n(t)| \geq N - 1 \right\} , \end{aligned}$$

which, by virtue of the property (2) of the sequence $\{\xi_n\}$, $n \geq 1$, leads to

$$\begin{aligned} \lim_{N\to\infty} \overline{\lim_{n\to\infty}} P\{\rho_T(\xi_n^N, \xi_n) \geq \varepsilon\} &\leq \lim_{N\to\infty} \overline{\lim_{n\to\infty}} P\left\{ \sup_{0 \leq t \leq T} |\xi_n(t)| \geq N - 1 \right\} \\ &= 0 . \end{aligned}$$

Assume now that for any $N > 0$ the conditions ($\forall t \geq 0$ and $\forall \varepsilon \in (0,1]$)

$$(A_N'): \qquad \int_0^t \int_{|x|>\varepsilon} \nu_n^N(ds, dx) \overset{P}{\to} 0 ,$$

$$(\sup B_N'): \qquad \sup_{0 \leq s \leq t} \left| B_n^N(s) - \int_0^s a_N(u, \xi_n^N) du \right| \overset{P}{\to} 0 ,$$

$$\begin{aligned} (C_N'): \qquad & \langle \xi_n^{Nc} \rangle_t + \int_0^t \int_{|x| \leq \varepsilon} x^2 \nu_n^N(ds, dx) \\ & - \sum_{0 \leq s \leq t} \left(\int_{|x| \leq \varepsilon} x \, \nu_n^N(\{s\}, dx) \right)^2 - \int_0^t b_N^2(s, \xi_n^N) ds \overset{P}{\to} 0 \end{aligned}$$

hold, where $(B_n^N, \langle \xi_n^{Nc} \rangle, \nu_n^N)$ is the triplet of predictable characteristics of the semimartingale ξ_n^N. Then from Theorem 1 of [6] for any $N > 0$ there follows the convergence

$$\xi_n^N \overset{\mathcal{D}}{\to} \xi_0^N \qquad (n\to\infty) ,$$

which, in turn, by Theorem 4.2 of [8] and by the relations (10) and (11) leads to

$$\xi_n \to \xi_0 , \qquad n\to\infty .$$

Using the obvious equalities (see, e.g., [9])

$$B_n^N(t) = B_n(t\wedge\tau_N(\xi_n)) + \int_{t\wedge\tau_n(\xi_n)}^{t} a_N(s,\xi_n^N)\,ds ,$$

$$\langle\xi_n^{Nc}\rangle_t = \langle\xi_n^c\rangle_{t\wedge\tau_N(\xi_n)} + \int_{t\wedge\tau_N(\xi_n)}^{t} b_N^2(s,\xi_n^N)\,ds ,$$

$$\nu_n^N([0,t]\times B) = \nu_n([0, t\wedge\tau_N(\xi_n)]\times B) ,$$

conditions (A_N'), $(\sup B_N')$ and (C_N') can be written in the following form: for any $t > 0$ and $\varepsilon \in (0,1]$

(A_N): $$A^n(t\wedge\tau_N(\xi_n)) \overset{P}{\to} 0 ,$$

$(\sup B_N)$: $$\sup_{0\le s\le t\wedge\tau_N(\xi_n)} \left| B_s^n - \int_0^s a(u,\xi_n)du \right| \overset{P}{\to} 0 , \tag{12}$$

(C_N): $$C_\varepsilon(t\wedge\tau_N(\xi_n)) - \int_0^{t\wedge\tau_N(\varepsilon_n)} b^2(u,\xi_n)du \overset{P}{\to} 0 .$$

Hence, if for any $N > 0$ conditions (A_N), $(\sup B_N)$ and (C_N) are satisfied, then $\xi_n \to \xi_0$, $n\to\infty$; and to prove the Theorem we need only to note that when conditions (A), (sup B), (sup C) are satisfied conditions (A_N), $(\sup B_N)$ and (C_N) are also satisfied for all $N > 0$. Indeed, by Lemma 2.1 of [6] condition (A) is equivalent to

(sup A): $$\sup_{0 \le s \le t} |A_n^\varepsilon(t)| \xrightarrow{P} 0 \;;$$

and since

$$|A_n^\varepsilon(t \wedge \tau_N(\xi_n))| \le \sup_{0 \le s \le t} |A_n^\varepsilon(s)| \;,$$

then (A) => (A_N) for all $N > 0$. The validity of the implications

$$(\sup B) \Rightarrow (\sup B_N) \quad \text{for all} \quad N > 0 \;,$$

$$(\sup B) \Rightarrow (C_N) \quad \text{for all} \quad N > 0$$

is also obvious. ///

REMARK. As it follows from the proof, the assertion of Theorem 1 holds when conditions (A), (sup B), (C) are replaced by local conditions (A_N), (sup B_N), (C_N), $N > 0$.

Proof of Theorem 2. Let $\tau(x) = \{\tau_t(x),\ t \ge 0\}$, $x \in D$, be a process given by

$$\tau_t(x) = \inf\left\{s:\ \int_0^s b^2(u,x)du \ge t\right\}, \qquad t \ge 0 \;. \tag{13}$$

It can be easily verified that under the conditions of the Theorem $\tau(x)$ is a process of time change with respect to the filtration $\{\mathcal{D}_t\}$, $t \ge 0$, and satisfies the equation

$$\tau_t = \int_0^t [b(\tau_s,x)]^{-2}\, ds \;, \qquad t \ge 0 \;. \tag{14}$$

Furthermore, there exists $\tau_t^{-1}(x)\left\{\tau_{\tau_t^{-1}(x)}(x) = t\right\}$ and

$$\tau_t^{-1}(x) = \int_0^t b^2(s,x)\, ds , \qquad t \geq 0 .$$

Now we define the mapping $T^{\tau^{-1}} : x \in D \to T^{\tau^{-1}} x \in D$:

$$\left(T^{\tau^{-1}} x\right)(t) = x(\tau_t(x)) , \qquad t \geq 0 ,$$

and prove that:

- (a) the mapping $T^{\tau^{-1}}$ is one-to-one;
- (b) the inverse mapping has the form

$$T^{\Phi}: x \in D \to T^{\Phi}x \in D ,$$

$$(T^{\Phi}x)(t) = x(\Phi_t^{-1}(x)) , \qquad t \geq 0 ,$$

where $\Phi(x) = \{\Phi_t(x), t \geq 0\}$ is a solution of the equation (4);

- (c) the mapping T^{Φ} is continuous on C.

We prove first the assertion (a). Assume $x_1, x_2 \in D$, $x_1 \neq x_2$ and $T^{\tau^{-1}} x_1 = T^{\tau^{-1}} x_2 = y$. Then it can be easily found that $\tau(x_1)$ and $\tau(x_2)$ are solutions of the equation

$$\tau_t = \int_0^t [b(\tau_s, T^{\tau}y)]^{-2} ds ,$$

and since this equation (by condition (b)) has a unique solution, then $\tau_t(x_1) = \tau_t(x_2)$, $t \geq 0$. But the last equality is impossible since $x_1 \neq x_2$ and $x_1(\tau_t(x_1)) = x_2(\tau_t(x_2)) = x_2(\tau_t(x_1))$, $t \geq 0$. Hence the assumption $T^{\tau^{-1}} x_1 = T^{\tau^{-1}} x_2$ for $x_1 \neq x_2$ is not true.

In order to prove assertion (b), it is enough to show that if $y = T^{\tau^{-1}}x$, then $\tau_t(x) = \Phi_t(y)$, $t > 0$. Since $x = T^{\tau}y$ (i.e., $x(t) = y(\tau_t^{-1}(x), t \geq 0)$, then $\tau_t(x)$ satisfies the equation

$$\Phi_t = \int_0^t [b(\Phi_s, T^{\Phi}y)]^{-2}\, ds ,$$

which has a unique solution $\Phi(y)$ and hence $\Phi(y) = \tau(x)$.

Let us prove the continuity of the mapping T^{Φ} on C. Let $\rho(x_n, x) \to 0$, $n \to \infty$, $x_n \in D$, $n \geq 1$, $x \in C$. We state that $\rho(T^{\Phi}x_n, T^{\Phi}x) \to 0$, $n \to \infty$. Since the convergence with respect to Skorokhod's metric to the element x on the space C is equivalent to the uniform convergence in any finite interval $[0,T]$, $T > 0$, then it is sufficient to prove that for any $T > 0$

$$\sup_{0 \leq t \leq T} |T^{\Phi}x_n(t) - T^{\Phi}x(t)| \to 0 , \qquad n \to \infty . \tag{15}$$

We have

$$\sup_{0 \leq t \leq T} |T^{\Phi}x_n(t) - T^{\Phi}x(t)| = \sup_{0 \leq t \leq T} |x_n(\Phi_t^{-1}(x_n)) - x(\Phi_t^{-1}(x))|$$

$$\leq \sup_{0 \leq t \leq \Phi_T^{-1}(x_n)} |x_n(t) - x(t)| + \sup_{0 \leq t \leq T} |x(\Phi_t^{-1}(x_n)) - x(\Phi_t^{-1}(x))| .$$

By condition (c) for any $T > 0$

$$\sup_{0 \leq t \leq T} |\Phi_t^{-1}(x_n) - \Phi_t^{-1}(x)| \to 0 , \qquad n \to \infty .$$

Then for any $\varepsilon_1 > 0$ (say $\varepsilon_1 = 1$) we can find an integer n_1 such that for every $n \geq n_1$

$$\sup_{0\le t\le T} |\Phi_t^{-1}(x_n) - \Phi_t^{-1}(x)| \le \varepsilon_1 .$$

Then for all $n \ge n_1$ we have

$$\sup_{0\le t\le T} |T^\Phi x_n(t) - T^\Phi x(t)| \le \sup_{0\le t\le \Phi_T^{-1}(x)+1} |x_n(t) - x(t)|$$

$$+ \sup_{0\le t\le T} |x(\Phi_t^{-1}(x_n)) - x(\Phi_t^{-1}(x))| .$$

Fix now some $\varepsilon > 0$ and show that an integer n_ε can be found such that for all $n \ge n_\varepsilon$ the inequality

$$\sup_{0\le t\le T} |T^\Phi x_n(t) - T^\Phi x(t)| \le \varepsilon \tag{16}$$

holds. Since for all $n \ge n_1$ and $t \in [0,T]$ the following relations are true:

$$\Phi_t^{-1}(x_n) \in [0, \Phi_t^{-1}(x_n)] \subset [0, \Phi_T^{-1}(x) + 1]$$

and

$$\Phi_t^{-1}(x) \in [0, \Phi_T^{-1}(x) + 1] ,$$

then in view of the uniform continuity of the function $x(t)$, $t \in [0,T]$, a number $\delta_\varepsilon > 0$ can be found such that the inequality

$$\sup_{\substack{t,s\in[0,\Phi_t^{-1}(x)+1] \\ |t-s|\le\delta_\varepsilon}} |x(t) - x(s)| < \frac{\varepsilon}{2}$$

is true. If we choose an integer n_2 such that for all $n \ge n_2$ the inequality

$$\sup_{0 \le t \le T} |\Phi_t^{-1}(x_n) - \Phi_t^{-1}(x)| < \varepsilon$$

holds, then for all $n \ge \max(n_1, n_2)$ we have

$$\sup_{0 \le t \le T} |T^{\Phi}x_n(t) - T^{\Phi}x(t)| \le \sup_{0 \le t \le \Phi_T^{-1}+1} |x_n(t) - x(t)| + \frac{\varepsilon}{2} .$$

Finally, if $n \ge \max(n_1, n_2, n_3)$ where n_3 is given by the condition: for all $n \ge n_3$

$$\sup_{0 \le t \le \Phi_T^{-1}+1} |x_n(t) - x(t)| < \frac{\varepsilon}{2} ,$$

then the inequality (16) holds and thus the continuity of the mapping T^{Φ} is proved.

Now instead of the initial sequence of semimartingales $\{\xi_n\}$, $n \ge 1$, we consider a new sequence of the processes $\eta_n = \{\eta_n(t), t \ge 0\}$, $n \ge 1$, defined by the relation

$$\eta_n(t) = \xi_n(\tau_t(\xi_n)) , \qquad t \ge 0 .$$

It is well known (see e.g., [9, Chapter 10]) that the process

$$\eta_n = \{\eta_n(t), \tilde{F}_t^n = F^n_{\tau_t(\xi_n)}, t \ge 0\}$$

is a semimartingale with canonical representation

$$\eta_n(t) = \tilde{B}_n(t) + \eta_n^c(t) + \int_0^t \int_{|x|>1} x \, \tau \mu_n(ds, dx)$$

$$+ \int_0^t \int_{|x|\le 1} x(\tau\mu_n - \tau\nu_n)(ds, dx) ,$$

where $\tilde{B}_n(t) = B_n(\tau_t(\xi_n))$, $\eta_n^c(t) = \xi_n^c(\tau_t(\xi_n))$, an integer-valued random measure $\tau\mu_n$ is defined by the relation

$$\tau\mu_n([0,t]\times B) = \mu_n\Big([0, \tau_t(\xi_n)]\times B\Big) ,$$

$\tau\nu_n$ being the compensator of the measure $\tau\mu_n$.

Define the process $\eta_0 = \{\eta_0(t), t \geq 0\}$ in the following way:

$$\eta_0(t) = \xi_0(\tau_t(\xi_0)) , \qquad t \geq 0 .$$

It is clear that $\eta_0 = \{\eta_0(t), \tilde{F}_t^0 = F^0_{\tau_t(\xi_0)}, t \geq 0\}$ is a Wiener process.

By virtue of Theorem 1 of [5] conditions (A_τ), $(\sup B_\tau)$ and (C_τ) are sufficient for weak convergence of a sequence of semi-martingales η_n, $n \geq 1$ to a Wiener process η_0, i.e.,

$$(A_\tau),\ (\sup B_\tau),\ (C_\tau) \Rightarrow \eta_n \overset{D}{\to} \eta_0 , \qquad n \to \infty . \quad (17)$$

But since $\xi_n = T^\Phi \eta_n$, $n \geq 0$ (see assertion (b)), the measure P^{η_0} is concentrated on the space C and the mapping $T^\Phi : x \in D \to T^\Phi x \in D$ is continuous with respect to x on C, then it follows from Theorem 5.1 if [8] and the relation (17) that

$$\xi_n \overset{D}{\to} \xi_0 , \qquad n \to \infty \quad . \qquad ///$$

Proof of Theorem 3. First consider the case when the functional $b(t,x)$ satisfies condition (E). If we prove that conditions (D) and (E) imply condition (c) of Theorem 2, then the assertion of Theorem 3 will be a direct consequence of Theorem 2.

Let $x \in C$ and x_n, $n \geq 1$, be the sequence of elements from D converging to x. Let us show now that for any $T > 0$

$$\sup_{0 \le t \le T} |\Phi_t^{-1}(x_n) - \Phi_t^{-1}(x)| \to 0 , \qquad n \to \infty . \tag{18}$$

By virtue of condition (E) the sequence $\Phi_T^{-1}(x_n) = \{\Phi_t^{-1}(x_n),\ t \le T\}$, $n \ge 1$, is relatively compact. Then it is sufficient to prove that it has a unique limit point $\Phi_T^{-1}(x) = \{\Phi_t^{-1}(x),\ t \le T\}$.

Let $\{\Phi_T^{-1}(x_{n'})\}$ be a converging subsequence of the sequence $\{\Phi_T^{-1}(x_n)\}$ and let $\tilde{\Phi}_T^{-1}$ be its limit. For any n' we have

$$\Phi_t^{-1}(x_{n'}) = \int_0^t b^2(s, T^{\Phi} x_{n'})\, ds , \qquad t \le T .$$

Passing to the limit in the last relation as $n' \to \infty$ and taking into account conditions (D) and (E), by the Lebesgue theorem on majorized convergence we have

$$\tilde{\Phi}_t^{-1} = \int_0^t b^2(s, T^{\tilde{\Phi}} x)\, ds , \qquad t \le T ,$$

or

$$\tilde{\Phi}_t = \int_0^t [b(\tilde{\Phi}_s,\ T^{\tilde{\Phi}} x)]^{-2}\, ds , \qquad t \le T .$$

But the last equation by virtue of condition (b) has a unique solution $\Phi_T(x)$ and hence $\tilde{\Phi}_T^{-1} = \Phi_T^{-1}(x)$, as required.

Note that when condition (E) is satisfied $\Phi(x) = \{\Phi_t(x), t \ge 0\}$ is also continuous with respect to x mapping onto C. (Recall that $\Phi(x)$ is a solution of the equation (4)).

Indeed, if we assume that $\rho(x_n, x) \to 0$ as $n \to \infty$, but $\lim_{n\to\infty} \Phi_t(x_n) \ne \Phi_t(x)$, we can then find $\varepsilon > 0$ and a subsequence $\{x_{n'}\}$ of the sequence $\{x_n\}$, such that

$$\Phi_t(x_{n'}) > \Phi_t(x) + \varepsilon \qquad (\text{or} \quad \Phi_t(x_{n'}) < \Phi_t(x) - \varepsilon) .$$

The last inequality by the definition

$$\Phi_t(x) = \inf \left\{ s: \int_0^s b^2(u, T^\Phi x)du \geq t \right\} , \qquad t \geq 0, \quad (19)$$

is equivalent to

$$\int_0^{\Phi_t(x)+\varepsilon} b^2(s, T^\Phi x_{n'})ds < t \qquad \left(\text{or} \int_0^{\Phi_t(x)-\varepsilon} b^2(s, T^\Phi x_{n'})ds \geq t \right) ,$$

which together with the continuity of the mapping T^Φ (see Theorem 2) and passage to the limit as $n' \to \infty$ yield

$$\int_0^{\Phi_t(x)+\varepsilon} b^2(s, T^\Phi x)ds < t \qquad \left(\int_0^{\Phi_t(x)-\varepsilon} b^2(s, T^\Phi x)ds \geq t \right) ,$$

which contradicts (19).

Hence, for any $t \geq 0$ $\lim_{n\to\infty} \Phi_t(x_n) = \Phi_t(x)$ if $\lim_{n\to\infty} \rho(x_n, x) = 0$, $x_n \in D$, $x \in C$ and the continuity of the mapping $\Phi(x)$ (i.e., $\lim_{n\to\infty} \sup_{0\leq t\leq T} |\Phi_t(x_n) - \Phi_t(x)|$ for $\forall T > 0$) follows from Lemma 2.1 of [6].

Now return to the case when the functional $b(t,x)$ satisfies condition (E_L). We need to prove that the implication

$$(D), (E_L) \Rightarrow (c)$$

is true.

First we state some facts we shall need in what follows.

Let $g_N(t,x)$ and $b_N(t,x)$, $t \geq 0$, $x \in D$, $N > 0$, be the functionals introduced in the proof of Theorem 1. It can be easily verified that if Φ_t, $t \geq 0$ is a time-change process, i.e., with $\Phi_S = \infty$, then

$$\tau_N(T^\Phi x) = \Phi_{\tau_N^\Phi(x)} , \tag{20}$$

$$g_N(t, T^\Phi x) = T^\Phi g_N^\Phi(\cdot, x)(t) , \tag{21}$$

where

$$\tau_N^\Phi(x) = \inf \{s: \|x\|_s + f(\Phi_s) \geq N\} ,$$

$$g_N^\Phi(t,x) = \begin{cases} x(t) , & t < \tau_N^\Phi(x) , \\ [x(\tau_N^\Phi(x))]_{-N}^N , & t \geq \tau_N^\Phi(x) . \end{cases}$$

Put $\tilde{\tau}_N(x) = \inf \{s: \|x\|_s \geq N\}$, $x \in D$, $N > 0$. It is obvious that

$$\tilde{\tau}_{N-1}(x) \leq \tau_N^\Phi(x) \leq \tilde{\tau}_N(x) . \tag{22}$$

Note the following property of the functional $\tilde{\tau}_N(x)$: if $\lim_{n\to\infty} \rho(x_n, x) = 0$, $x_0 \in C$, $x_n \in D$, $n \geq 1$, and an integer N_T is such that

$$\tilde{\tau}_{N_T}(x_0) > T , \tag{23}$$

we can find a number n_T such that the inequality

$$\tilde{\tau}_N(x_n) > T$$

holds for all $n \geq n_T$ and $N \geq N_T$.

Assume the reverse. Then a subsequence $\{x_{n'}\}$ will exist for which

$$\tilde{\tau}_N(x_{n'}) \leq T .$$

The last inequality, by the definition of $\tilde{\tau}_N(x)$, is equivalent to

$$\sup_{0\le s\le T} |x_{n'}(s)| > N_T ,$$

which as $n' \to \infty$ leads to

$$\sup_{0\le s\le T} |x_0(s)| > N_T \Rightarrow \tilde{\tau}_{N_T}(x_0) > T ,$$

and thus we have arrived at a contradiction.

Let $x \in D$, $N > 0$. Consider the equation

$$\Phi_t = \int_0^t [b_N(\Phi_s, T^{\Phi}x)]^{-2}\, ds .$$

By virtue of (21) and condition (b) this equation has a unique solution $\Phi_N(x) = \{\Phi_N(t,x),\ t \ge 0\}$, and the mapping $T^{\Phi_N}: x \in D \to T^{\Phi_N}x \in D$ is continuous with respect to x on C since the functional $b_N(t,x)$ satisfies conditions (D) and (E). It can be easily verified that

$$\Phi_N(t,x) = \Phi_t(x) \quad \text{for} \quad t < \tau_N^{\Phi}(x) , \tag{24}$$

$$\Phi_N^{-1}(t,x) = \Phi_t^{-1}(x) \quad \text{for} \quad t < \Phi(\tau_N^{\Phi}(x),x) = \tau_N(T^{\Phi}x) , \tag{25}$$

where $\Phi(t,x)$, $t \ge 0$, is a solution of the equation

$$\Phi_t = \int_0^t [b(\Phi_s, T^{\Phi}x)]^{-2}\, ds ,$$

and $\tau_N^{\Phi}(x) = \inf \{s: ||x||_s + f(\Phi_s)(x)) \ge N\}$.

Now let us prove the continuity of the mapping

$$x \to T^{\Phi}x , \qquad T^{\Phi}x(t) = x(\Phi_t^{-1}) , \qquad t \ge 0 .$$

Let $\rho(x_n, x_0) \to 0$, $n \to \infty$, $x_n \in D$, $x_0 \in C$. It can be proved that for any $T > 0$

$$\sup_{t \le T} |\Phi_y^{-1}(x_n) - \Phi_t^{-1}(x_0)| \to 0, \quad n \to \infty. \tag{26}$$

Fix an arbitrary $T > 0$. Since $\lim_{n\to\infty} \tilde{\tau}_{N-1}(x_0) = \infty$, we can find a number $N_T > 0$ and $n_T > 0$ such that

$$\tilde{\tau}_{N-1}(x_0) > \Phi^{-1}(T, x_0), \tag{27}$$

$$\tilde{\tau}_{N-1}(x_n) > \Phi^{-1}(T, x_0) \tag{28}$$

for all $n \ge n_T$ and $N \ge N_T$. It can be concluded from these inequalities that for all $n \ge n_T$ and $N \ge N_T$ the inequalities

$$\begin{aligned} \tau_N^{\Phi}(x_0) &> \Phi^{-1}(T, x_0), \\ \tau_N(x_n) &> \Phi^{-1}(T, x_0) \end{aligned} \tag{29}$$

will be simultaneously satisfied, which in turn leads to

$$\Phi(\tau_N^{\Phi}(x_n), x_n) > \Phi(\Phi^{-1}(T, x_0), x_n)$$

and

$$\Phi(\Phi^{-1}(T, x_0), x_n) = \Phi_N(\Phi^{-1}(T, x_0), x_n).$$

These relations by virtue of the equality

$$\begin{aligned} \lim_{N\to\infty} \Phi_N(\Phi^{-1}(T, x_0), x_n) &= \Phi_N(\Phi^{-1}(T, x_0), x_0) \\ &= \Phi(\Phi^{-1}(T, x_0), x_0) = T \end{aligned}$$

imply that for all $N \ge N_T$

$$\liminf_{n \geq n_T} \Phi(\tau_N^\Phi(x_n)\, x_n) \quad > \quad T \quad .$$

The last inequality for an arbitrary $\varepsilon > 0$ enables us to choose a number n_ε such that for all $n \geq \max(n_T, n_\varepsilon)$

$$\Phi(\tau_N^\Phi(x_n),\ x_n) \quad > \quad T - \varepsilon \quad . \tag{30}$$

Hence by the inequalities (29), (30) and the formulas (24), (25) for any $T > 0$ and $\varepsilon > 0$ one can find N_T and $n_{\varepsilon,T}$ such that for all $N \geq N_T$ and $n \geq n_{\varepsilon,T}$ the equalities

$$\Phi^{-1}(t,x_n) \quad = \quad \Phi_N^{-1}(t,x_n) \quad , \qquad t \leq T < \varepsilon \ ,$$

$$\Phi^{-1}(t,x_0) \quad = \quad \Phi_0^{-1}(t,x_0) \quad , \qquad t \leq T$$

hold and hence

$$\sup_{0 \leq s \leq T-\varepsilon} |\Phi^{-1}(t,x_n) - \Phi^{-1}(t,x_0)| \quad = \quad \sup_{0 \leq s \leq T-\varepsilon} |\Phi_N^{-1}(t,x_n) - \Phi_N^{-1}(t,x_0)| \ .$$

But since $\sup_{0 \leq s \leq T} |\Phi_N^{-1}(t,x_n) - \Phi_N^{-1}(t,x_0)| \to 0, \quad n \to \infty,$ then it follows from the last equality that

$$\sup_{0 \leq s \leq T-\varepsilon} |\Phi^{-1}(t,x_n) - \Phi^{-1}(t,x_0)| \quad = \quad 0 \quad . \qquad ///$$

<u>EXAMPLE</u>. Let $b(s,x) = \beta(x(s))$, where $\beta^2(y)$, $y \in R^1$ is continuous. By the condition of the existence of a unique weak solution of the equation

$$d\xi_t \quad = \quad b(t,\xi)\, dw_t \ , \qquad \xi_0 = 0 \quad ,$$

it follows that $\beta^2(y) > 0$ for all $y \in R^1$ (see [10]). Let us make sure that the functional $b(t,x)$ satisfies all the conditions of Theorem 2. Condition (a) holds automatically.

As for condition (b), note that in the given case for a fixed $x \in D$ the equation (4) has the form

$$\Phi_t(x) = \int_0^t [\beta(x(s))]^{-2} ds , \qquad t \geq 0 ,$$

and since for any $x \in D$ $\sup_{0 \leq s \leq t} |x(s)| < \infty$ for all $t > 0$ and $\beta^2(y)$ is a continuous function and $\beta^2(y) > 0$, then for any $t > 0$ a constant C_t can be found (generally speaking, depending on $x \in D$) such that $\beta^2(x(s)) > C$ for all $s \leq t$ and consequently

$$\Phi_t(x) = \int_0^t [\beta(x(s))]^{-2} ds \leq C_t \cdot t < \infty ,$$

hence condition (b) is satisfied.

The fact that condition (c) is satisfied follows from the local boundedness and continuity of the functional $b^2(s,x)$ (see the proof of Theorem 3: (D), (E_L) => (C)).

The author is grateful to R.J. Chitashvili for his interest in this paper.

REFERENCES

[1] Gikhman, I.I., and Skorokhod, A.V. *Stochastic Differential Equations*. Berlin Heidelberg New York: Springer-Verlag Inc. 1972.

[2] Skorokhod, A.V. *Studies in the Theory of Random Processes*. Reading, Mass.: Addison-Wesley Pub. Co., 1965.

[3] Grigelionis, B.I., and Mikulevichus, P. "Weak Convergence of Semimartingales." *Lithuanian Math. Journal*, vol.21, no.3 (1981): 213-224.

[4] Grigelionis, B.I., Kubilyus, K., and Mikulyavichyus, R.A. "The Martingale Approach to Functional Limit Theorems." *Russian Math. Surveys*, vol.37, no.6 (1982): 41-54.

[5] Liptser, R.Sh., and Shiryayev, A.N. "A Functional Central Limit Theorem for Semimartingales." *Theory Probab. Applications*, vol.25, no.4 (1980): 667-688.

[6] Liptser, R.S., and Shiryaev, A.N. "Weak Convergence of Semimartingales to the Diffusion-Type Process." *Math. USSR Sbornik*, vol.49, no.1 (1984): 171-195.

[7] Butov, A.A. "On the Problem of Weak Convergence of Semimartingales to the Diffusion-Type Process." *Russian Math. Surveys*, vol.38, no.5 (1983): 135-136.

[8] Billingsley, P. *Convergence of Probability Measures*.New York: John Wiley, 1968.

[9] Jacod, J. Calcul stochastique et problèmes des martingales. *Lecture Notes in Math.*, vol.714. Berlin Heidelberg New York: Springer-Verlag Inc., 1978.

[10] Engelebert, H.-J., and Schmidt, W. "On Solutions of One-Dimensional Stochastic Differential Equations Without Drift." Unpublished.

V.A. LEBEDEV

ON INFINITE-DIMENSIONAL STOCHASTIC INTEGRALS

Our objective in this paper is to connect two approaches to infinite-dimensional stochastic integration. One approach, developed in [1], defines the stochastic integral with respect to an L^0-valued random measure by generalizing the classical procedure of extension of the integral from some integration lattice on which it is defined in a natural way. In the other approach, developed in [2] as well as in several previous papers, the extension of the stochastic integral from the original integration lattice is carried out on the basis of prelocal L^2-domination. In our paper, we show first that under broad assumptions the stochastic integral with respect to the given random measure in the sense of [1] is the stochastic integral in the sense of [2]. On the other hand, the converse does not always hold, but the stochastic integral in the sense of [2] can be represented as the stochastic integral in the sense of [1] with respect to another random measure. Such reduction is far from trivial, however, the theory of the integral with respect to an L^0-valued measure in the sense of [1] appears to be simpler to obtain general results on stochastic differential equations with such measures.

*

Let $(\Omega, \mathcal{F}, \underline{\mathcal{F}}, P)$ be a stochastic basis consisting of a complete probability space $(\Omega, \mathcal{F}, P)$ and a filtration $\underline{\mathcal{F}} = (\mathcal{F}_t)_{t \in R_+}$ satisfying usual conditions. We denote by $\mathcal{S}$ the space of all semimartingales with the topology of Emery [3], by $\mathcal{S}^p$ for $p \geq 1$ the space of semimartingales, belonging to $\mathcal{H}^p$ from [3] on every finite interval, with the corresponding topology, by $\mathcal{V}$ the space of all cadlag adapted processes with finite variation over finite intervals, by $\mathcal{M}_{loc}$ the space of all local martingales.

Let be for $p \geq 1$ $\mathcal{V}^p = \mathcal{S}^p \cap \mathcal{V}$ and $\mathcal{M}^p = \mathcal{S}^p \cap \mathcal{M}_{loc}$ with the relative topology of $\mathcal{S}^p$, $\mathcal{L}^c$ the space of all continuous local martingales starting from 0 at $t = 0$, $\mathcal{L}_{loc}$ and $\mathcal{L}^p$ the spaces of processes respectively from $\mathcal{M}_{loc}$ and $\mathcal{M}^p$ starting from 0 at $t = 0$. Let $X^c \in \mathcal{L}^c$ denote the continuous local martingale part of $X \in \mathcal{S}$.

If X is a cadlag process then X_- is the process of left-hand limits (and $X_{0-} = 0$), and $\Delta X = X - X_-$ (hence $\Delta X_0 = X_0$). If τ is a stopping time then X^τ (respectively $X^{\tau-}$) denotes the process X stopped at τ (respectively before τ).

The optional and the predictable σ-algebras on $\Omega \times R_+$ are denoted respectively by $\mathcal{O}$ and $\mathcal{P}$, sometimes $\mathcal{P}$ will denote also the set of all predictable processes. $b\mathcal{P}$ (respectively $b\mathcal{O}, \ldots$) denotes the set of all bounded functions which are $\mathcal{P}$-measurable (respectively $\mathcal{O}$-measurable, ...). If $X \in \mathcal{S}$ and H is a predictable process which is integrable with respect to X then we denote by $H \cdot X$ the stochastic integral process.

We endow the space $L^p(\Omega, \mathcal{F}, P)$ with the usual norm

$||Z||_p = (E|Z|^p)^{1/p}$ for $p \geq 1$ and with the quasi-norm $||Z||_p = E(|Z| \wedge |Z|^p)$ for $0 \leq p < 1$ making $L^p(\Omega, \mathcal{F}, P)$ an F-space and for $p \geq 1$ a Banach space.

Let $(E, \mathcal{E})$ be a measurable space, $\tilde{\Omega} = \Omega \times R_+ \times E$, $\tilde{\mathcal{P}} = \mathcal{P} \otimes \mathcal{E}$, $\tilde{\mathcal{O}} = \mathcal{O} \otimes \mathcal{E}$.

The main concept of the present paper, as also of [1], is a σ-finite L^p-valued random measure on $(\tilde{\Omega}, \tilde{\mathcal{P}})$ that is a family $\theta = (\theta_t)_{t \in R_+}$ that satisfies the following:

•(i) for each $t \geq 0$, θ_t is a σ-finite measure on $(\tilde{\Omega}, \tilde{\mathcal{P}})$ with values in $L^p(\Omega, \mathcal{F}_t, P)$ (which means that there is a strictly positive $\tilde{\mathcal{P}}$-measurable function V such that, if $\tilde{\mathcal{P}}_V = \{\phi\colon \tilde{\mathcal{P}}\text{-measurable, } \phi/V \text{ bounded}\}$, we have:

•(i.1) θ_t is a linear mapping from $\tilde{\mathcal{P}}_V$ into $L^p(\Omega, \mathcal{F}_t, P)$,

•(i.2) if (ϕ_n) is a sequence in $\tilde{\mathcal{P}}_V$ with $|\phi_n| \leq V$ converging pointwise to 0, then $\theta_t(\phi_n) \to 0$ in L^p);

•(ii) $\theta_s(\phi) = \theta_t(\phi 1_{[0,s]})$ for all $\phi \in \tilde{\mathcal{P}}_V$, $s \leq t$;

•(iii) $\theta_t(\phi 1_{A \times I \times E}) = 1_A \theta_t(\phi 1_{\Omega \times I \times E})$ for all $\phi \in \tilde{\mathcal{P}}_V$, $t \in R_+$, if $A \in \mathcal{F}_0$ and $I = R_+$, or if $A \in \mathcal{F}_s$ and $I =]s,s']$ with $s < s'$.

In particular, when $V \equiv 1$, θ is called a finite L^p-valued random measure.

We shall denote by $\tilde{\mathcal{S}}^p_\sigma$ (respectively $\tilde{\mathcal{S}}^p$) the space of all σ-finite (respectively finite) L^p-valued random measures on $(\tilde{\Omega}, \tilde{\mathcal{P}})$.

<u>Example</u>. Let $E = \{1\}$, that is, it consists of one point so that we consider random measures on $(\Omega \times R_+, \mathcal{P})$. Then by Theorem 1.1 of [1] for every $\theta \in \tilde{\mathcal{S}}^0$ there exists and is unique up to

P-indistinguishability a semimartingale X such that for every $H \in b\mathcal{P}$ at each $t \in R_+$

$$\theta_t(H) = H \cdot X_t \quad . \tag{1}$$

Conversely, for every semimartingale X (1) defines a finite L^0-valued random measure θ on $(\Omega \times R_+, \mathcal{P})$ so that we have a bijective mapping between $\mathcal{S}$ and $\tilde{\mathcal{S}}^0$ on $(\Omega \times R_+, \mathcal{P})$. As far as σ-finite L^0-valued random measures on $(\Omega \times R_+, \mathcal{P})$ are concerned, we shall call them formal semimartingales. ///

We put for $\theta \in \tilde{\mathcal{S}}^p_\sigma$ on $(\tilde{\Omega}, \tilde{\mathcal{P}})$

$$\|\phi\|_{L^{1,p}(\theta)} = \sum_{n=1}^{\infty} 2^{-n}\Big(1 \wedge \sup_{\psi \in \tilde{\mathcal{P}}_V, |\psi| \leq |\phi|} \|\theta_n(\psi)\|_p\Big)$$

and denote by $L^{1,p}(\theta)$ the set of all $\tilde{\mathcal{P}}$-measurable functions ϕ for each of which there is a sequence $(\phi_n) \subset \tilde{\mathcal{P}}_V$ with $\|\phi_n - \phi\|_{L^{1,p}(\theta)} \to 0$. Then for each $t \in R_+$, $\theta_t(\phi_n)$ tends in $L^p(\Omega, \mathcal{F}_t, P)$ to a limit which does not depend on the choice of (ϕ_n) for the given ϕ, and is denoted by $\theta_t(\phi)$. Thus we have an extension of θ to $L^{1,p}(\theta)$ and every $\phi \in L^{1,p}(\theta)$ satisfies properties (ii) and (iii). Notice that definitions of $\|\cdot\|_{L^{1,p}(\theta)}$ and $L^{1,p}(\theta)$ do not depend on the choice of V and in the definition of θ we can take, instead, any strictly positive function V' belonging to $L^{1,q}(\theta)$. Finally, if $p \geqslant q$ and, furthermore, $\theta \in \tilde{\mathcal{S}}^q_\sigma$, then $\theta \in \tilde{\mathcal{S}}^p_\sigma$ and $L^{1,q}(\theta) \subset L^{1,p}(\theta)$.

Let $\theta \in \tilde{\mathcal{S}}^p_\sigma$ and $\phi \in L^{1,p}(\theta)$. Then the equality

$$(\phi \times \theta)_t(H) = \theta_t[\phi(H \otimes 1)]$$

for $H \in bP$ defines the family $\phi \times \theta$ as a finite L^p-valued random measure on $(\Omega \times R_+, P)$ for which by the Example there is a semimartingale denoted also by $\phi \times \theta$ and called the stochastic integral process for ϕ with respect to θ and by (1)

$$\phi \times \theta_t = (\phi \times \theta)_t(1) = \theta_t(\phi) .$$

Notice that in concrete applications it is, as a rule, difficult to define $\theta \in \tilde{S}^p_\sigma$ on the whole $\tilde{P}_V$ at once. Usually it is defined first on some set of simple functions which is dense in $\tilde{P}_V$ and hence $L^{1,p}(\theta)$ in the corresponding sense, as for example, step functions for integration with respect to semimartingales. Such sets usually fit the following general concept of an integration lattice.

Thus, any integration lattice on $(\tilde{\Omega}, \tilde{P})$ is the vector lattice T of functions on $\tilde{\Omega}$ with:

- (i) T is stable under the mapping: $\phi \to 1 \wedge \phi$;
- (ii) there is a sequence $(\phi_n) \subset T$ such that $\sup_{n \in \underline{N}} \phi_n = 1$;
- (iii) T generates the σ-algebra $\tilde{P}$;
- (iv) T is stable under mappings $\phi \to \phi 1_{A \times I \times E}$, for $A \in F_0$ and $I = R_+$, and for $A \in F_s$ and $I =]s,s']$ with $s < s'$.

The measure θ itself must be defined on T so that it would satisfy conditions (i) - (iii) of the definition of a σ-finite L^p-valued random measure on $(\tilde{\Omega}, \tilde{P})$ with T instead of $\tilde{P}_V$ and with some $\phi \in T$ instead of V in (i.2). Then by Theorem 2.25 of [1] it is extended uniquely as a σ-finite L^p-valued random measure to the corresponding $L^{1,p}(\theta)$.

We can define integrals with respect to $\theta \in \tilde{S}^p_\sigma$ for a larger class of $\tilde{P}$-measurable functions than $L^{1,p}(\theta)$. We introduce the set $L^p_\sigma(\theta) = \{\phi$: $\tilde{P}$-measurable, and there is a strictly positive predictable process K such that $(K \otimes 1)\phi \in L^{1,p}(\theta)\}$. Then the equality

$$(\phi \times \theta)_t(H) = \theta_t [(H \otimes 1)\phi]$$

for predictable H with bounded H/K defines $\phi \times \theta$ as a σ-finite L^p-valued random measure on $(\Omega \times R_+, P)$ that is a formal semimartingale. Finally we select the set $\hat{L}^p(\theta) \subset L^p_\sigma(\theta)$ of such ϕ for which $\phi \times \theta$ is a finite L^p-valued measure that is a semimartingale. Obviously, $L^{1,p}(\theta) \subset \hat{L}^p(\theta)$ and $\hat{L}^q(\theta) \subset \hat{L}^p(\theta)$ for $p \le q$.

We select also the following classes of random measures:

$$PV - \tilde{S}^p_\sigma = \{\theta \in \tilde{S}^p_\sigma : \exists V \in L^{1,p}(\theta),\ V > 0 \text{ and } \phi\times\theta \in P \cap V \text{ for all } \phi \in \tilde{P}_V\},$$

$$L^c - \tilde{S}^p_\sigma = \{\theta \in \tilde{S}^p_\sigma : \exists V \in L^{1,p}(\theta),\ V > 0 \text{ and } \phi\times\theta \in L^c \text{ for all } \phi \in \tilde{P}_V\},$$

$$M - \tilde{S}^p_\sigma = \{\theta \in \tilde{S}^p_\sigma : \exists V \in L^{1,p}(\theta),\ V > 0 \text{ and } \phi\times\theta \in M_{loc} \text{ for all } \phi \in \tilde{P}_V\}.$$

A certain natural extension of the measure θ to some larger σ-algebra than $\tilde{P}$ will be useful. Thus, let $\bar{O}$ be the σ-algebra on $\Omega \times R_+ \times E$ generated by $\tilde{P}$ and sets of the form $[\![\tau]\!] \times B$, where τ is a stopping time and $B \in E$. In particular, we have $\bar{O} = \tilde{O}$ if E is separable. Consider the set $\bar{T}$ of functions on $\tilde{\Omega}$ of the form $\phi = \phi_0 + \sum_{k=1}^{\infty} 1_{[\![\tau_k]\!]} \phi_k$, where functions ϕ_k, $k = 0,\dots,r$, belong to $L^{1,p}(\theta)$ and τ_k, $k = 1,\dots,r$, are stopping times. Obviously it is an integration lattice on $(\tilde{\Omega}, \bar{O})$, and if we define

$$\theta_t(\phi) = \phi \times \theta_t = \phi_0 \times \theta_t + \sum_{k=1}^{2} \Delta(\phi_k \times \theta)_{\tau_k} 1_{\tau_k \leq t} ,$$

then by Proposition 4.18 of [1] this does not depend on the choice of ϕ_k and τ_k, satisfies conditions of the definition of the measure θ on an integration lattice, and allows to define integrals with respect to θ for functions from the spaces $\bar{L}^{1,p}(\theta)$ and $\hat{\bar{L}}^p(\theta)$ defined analogously to $L^{1,p}(\theta)$ and $\hat{L}^p(\theta)$ respectively with $\bar{\mathcal{O}}$ instead of $\tilde{\mathcal{P}}$.

Finally if W is an $\bar{\mathcal{O}}$-measurable function then we define the measure $W \cdot \theta \in \tilde{\mathcal{S}}^p_\sigma$ by putting

$$W \cdot \theta_t(\phi) = \theta_t(W\phi)$$

if $W\phi \in \bar{L}^{1,p}(\theta)$.

First we prove the following lemmas.

<u>LEMMA 1</u>. Let $\phi \in \hat{L}^p(\theta_1)$ and $\phi \in \hat{L}^p(\theta_2)$ for some p. Then for every $\alpha_1, \alpha_2 \in R$ $\phi \in \hat{L}^p(\alpha_1\theta_1 + \alpha_2\theta_2)$ and

$$\phi \times (\alpha_1\theta_1 + \alpha_2\theta_2) = \alpha_1(\phi \times \theta_1) + \alpha_2(\phi \times \theta_2) \quad . \tag{2}$$

<u>Proof</u>. First of all $\phi \in L^p_\sigma(\theta_1)$ and $\phi \in L^p_\sigma(\theta_2)$. By definition there exist strictly positive predictable processes K_1 and K_2 such that $(K_1 \otimes 1)\phi \in L^{1,p}(\theta_1)$ and $(K_2 \otimes 1)\phi \in L^{1,p}(\theta_2)$. Now if $K = K_1 \wedge K_2$, then $(K \otimes 1)\phi \in L^{1,p}(\theta_1)$ and $(K \otimes 1)\phi \in L^{1,p}(\theta_2)$ and hence $(K \otimes 1)\phi \in L^{1,p}(\alpha_1\theta_1 + \alpha_2\theta_2)$, so that $\phi \in L^p_\sigma(\alpha_1\theta_1 + \alpha_2\theta_2)$ and (2) is valid in the sense of formal semimartingales. At last, since $\phi \in \hat{L}^p(\theta_1)$ and $\phi \in \hat{L}^p(\theta_2)$ the right-hand member of (2) is a semimartingale from $\mathcal{S}^p$ that is a finite L^p-valued random measure on $(\Omega \times R_+, \mathcal{P})$, (2) is a usual semimartingale equality,

and $\phi \in \hat{L}^p(\alpha_1\theta_1 + \alpha_2\theta_2)$. ///

<u>LEMMA 2</u>. Let there be a P-a.s. increasing to infinity sequence of stopping times $(\tau_n)_{n\in\underline{N}}$. Then for $\phi \in \hat{L}^0(\theta)$ it is necessary and sufficient that for all $n \in \underline{N}$ either $\phi \in \hat{L}^0(1_{[\![0,\tau_n]\!]\times E}\cdot\theta)$ or $\phi \in \hat{L}^0(1_{[\![0,\tau_n[\![\times E}\cdot\theta)$.

<u>Proof</u>. Necessity. Let $\phi \in \hat{L}^0(\theta)$. Then $\phi \in L^0_\sigma(\theta)$ and hence for every n, $\phi 1_{[\![0,\tau_n]\!]\times E} \in L^0_\sigma(\theta)$ with the same K in the definition, and

$$\phi\, 1_{[\![0,\tau_n]\!]\times E} \times \theta = 1_{[\![0,\tau_n]\!]\times E} \times (\phi\cdot\theta) = \phi \times (1_{[\![0,\tau_n]\!]\times E}\cdot\theta), \tag{3}$$

but $\phi\times\theta \in \mathcal{S}$ and the middle member of (3) belongs to $\mathcal{S}$ so that

$$\phi \in \hat{L}^0(1_{[\![0,\tau_n]\!]\times E}\cdot\theta) .$$

Now we apply Proposition 4.18 of [1] with $D = [\![\tau_n]\!]$ and obtain that

$$\phi \in \hat{L}^0(1_{[\![\tau_n]\!]\times E}\cdot\theta)$$

and by Lemma 1 that

$$\phi \in \hat{L}^0(1_{[\![0,\tau_n[\![\times E}\cdot\theta) .$$

Furthermore,

$$\phi \times (1_{[\![0,\tau_n]\!]\times E}\cdot\theta) \quad \text{and} \quad \phi \times (1_{[\![0,\tau_n[\![\times E}\cdot\theta)$$

coincide with $\phi\times\theta$ stopped respectively at τ_n and before it.

Sufficiency. If $n_1 < n_2$, then by the proof above the process

$$\phi \times (1_{[\![0,\tau_{n_1}]\!]\times E}\cdot\theta) \quad (\text{respectively, } \phi \times (1_{[\![0,\tau_{n_1}[\![\times E}\cdot\theta)\ )$$

coincides with the process

$$\phi \times (1_{[\![0,\tau_{n_2}]\!]\times E} \cdot \theta) \quad (\text{respectively} \quad \phi \times (1_{[\![0,\tau_{n_2}[\![\times E} \cdot \theta) \quad)$$

stopped at τ_{n_1} (respectively before τ_{n_1}). Therefore there exists $X \in S$ such that for each $n \in \underline{N}$

$$X^{\tau_n} = \phi \times (1_{[\![0,\tau_n]\!]\times E} \cdot \theta) \quad (\text{respectively} \quad X^{\tau_n-} = \phi \times (1_{[\![0,\tau_n[\![\times E} \cdot \theta)).$$

We must prove that $X = \phi \times \theta$. First of all, reasoning analogous to [4] shows that there exists a strictly positive predictable process K such that for all $n \in \underline{N}$

$$(K \otimes 1)\phi \in L^{1,0}(1_{[\![0,\tau_n]\!]\times E} \cdot \theta)$$

$$(\text{respectively} \quad (K \otimes 1)\phi \in L^{1,0}(1_{[\![0,\tau_n[\![\times E} \cdot \theta) \quad),$$

and let V be a strictly positive function belonging to $L^{1,0}(\theta)$. Then, if $\psi \in L^{1,0}(\theta)$, analogously to the proof of necessity for all $n \in \underline{N}$

$$\psi \in L^{1,0}(1_{[\![0,\tau_n]\!]\times E} \cdot \theta) \quad (\text{respectively} \quad \psi \in L^{1,0}(1_{[\![0,\tau_n[\![\times E} \cdot \theta) \;)$$

and

$$\psi \times (1_{[\![0,\tau_n]\!]\times E} \cdot \theta) = (\psi \times \theta)^{\tau_n}$$

$$(\text{respectively} \quad \psi \times (1_{[\![0,\tau_n[\![\times E} \cdot \theta) = (\psi \times \phi)^{\tau_n-} \quad).$$

Now let $V' = V + (K\otimes 1)|\phi|$. Define the L^0-valued measure θ' so that for $\psi \in \tilde{P}_{V'}$

$$\psi \times \theta' = \psi \times (1_{[\![0,\tau_1]\!]\times E} \cdot \theta) + \sum_{i=2}^{\infty} \psi \times (1_{]\!]\tau_{i-1},\tau_i]\!]\times E} \cdot \theta)$$

(respectively,

$$\psi \times \theta' = \psi \times (1_{[\![0,\tau_1[\![\times E} \cdot \theta) + \sum_{i=2}^{\infty} \psi \times (1_{[\![\tau_{i-1},\tau_i[\![\times E} \cdot \theta)) .$$

Then $\psi \times \theta' = \psi \times \theta$ for $\psi \in \tilde{P}_{V'} \cap L^{1,0}(\theta)$ so that $\theta' = \theta$ and $\tilde{P}_{V'} \subset L^{1,0}(\theta)$. Therefore $(K \otimes 1)\phi \in L^{1,0}(\theta)$ and $\phi \in L^0_\sigma$ so that $\phi \times \theta$ is defined as a formal semimartingale. Now if $H \in P$ is such that H/K is bounded then for every $t \in R_+$

$$(\phi \times \theta)_t(H) = \theta_t[(H \otimes 1)\phi]$$

and by construction

$$H \cdot X_t = \theta_t[(H \otimes 1)\phi]$$

so that $\phi \times \theta = X$ and $\phi \in \hat{L}^0(\theta)$. ///

Now we need the following characterization of separable subsets of H^2.

<u>LEMMA 3</u>. Let U be a set of semimartingales which is separable in H^2. Then there exists a sequence of semimartingales $X = (X^n)_{n \in \underline{N}}$ with an increasing $X^1 \in P \cap V^2$, $X^n \in L^2$ for $n \geq 2$ and $\langle X^n, X^m \rangle \equiv 0$ for $n \neq m$, such that for every $Y \in U$ $Y = H \times X$ with $H = (H^1, \ldots) \in L^{1,2}(X)$, H^n being unique up to null sets respectively in measures $P \times dX^1$ and $P \times d\langle X^n \rangle$ for $n \geq 2$.

<u>Proof</u>. For every $X \in H^2$ consider its canonical decomposition $X = A + M$, where $A \in P \cap V^2$ and $M \in L^2$. Then the mappings from H^2 respectively into $P \cap V^2$ and L^2, which are induced by it,

are continuous, and corresponding sets U_V and U_L are also separable. For some countable dense subset $S \subset U_V$ there exists a predictable bounded increasing process X^1 such that every $A \in S$ is P-a.s. absolutely continuous with respect to X^1. But then every $A \in U_V$ is also P-a.s. absolutely continuous with respect to X^1 and is represented as $A = H^1 \cdot X^1$ with a predictable function H^1 which is obviously unique up to null sets in the measure $P \times dX^1$.

Now for a countable dense subset of U_L the sub-σ-algebra $\mathcal{G} \subset \mathcal{F}$, generated by it, is separable and every process from U_L is measurable with respect to the completion of $\mathcal{G}$ by P-null sets from $\mathcal{F}$. There exists a sequence of reciprocally orthogonal bounded martingales X^i, $i \geq 2$, such that for stochastic integrals of the form $H^{ij} \cdot X^i$ with $i \geq 2$, $j \in \underline{N}$ and some $\mathcal{P}$-measurable $H^{ij} \in L^{1,2}(X^i)$ values $H^{ij} \cdot X^i_\infty$ constitute a complete system in the subspace of random variables ξ from $L^2(\mathcal{G})$ for which P-a.s. $E[\xi \mid \mathcal{F}_0] = 0$. Now let H^i for $i \geq 2$ be predictable versions of derivatives $d\langle M, X^i\rangle / d\langle X^i\rangle$. Then the stochastic integral $H^i \cdot X^i$ is defined for each $n \in \underline{N}$. Now we prove that $H \in L^{1,2}(X)$. Consider finite-dimensional predictable functions K that are such that for each of them there exists $N \in \underline{N}$ with $K^i \equiv 0$ for $i > N$, and let $K^i \cdot X^i$ be defined for all $i \leq N$ and be elements of $\mathcal{H}^2$. Then the process $|K^1| \cdot X^1$ is an element of $\mathcal{V}^2$ and the process $\sum_{i=2}^{N} (K^i)^2 \cdot \langle X^i\rangle$ is an element of $\mathcal{V}^1$, and functions of the form K constitute an integration lattice.

Now let K be infinite-dimensional and such that, as before,

$|K^1| \cdot X^1$ is an element of V^2 and $\sum_{i=2}^{N} (K^i)^2 \cdot \langle X^i \rangle$ is an element of V^1. Then for any sequence $(L_n)_{n \in \underline{N}}$ of finite-dimensional functions with $|L_n| \leq |K|$ converging to K, the sequence of stochastic integrals $L_n \times X$ is fundamental in H^2 and has a limit which by mixing does not depend on the choice of a sequence so that $K \in L^{1,2}(X)$ and $K \times X$ is defined as this limit. Now

$$\left\langle M, \sum_{i=2}^{N} H^i \cdot X^i \right\rangle = \sum_{i=2}^{N} H^i \cdot \langle M, X^i \rangle$$

$$= \sum_{i=2}^{N} (H^i)^2 \cdot \langle X^i \rangle$$

$$= \left\langle \sum_{i=2}^{N} H^i \cdot X^i \right\rangle ,$$

which means that $\sum_{i=2}^{N} H^i \cdot X^i$ is the orthogonal projection of M to the stable subspace generated by all X^i with $i \leq N$. Therefore for every $N \in \underline{N}$

$$\sum_{i=2}^{N} (H^i)^2 \cdot \langle X^i \rangle \preceq \langle M \rangle ,$$

where $\preceq$ denotes the strong majorization (in the sense of the measure generated by increments) by the right-hand term for increasing processes, and hence

$$\sum_{i=2}^{N} (H^i)^2 \cdot \langle X^i \rangle \preceq \langle M \rangle$$

by the preceding $H \in L^{1,2}(X)$ and the stochastic integral $H \times X$ is defined. Put $\tilde{H} = H - (H^1, 0, \ldots)$ and prove that $M = \tilde{H} \times X$. Indeed, $M - \tilde{H} \times X \in L^2$ is orthogonal to each martingale X^i,

$i \geq 2$, and by completeness for every $\varepsilon > 0$ there exists a martingale $K \times X$ with a finite-dimensional K and $K^1 \equiv 0$ such that $E[M - (\tilde{H}+K) \times X]_\infty^2 < \varepsilon$ but

$$E[M - (\tilde{H}+K) \times X]_\infty^2 = E\langle M - (\tilde{H}+K) \times X\rangle_\infty < \varepsilon$$

so that $E\langle M - \tilde{H} \times X\rangle_\infty < \varepsilon$ and by arbitrary smallness $E\langle M - \tilde{H} \times X\rangle_\infty = 0$ and $M = \tilde{H} \times X$.

Finally if $K \in L^{1,2}(X)$ and $K \times X = Y$, then $K^1 \cdot X^1 = A$ and $\tilde{K} \times X = M$, where $Y = A + M$ is the canonical decomposition and by the proof above $K^1 = H^1$, $P \times dX^1$ - a.e. and $\tilde{K} \times X = \tilde{H} \times X$. But then for every $i \geq 2$ $\langle \tilde{K} \times X, X^i\rangle = \langle \tilde{H} \times X, X^i\rangle$ that is $K^i \cdot \langle X^i\rangle = H^i \cdot \langle X^i\rangle$ and H^i for $i \geq 2$ are unique up to null sets respectively in the measure $P \times d\langle X^i\rangle$. ///

We shall need the following construction generating the operator $[\cdot,\cdot]$ for semimartingales. Indeed, let $\theta^{(1)}$ and $\theta^{(2)}$ be L^0-valued measures respectively on $(\Omega \times R_+ \times E^{(1)}, \mathcal{P} \otimes \mathcal{E}^{(1)})$ and $(\Omega \times R_+ \times E^{(2)}, \mathcal{P} \otimes \mathcal{E}^{(2)})$. Assume both finite to begin with. Consider the set of functions of the form

$$f(\omega,t,u_1,u_2) = \sum_{i=1}^{n} \sum_{j=1}^{m} f_{ij}(\omega,t)\, 1_{B_i^{(1)}}(u_1)\, 1_{B_j^{(2)}}(u_2) \tag{4}$$

for bounded $\mathcal{P}$-measurable functions f_{ij} on $\Omega \times R_+$ and respectively $\mathcal{E}^{(1)}$- and $\mathcal{E}^{(2)}$-measurable sets

$$B_i^{(1)} \text{ and } B_j^{(2)}, \qquad i = 1,\dots,n, \qquad j = 1,\dots,m,$$

$$B_i^{(1)} \cap B_k^{(1)} = \emptyset \quad \text{and} \quad B_j^{(2)} \cap B_\ell^{(2)} = \emptyset$$

$$\text{for } i \neq k \text{ and } j \neq \ell,$$

$$\bigcup_{i=1}^{n} B_i^{(1)} = E^{(1)} , \qquad \bigcup_{j=1}^{m} B_j^{(2)} = E^{(2)} . \tag{5}$$

It is an integration lattice on $(\Omega \times R_+ \times E^{(1)} \times E^{(2)}, \ \mathcal{P} \otimes \mathcal{E}^{(1)} \otimes \mathcal{E}^{(2)})$ and for f of the form (4) we define

$$[\theta^{(1)},\theta^{(2)}]_t(f) = \sum_{i=1}^{n} \sum_{j=1}^{m} \left| 1_{B_i^{(1)}} \times \theta^{(1)}, \ 1_{B_j^{(2)}} \times \theta^{(2)} \right|_t (f_{ij}) . \tag{6}$$

By passing to a common fine division we can see easily that (6) does not depend on the choice of the division (5) and by Theorem 2.25 of [1] defines $[\theta^{(1)},\theta^{(2)}]$ as an L^0-valued finite measure on $(\Omega \times R_+ \times E^{(1)} \times E^{(2)}, \ \mathcal{P} \otimes \mathcal{E}^{(1)} \otimes \mathcal{E}^{(2)})$.

In the general case let there exist strictly positive functions $V^{(1)}$ and $V^{(2)}$, respectively, $\mathcal{P} \otimes \mathcal{E}^{(1)}$- and $\mathcal{P} \otimes \mathcal{E}^{(2)}$-measurable, such that the measures $V^{(1)} \cdot \theta^{(1)}$ and $V^{(2)} \cdot \theta^{(2)}$ are finite. Then we define

$$[\theta^{(1)},\theta^{(2)}]_t(f) = [V^{(1)} \cdot \theta^{(1)}, \ V^{(2)} \cdot \theta^{(2)}]_t \left(\frac{f}{V^{(1)}(\cdot,u_1)V^{(2)}(\cdot,u_2)} \right)$$

with $u_1 \in E_1$ and $u_2 \in E_2$, and obviously this does not depend on the choice of functions $V^{(1)}$ and $V^{(2)}$ and defines $[\theta^{(1)},\theta^{(2)}]$ as an L^0-valued σ-finite measure on $(\Omega \times R_+ \times E^{(1)} \times E^{(2)}, \ \mathcal{P} \otimes \mathcal{E}^{(1)} \otimes \mathcal{E}^{(2)})$. If $\theta^{(1)} \in \tilde{S}_\sigma^{p_1}$ and $\theta^{(2)} \in \tilde{S}_\sigma^{p_2}$ for $p_1 > 0$ and $p_2 > 0$ then $[\theta^{(1)},\theta^{(2)}] \in \tilde{S}_\sigma^{p}$ for $p = p_1 p_2 / (p_1+p_2)$. In particular, if $E^{(1)} = E^{(2)}$ and $\mathcal{E}^{(1)} = \mathcal{E}^{(2)}$, then we can define the measure $[\theta] = [\theta,\theta]$. If $\theta^{(1)} \in M - \tilde{S}_\sigma^{p_1}$ and $\theta^{(2)} \in M - \tilde{S}_\sigma^{p_2}$ with $p = p_1 p_2 / (p_1+p_2) \geq 1$, then by Theorem 4.14 of [1] $[\theta^{(1)},\theta^{(2)}] \in \tilde{S}^p_\sigma$ has the compensator $\langle\theta^{(1)},\theta^{(2)}\rangle \in PV - \tilde{S}^p_\sigma$.

Henceforth we shall assume permanently that $(E,\mathcal{E})$ is a Lusin space with its Borel σ-algebra. For this case with use of Lemma 3 we can obtain the following strengthening of Theorem 4.14 of [1] for $\theta \in \tilde{S}^2_\sigma$.

<u>THEOREM 1</u>. Let $\theta \in \tilde{S}^2_\sigma$. Then there are sequences of semimartingales X^i and measures $\theta^{(i)} \in \tilde{S}^2_\sigma$ which possess the following properties:

• (i) X^1 is increasing and belongs to $\mathcal{P} \cap \mathcal{V}^2$, $X^i \in \mathcal{L}^2$ for $i \geq 2$ and $\langle x^i, x^j\rangle \equiv 0$ for $i \neq j$;

• (ii) $\theta^{(1)} \in PV - \tilde{S}^2_\sigma$ and $\theta^{(i)} \in M - \tilde{S}^2_\sigma$ for $i \geq 2$, $L^{1,2}(\theta) \subset \bigcap_{i=1}^{\infty} L^{1,2}(\theta^{(i)})$ and $\hat{L}^2(\theta) \subset \bigcap_{i=1}^{\infty} \hat{L}^2(\theta^{(i)})$, and besides that for every $\phi \in \hat{L}^2(\theta)$ and $i \in \underline{N}$, $\phi \times \theta^{(i)}$ belongs to the stable space generated by X^i;

• (iii) for every $\phi \in \hat{L}^2(\theta)$, $\phi \times \theta = \sum_{i=1}^{\infty} \phi \times \theta^{(i)}$, the series converging unconditionally in $\mathcal{S}^2$.

Besides that there exist regular transition (signed) measures ρ^i from $(\Omega \times R_+, \mathcal{P})$ to $(E,\mathcal{E})$ such that for $\phi \in \hat{L}^2(\theta)$ its sections are ρ^i-integrable almost everywhere respectively in the measures $P \times dX^1$ and $P \times d\langle X^i\rangle$ for $i \geq 2$, and moreover

$$\phi \times \theta^{(i)} = \int_E \phi\rho^i_s(\omega,du)\cdot X^i .$$

<u>Proof</u>. By Proposition 4.23 of [1] the set of processes of the form $\phi \times \theta$ for $\phi \in L^2(\theta)$, which belong to $\mathcal{S}^2$, is separable in $\mathcal{S}^2$, and by Lemma 3 there exist the required processes X^i, and projections of the given process to stable subspaces generated by X^i, which are defined in the proof of the Lemma, are continuous

linear mappings from S^2 into itself so that for $\phi \in L^{1,2}(\theta)$ we can define $\phi \times \theta^{(i)}$ as corresponding projections and $L^{1,2}(\theta) \subset L^{1,2}(\theta^{(i)})$ for every $i \in \underline{N}$, and for $\phi \in \hat{L}^2(\theta)$ we have also $\phi \in \hat{L}^2(\theta^{(i)})$ and the series $\phi \times \theta = \sum_{i=1}^{\infty} \phi \times \theta^{(i)}$ converges unconditionally in S^2 because all its terms except the first one are reciprocally orthogonal martingales.

Now we have $\theta^{(1)} \in PV - \tilde{S}^2_\sigma$ and apply Theorem 4.10 of [1] to it. By this theorem $\theta^{(1)}$ coincides in a natural way with a predictable strict random measure in its restriction to $(\Omega \times R_+ \times E, P \otimes E)$, which we shall denote also by $\theta^{(1)}$. Let V be a strictly positive function from $L^{1,2}(\theta^{(1)})$. By Theorem 4.10 of [1] the measure $\theta^{(1)}$ has the predictable variation $|\theta^{(1)}| \in \tilde{S}^0_\sigma$ with respect to which V is also integrable so that $V \times |\theta^{(1)}|$ is a predictable increasing process, which is P-a.s. equivalent to X^1 in the sense of the measure generated by increments, so that $|\theta^{(1)}| \in \tilde{S}^2_\sigma$ and $\theta^{(1)}$ admits the factorization $\theta^{(1)}(ds \times du) = \rho^1_s(du)\, dX^1_s$ with a signed transition measure ρ^1 from $(\Omega \times R_+, P)$ to (E, E) and so for $\phi \in \hat{L}^2(\theta^{(1)})$

$$\phi \times \theta^{(1)} = \int_E \phi \rho^1_s(du) \cdot X^1 .$$

For $i \geq 2$ we have $\langle \theta^{(i)}, X^i \rangle \in PV - \tilde{S}^1_\sigma$ and the factorization $\langle \theta^{(i)}, X^i \rangle (ds \times du) = \rho^i_s(du)\, d\langle X^i \rangle_s$ so that for $\phi \in \hat{L}^2(\theta^{(i)})$

$$\phi \times \langle \theta^{(i)}, X^i \rangle = \int_E \phi \rho^i_s(du) \cdot \langle X^i \rangle$$

and

$$\phi \times \theta^{(i)} = \int_E \phi \rho^i_s(du) \cdot X^i$$

because $\phi \times \theta^{(i)}$ belongs to the stable subspace generated by X^i.

///

REMARK 1. Let analogously to [2] there be a Λ-space $\Pi = \Lambda(\tilde{P}, A, \tilde{A}, \lambda)$ of $\tilde{P}$-measurable functions associated with positive increasing adapted processes A and $\tilde{A}$ and with the mapping λ from Π into the set of positive adapted processes, that is a vector space of $\tilde{P}$-measurable functions which contains some integration lattice T and for which there exists a sequence $(\tau_n)_{n\in\underline{N}}$ of stopping times with P - a.s. $\lim_{n\to\infty} \tau_n = \infty$ such that for every $n \in \underline{N}$ and $\phi \in \Pi$

$$E\left(\tilde{A}_{\tau_n^-} \int_{[0,\tau_n[} \lambda_s(\phi)\, dA_s\right) < \infty \tag{7}$$

and T is dense in Π for seminorms given by the left-hand member of (7) for $n \in \underline{N}$. Let this Λ-space be *-associated with some measure $\theta \in \tilde{S}^2_\sigma$ which means that it is defined on T in the natural way and such that for $\phi \in T$

$$E\left(\sup_{0\le s<\tau} |\phi \times \theta_s|^2\right) \le E\left(\tilde{A}_{\tau-} \int_{[0,\tau[} \lambda_s(\phi)\, dA_s\right) . \tag{8}$$

Then the integral $\phi \times \theta$ defined on T is extended uniquely up to P-indistinguishability to Π with keeping inequality (8) and is called the Λ-stochastic integral on Π with respect to θ, and on $\Pi \cap \hat{L}^2(\theta)$ it coincides obviously with the usual integral with respect to θ. Now let the operator λ be such that for any P-measurable process K on $\Omega \times R_+$ with $|K| \le 1$, $P \times dA$ - a.e. $\lambda_t((K \otimes 1)\phi) \le \lambda_t(\phi)$ and $\lim_{n\to\infty} \lambda_t(\phi/n) = 0$. Then for $\phi \in \Pi$ with

$$E\left(\tilde{A}_{t-} \int_{[0,t[} \lambda_s(\phi)\, dA_s\right) < \infty$$

for every $t \in R_+$, $\phi \times \theta$ as the Λ-stochastic integral is a limit of some sequence of integrals with respect to θ of functions from T in S^2. By [3] and [5] for each of $\theta^{(i)}$ ϕ belongs to the corresponding Λ-space with cA instead of A, where c is some absolute positive constant so that $\phi \times \theta^{(i)}$ are defined as Λ-stochastic integrals and

$$\phi \times \theta = \sum_{i=1}^{\infty} \phi \times \theta^{(i)} = \frac{d(\phi\times\theta^{(1)})}{dX^1}\cdot X^1 + \sum_{i=2}^{\infty} \frac{d\langle \phi\times\theta^{(i)}, X^i\rangle}{d\langle X^i\rangle}\cdot\langle X^i\rangle , \tag{9}$$

the series converging also unconditionally in S^2. However, we can deduce an example to show that unlike the conditions of Theorem 1, the derivatives in (9) are not always (a.e. in proper measures) absolutely convergent integrals in transition measures ρ^i. Let $E = \underline{N}$. As θ we take the sequence $Y = (Y^n)_{n\in\underline{N}}$ of semimartingales with

$$Y_t^n = \begin{cases} 0 , & t < 1 , \\ 1 , & t \geq 1 . \end{cases}$$

Then we have

$$L^{1,0}(Y) = \hat{L}^0(Y) = \left\{\phi = (\phi^1, \ldots): \phi^n \in P,\ \sum_{n=1}^{\infty} |\phi_1^n| < \infty \quad P\text{-a.s.}\right\} ,$$

$$L^{1,2}(Y) = \hat{L}^2(Y) = \left\{\phi = (\phi^1, \ldots): \phi^n \in P,\ E\left(\sum_{n=1}^{\infty} |\phi_1^n|\right)^2 < \infty\right\},$$

and for $\phi \in \hat{L}^0(Y)$

$$\phi \times Y_t = \begin{cases} 0 , & t < 1 , \\ \sum_{n=1}^{\infty} \phi_1^n , & t \geq 1 , \end{cases} \tag{10}$$

the series converging P-a.s. absolutely. Now consider the Λ-space Π with

$$A = \begin{cases} 0, & t < 1 , \\ 1, & t \geq 1 , \end{cases} \qquad \tilde{A} \equiv 1 \qquad \text{and} \qquad \Lambda_1(\phi) = \left(\sum_{n=1}^{\infty} \phi_1^n \right)^2 ,$$

where the series converges unconditionally in $L^2(\Omega, F_{1-}, P)$. For $\phi \in L^{1,2}(Y)$ we have $E(\phi \times Y_t)^2 = 0$ at $t < 1$ and

$$E(\phi \times Y_t)^2 = E\left(\sum_{n=1}^{\infty} \phi_1^n \right)^2$$

at $t \geq 1$, that is, $L^{1,2}(Y) \subset \Pi$ and Π is *-associated with Y, and for $\phi \in \Pi$ we can define the Λ-stochastic integral $\phi \times Y$ by (10) where the series converges unconditionally in $L^2(\Omega, F_{1-}, P)$ but not always P-a.s. absolutely if the σ-algebra F_{1-} is rich enough. In terms of Theorem 1 we have $Y^{(1)} = Y$ and $Y^{(i)} \equiv 0$ for $i \geq 2$, we can take

$$X^1 = \begin{cases} 0 , & t < 1 , \\ 1 , & t \geq 1 , \end{cases}$$

and the measure $\rho_1^1(\cdot)$ on $\underline{N}$ equals 1 at each point so that the integral with respect to it is reduced to the sum of the series, but in the representation (9)

$$\left.\frac{d(\phi\times Y^{(1)})}{dX^1}\right|_{t=1} = \sum_{n=1}^{\infty} \phi_1^n ,$$

the series converging not always absolutely. ///

As we have noticed above, $\tilde{S}_\sigma^q \subset \tilde{S}_\sigma^p$ for $p \le q$, and for $\theta \in \tilde{S}_\sigma^q$, $L^{1,q}(\theta) \subset L^{1,p}(\theta)$ and $\hat{L}^q(\theta) \subset \hat{L}^p(\theta)$. The following result, which is converse to the given one in some sense, plays an important role in the sequel, and is interesting in itself.

THEOREM 2. Let $\theta \in \tilde{S}_\sigma^0$. Then θ is an L^2-valued measure pre-locally, that is, for all functions $\phi \in \hat{L}^0(\theta)$ there exists a P-a.s. increasing to the infinite sequence $(\tau_{(n)})_{n\in\underline{N}}$ of stopping times such that $\phi \in \hat{L}^2(1_{[\![0,\tau_{(n)}[\![\times E} \cdot \theta)$ for all $n \in \underline{N}$.

Proof. Let V be a strictly positive $\mathcal{P}\otimes\mathcal{E}$-measurable function belonging to $L^{1,0}(\theta)$ for $\theta \in \tilde{S}_\sigma^0$. It is not difficult to prove that there exist strictly positive constants a_k, $k \in \underline{N}$, such that the function

$$\tilde{V} = \left\{a_1 \, 1_{[0,1]\times E} + \sum_{k=2}^{\infty} a_k \, 1_{]k-1,k]\times E}\right\} V$$

belongs to $L^{1,0}(\theta)$ on all $(\tilde{\Omega},\tilde{\mathcal{P}})$ at once or, as we shall say, belongs to $L^{1,0}(\theta)$ globally.

Now let $\phi \in \hat{L}^0(\theta)$. By the above there exists a strictly positive $\mathcal{P}$-measurable function K such that $(K\otimes 1)\phi$ belongs to $L^{1,0}(\theta)$ globally. Then we can apply Theorem 2.16 of [1] for $t = \infty$ so that there exists a probability measure Q on $(\Omega,\mathcal{F})$, which is equivalent to P with a bounded $Z_\infty = dQ/dP$ and such that $W = (K\otimes 1)|\phi| + \tilde{V}$ belongs to $L^{1,2}(\theta,Q)$ globally.

Now we define stopping times

$$\tilde{\tau}_{(n)} = \inf \{t \in R_+ : |(\phi \times \theta)|_t \geq n \wedge Z_t \leq \frac{1}{n}\} ,$$

where Z is the right-continuous martingale $Z_t = E[Z_\infty \mid F_t]$. Let $Z' = 1/Z$. Then for $t < \tilde{\tau}_{(n)}$ we have $Z' < n$ and hence $|\Delta Z'| < n$. Thus for each n to the measure θ we can apply Theorem 4.21 of [1] with optional stopping times $\tilde{\tau}_{(i)}$ for $i \geqslant n$.

Fix n and decompose $1_{[\![0,\tilde{\tau}_{(n)}[\![\times E} \cdot \theta$ as the sum of measures

$$\theta' \in M - \tilde{S}^0_\sigma \quad \text{with} \quad \psi \times \theta' \in L_{loc} \quad \text{for all} \quad \psi \in L^{1,0}(\theta')$$

and

$$\theta'' \in PV - \tilde{S}^0_\sigma \quad \text{with} \quad W \in L^{1,0}(\theta'') \quad .$$

In turn, by Theorem 4.13 of [1], θ' is decomposed as the sum of measures θ^c and θ^d with $\theta^c \in L^c - \tilde{S}^0_\sigma$, $L^{1,0}(\theta) \subset L^{1,0}(\theta^c)$ and $(\psi \times \theta)^c = \psi \times \theta^c$ for all $\psi \in L^{1,0}(\theta)$. Thus it suffices to prove that each of θ^c, θ^d and θ' belongs to $\tilde{S}^2_\sigma$ with $\phi \in L^2_\sigma(\tilde{\theta})$, where as $\tilde{\theta}$ we can take any of the measures θ^c, θ^d and θ''. By taking as K from the definition $\hat{L}^2(\tilde{\theta})$ the corresponding minimum we obtain hence that

$$1_{[\![0,\tilde{\tau}_{(n)}[\![\times E} \cdot \theta \in \tilde{S}^2_\sigma \quad .$$

In turn, $\phi \times (1_{[\![0,\tilde{\tau}_{(n)}[\![\times E} \cdot \theta)$, being bounded, belongs to S^2 locally, and there exists a stopping time $\sigma_{(n)}$ with $P(\sigma_{(n)} < \tilde{\tau}_{(n)}) \leq 2^{-n}$ such that

$$\phi \times (1_{([\![0,\tilde{\tau}_{(n)}[\![\cap [\![0,\sigma_{(n)}]\!])\times E} \cdot \theta) \in S^2 \quad .$$

Now let $\tau_{(n)} = \tilde{\tau}_{(n)} \wedge (\inf_{m \geq n} \sigma_{(m)})$. Then the sequence

$(\tau_{(m)})_{n\in\underline{N}}$ increases P - a.s. to infinity and

$$\phi \times (1_{[\![0,\tau_{(n)}[\![\times E} \cdot \theta) \in S^2$$

for every $n \in \underline{N}$ so that $\phi \in \hat{L}^2(1_{[\![0,\tau_{(n)}[\![\times E} \cdot \theta)$, as required.

So at first we consider the measure θ''. By Theorem 4.10 of [1] it has the predictable variation $|\theta''|$ and then $W \times |\theta''|$ is a predictable increasing process so that there exists a P-measurable strictly positive process L such that $(L \otimes 1)W \times |\theta''| \in V^2$ and thus $\theta'' \in \tilde{S}^2_\sigma$ and $\phi \in L^2_\sigma(\theta'')$ with $\tilde{K} = KL$ instead of K.

Now we pass to the measure θ^c. Since $\theta^c \in L^c - \tilde{S}^0_\sigma$, we have $[\theta^c, \theta^c] \in PV - \tilde{S}^0_\sigma$ and there exists a P-measurable strictly positive process $L \leq 1$ such that

$$(L \otimes 1)\, W(\omega,s,u_1)\, W(\omega,s,u_2) \times |[\theta^c,\theta^c]| \in V^1 . \tag{11}$$

By virtue of $L \leqslant 1$ we have

$$(\sqrt{L} \otimes 1)\psi \times \theta^c \in L^c \qquad \text{for} \quad |\psi| \leq W$$

and by (11) $(\sqrt{L} \otimes 1)\psi \times \theta^c \in L^2$ so that $\theta^c \in \tilde{S}^2_\sigma$ and $\phi \in L^2(\theta^c)$ with $\tilde{K} = K\sqrt{L}$ instead of K, and, moreover, we have $\langle \theta^c, \theta^c \rangle = [\theta^c, \theta^c]$.

Finally we study the measure θ^d. First of all, by Proposition 4.23 of [1] there exists a sequence $(\tau_i)_{i\in\underline{N}}$ of stopping times with disjoint graphs exhausting jumps of the measure θ^d. Let $|\psi| \leq W$. Then

$$E\left\{[\psi\times\theta^d,\psi\times\theta^d]_{\tilde{\tau}(n)}\right\} = E\left\{\psi(\omega,s,u_1)\ \psi(\omega,s,u_2)\times[\theta^d,\theta^d]_{\tilde{\tau}(n)}\right\}$$

$$= E\sum_{i=1}^{\infty}\psi(\omega,s,u_1)\ \psi(\omega,s,u_2)\times\left\{1_{[\![\tau_i]\!]\times E}\cdot[\theta^d,\theta^d]\right\}_{\tilde{\tau}(n)}$$

$$= \sum_{i=1}^{\infty}E\psi(\omega,s,u_1)\ \psi(\omega,s,u_2)\times\left\{1_{[\![\tau_i]\!]\times E}\cdot[\theta^d,\theta^d]\right\}_{\tilde{\tau}(n)}$$

$$= \sum_{i=1}^{\infty}E_Q\psi(\omega,s,u_1)\ \psi(\omega,s,u_2)\times\left\{1_{[\![\tau_i]\!]\times E}\cdot[\theta^d,\theta^d]\right\}_{\tilde{\tau}(n)}\ \frac{1}{Z_{\tau_i\wedge\tilde{\tau}(n)}}$$

$$= \sum_{i=1}^{\infty}E_Q\psi(\omega,s,u_1)\ \psi(\omega,s,u_2)\times\left\{1_{[\![\tau_i]\!]\times E}\cdot[\theta^d,\theta^d]\right\}_{\tilde{\tau}(n)}\ \frac{1_{\{\tau_i<\tilde{\tau}(n)\}}}{Z_{\tau_i}}$$

$$\leq n\,E_Q\left\{[\psi\times\theta^d,\psi\times\theta^d]_{\tilde{\tau}(n)}\right\}$$

so that $\psi\times\theta^d\in\mathcal{L}^2$, $\theta^d\in\tilde{\mathcal{S}}^2_\sigma$ and $\phi\in L^2(\theta^d)$ with the given K and, moreover, the compensator r $\langle\theta^d,\theta^d\rangle$ of $[\theta^d,\theta^d]$ is defined. ///

The following result is deduced easily from Theorem 2 and is relevant to the theory of stochastic differential equations with random measures.

COROLLARY 1. Let $\theta\in\tilde{\mathcal{S}}^0_\sigma$ and $\phi\in\hat{L}^0(\theta)$. Then there exists a discrete optional random set D_0 such that if D is any discrete optional random set with $D_0\subset D$, then the measure $\tilde{\theta}=1_{D^c\times E}\cdot\theta$ belongs to $\tilde{\mathcal{S}}^2_\sigma$ and $\phi\in L^2_\sigma(\tilde{\theta})$.

Proof. Apply Theorem 2 to θ and ϕ, and let $D_0=\bigcup_{n=1}^{\infty}[\![\tau(n)]\!]$. Since D is the union of graphs of stopping times also satisfying the conclusion of Theorem 2, we can believe that $D=D_0$. By applying Proposition 4.18 of [1] we can see easily that for every $n \in \underline{N}$

$$1_{([\![0,\tau_{(n)}[\![\cap D^c)\times E} \cdot \theta \in S^2$$

and

$$\phi \in \hat{L}^2(1_{([\![0,\tau_{(n)}[\![\cap D^c)\times E} \cdot \theta) \quad .$$

But

$$[\![0,\tau_{(n)}[\![\cap D^c = [\![0,\tau_{(n)}]\!] \cap D^c$$

and easily we obtain the required. ///

REMARK 2. Corollary 1 and Theorem 4.14 of [1] imply that in the case when $(E,\mathcal{E})$ is a Lusin space with its Borel σ-algebra, the set D_0 in the conditions of Theorem 4.21 of [1] can be such that measures θ' and θ'' with their sum $\tilde{\theta}$ belong to $\tilde{S}^2_\sigma$ and

$$V \in L^2_\sigma(\tilde{\theta}) \subset L^2_\sigma(\theta') \cap L^2_\sigma(\theta'') \quad . \qquad ///$$

Now we shall prove the following theorem.

THEOREM 3. Let $\theta \in \tilde{S}^2_\sigma$. Then for every $\phi \in \hat{L}^2(\theta)$ there exists a *-associated with θ, Λ-space in which the integral $\phi \times \theta$ in the sense of [1] is the Λ-stochastic integral.

Proof. By Theorem 4.14 of [1] we have $\theta = \theta' + \theta''$ with $\theta' \in M - \tilde{S}^2_\sigma$ and $\theta'' \in PV - \tilde{S}^2_\sigma$ and $\psi \times \theta' \in L^2$ for all $\psi \in L^{1,2}(\theta')$. At first we consider the measure θ''. Let there be a strictly positive function $V \in L^{1,2}(\theta'')$. By Theorem 4.10 of [1] θ'' has the predictable variation $|\theta''| \in \tilde{S}^2_\sigma$, and if $\psi \in L^{1,2}(\theta'')$, then $\psi \in \hat{L}^2(|\theta''|)$. Let $\hat{A} = V \times |\theta''|$. Then θ'' admits the factorization $\theta''(ds \times du) = \hat{\rho}''_s(du)\, d\tilde{A}_s$ with a regular transition measure $\hat{\rho}''$ from $(\Omega \times R_+, \mathcal{P})$ to $(E,\mathcal{E})$, and for $\phi \in \hat{L}^2(\theta)$ we have $\phi \times \theta'' = \int_E \phi\hat{\rho}''_s(du) \cdot \hat{A}$. Now let $A^{(1)}$ be a process equivalent to $\hat{A}$ (in the sense of the measure generated by increments) and such that P-a.s. for every $t \in R_+$

$$\left(\int_E \phi\rho''_s \,(du)\right)^2 \cdot A_t^{(1)} < \infty ,$$

where $\rho'' = \hat{\rho}''(A'/dA^{(1)})$. Then

$$E\left\{\sup_{0\le s\le\tau} |\phi\times\theta''_s|\right\}^2 \le E\left\{\left|\int_E \phi\rho''_s(du)\right| \cdot A_{\tau-}^{(1)}\right\}^2$$

$$\le E\left\{A_{\tau-}^{(1)}\left[\left(\int_E \phi\rho''_s(du)\right)^2 \cdot A_{\tau-}^{(1)}\right]\right\} ,$$

that is, ϕ belongs to the *-associated with θ'' Λ-space $\Lambda(\tilde{P},A^{(1)},A^{(1)},\lambda)$ with $\lambda_s(\phi) = \left(\int_E \phi\rho''_s(du)\right)^2$ for $\phi \in \hat{L}^2(\theta'')$ in which for $\phi \in L^{1,2}(\theta'')$ with $(\phi\rho''_s(du))^2 \cdot A_t^{(1)} < \infty$ P-a.s. for every $t \in R_+$ the integral $\phi\times\theta$ is defined in the usual way.

Now we pass to the measure θ'. First of all we have by [2] and [5] for $\phi \in \hat{L}^2(\theta')$

$$E\left(\sup_{0\le s<\tau} |\phi\times\theta'_s|\right)^2 \le 4\,E\,\phi(s,u_1)\,\phi(s,u_2) \times \langle\theta',\theta'\rangle_{\tau-}$$

$$+ 4\,E\,\phi(s,u_1)\,\phi(s,u_2) \times [\theta^j,\theta^j]_{\tau-},$$

where θ^j is the measure, which under integration with respect to it of $\psi \in L^{1,2}(\theta')$ gives the purely discontinuous accessible part of the martingale $\psi\times\theta'$. For $\langle\theta',\theta'\rangle$ analogously to preceding, we have the factorization

$$\langle\theta',\theta'\rangle(ds\times du_1\times du_2) = \rho'_s(du_1\times du_2)\,dA'_s$$

with a regular transition measure ρ' from $(\Omega\times R_t,\, P)$ to

$(E\times E,\ \mathcal{E}\otimes\mathcal{E})$ so that for $\phi\in\hat{L}^2(\theta')$

$$\phi(s,u_1)\ \phi(s,u_2)\times\langle\theta',\theta'\rangle = \int_{E\times E}\phi(s,u_1)\ \phi(s,u_2)\ \rho_s'(du_1\times du_2)\cdot A' .$$

Besides that by Proposition 4.23 of [1] there exists a sequence $(\tau_n)_{n\in\underline{N}}$ of predictable stopping times with disjoint graphs exhausting jumps of θ^j so that for $\phi\in\hat{L}^2(\theta^j)$

$$\phi(s,u_1)\ \phi(s,u_2)\times[\theta^j,\theta^j] = \sum_{n=1}^{\infty}\left(\phi(s,u)\times 1_{[\![\tau_n]\!]\times E}\cdot\theta^j\right)^2$$

so that there exist a predictable increasing process A^j and an $L^1(P\times dA^j)$ - valued measure ρ^j on $(\Omega\times R_+\times E\times E,\ \mathcal{P}\otimes\mathcal{E}\otimes\mathcal{E})$ such that for $\phi\in\hat{L}^2(\theta^j)$

$$\phi(s,u_1)\ \phi(s,u_2)\times[\theta^j,\theta^j] = \int_{E\times E}\phi(s,u_1)\ \phi(s,u_2)\ \rho_s^j(du_1\times du_2)\cdot A^j .$$

Now let $A^{(2)}=A'+A^j$, $b'=dA'/dA^{(2)}$, $B^j=dA^j/dA^{(2)}$. Then

$$E\left(\sup_{0\le s<\tau}|\phi\times\theta_s'|\right)^2$$

$$\le E\left[\int_{E\times E}\phi(s,u_1)\ \phi(s,u_2)\left(4b_s'\rho_s'+4b_s^j\rho_s^j\right)(du_1\times du_2)\cdot A_{\tau-}^{(2)}\right],$$

that is, ϕ belongs to the *-associated with θ' Λ-space $\Lambda(\tilde{\mathcal{P}},A^{(2)},1,\mu)$ with

$$\mu_s(\phi) = \int_{E\times E}\phi(s,u_1)\ \phi(s,u_2)\left(4b_s'\rho_s'+4b_s^j\rho_s^j\right)(du_1\times du_2)$$

for $\phi\in\hat{L}^2(\theta')$ in which for $\phi\in L^{1,2}(\theta')$ the integral $\phi\times\theta$ is defined in the usual way.

At last

$$E\left(\sup_{0\le s<\tau}|\phi\times\theta_s|\right)^2 \le 2E\left(\sup_{0\le s<\tau}|\phi\times\theta''_s|\right)^2 + 2E\left(\sup_{0\le s<\tau}|\phi\times\theta'_s|\right)^2$$

$$\le 2E\left[A^{(1)}_{\tau-}\left(\lambda_s(\phi)\cdot A^{(1)}_{\tau-}\right)\right] + 2E\left(\mu_s(\phi)\cdot A^{(2)}_{\tau-}\right)$$

$$\le E\left[\left(A^{(1)}_{\tau-}\vee 1\right)\left\{\left(2b^{(1)}_s\lambda_s(\phi)+2b^{(2)}_s\mu_s(\phi)\right)\left(A^{(1)}_{\tau-}+A^{(2)}_{\tau-}\right)\right\}\right],$$

where

$$b^{(1)} = \frac{dA^{(1)}}{dA^{(1)}+dA^{(2)}} \quad\text{and}\quad b^{(2)} = \frac{dA^{(2)}}{dA^{(1)}+dA^{(2)}},$$

that is, ϕ belongs to the *-associated with θ Λ-space $\Lambda(\tilde{P},\ A^{(1)}+A^{(2)},\ A^{(1)}\vee 1,\ 2b^{(1)}\lambda+2b^{(2)}\mu)$. ///

<u>COROLLARY 2</u>. Let $\theta\in\tilde{S}^0_\sigma$. Then for every $\phi\in\hat{L}^0(\theta)$ there exists an *-associated with θ Λ-space to which ϕ belongs prelocally and the integral $\phi\times\theta$ in the sense of [1] is the Λ-stochastic integral.

<u>Proof</u>. Let according to Theorem 1 $(\tau_n)_{n\in\underline{N}}$ be an increasing sequence of stopping times with P - a.s. $\lim_{n\to\infty}\tau_n=\infty$ such that for every $n\in\underline{N}$ $\phi\in\hat{L}^2(1_{[\![0,\tau_n[\![\times E}\cdot\theta)$. By Theorem 2 for each $n\in\underline{N}$ ϕ belongs to some *-associated with $1_{[\![\tau_{n-1},\tau_n[\![\times E}\cdot\theta$ for $\tau_0=0$ Λ-space $\Lambda(\tilde{P},\ A^{(n)},\ \tilde{A}^{(n)},\ \lambda^{(n)})$. Now if

$$\lambda_t = \sum_{n=1}^{\infty}\lambda^{(n)}_t\,1_{[\tau_{n-1},\tau_n[}(t),$$

$$A_t = A^{(1)}_0 + \sum_{k=1}^{n-1}\left(A^{(k)}_{\tau_k}-A^{(k)}_{\tau_{k-1}}\right) + \left(A^{(n)}_t + A^{(n)}_{\tau_{n-1}}\right)$$

for $t\in[\tau_{n-1},\tau_n[$, $\tilde{A}_t=\tilde{A}^{(1)}_t$ for $t\in[0,\tau_1[$ and

$\tilde{A}_t = \tilde{A}^{(n-1)}_{\tau_{n-1}} \vee A^{(n)}_t$ for $n \geq 2$ for $t \in [\tau_{n-1}, \tau_n]$, then ϕ belongs to the obtained Λ-space prelocally. ///

REFERENCES

[1] Bichteler, K., and Jacod, J. "Random Measures and Stochastic Integration." *Lecture Notes in Control and Information Sci.*, pp. 1-18, vol. 49. Berlin Heidelberg New York: Springer-Verlag Inc., 1983.

[2] Métivier, M. "Stability Theorems for Stochastic Differential Equations Driven by Random Measures and Semimartingales." *Journal of Integral Equations*, vol.3, no.2 (1981): 109-133.

[3] Emery, M. "Une topologie sur l'espace des semimartingales." *Lecture Notes in Math.*, pp. 260-280, vol.721. Berlin Heidelberg New York: Springer-Verlag Inc., 1979.

[4] Dellacherie, C. "Quelques applications du lemme de Borel-Cantelli à la théorie des semimartingales. *Lecture Notes in Math.*, pp. 742-745, vol.649. Berlin Heidelberg New York: Springer-Verlag Inc., 1978.

[5] Lenglart, E., Lépingle, D., and Pratelli, M. "Présentation unifiée de certaines inégalités de la théorie des martingales. *Lecture Notes in Math.*, pp. 26-48, vol.784. Berlin Heidelberg New York: Springer-Verlag Inc., 1980.

R.Sh. LIPTSER

ON A FUNCTIONAL LIMIT THEOREM FOR FINITE STATE SPACE MARKOV PROCESSES

1

Let $X^n = (X_t^n)$ be a Markov process (for each $n \geq 1$) with finite state space $J = (a_1,...,a_k)$ and intensity matrix of transition probabilities $\Lambda^n = n\Lambda$ ($\Lambda = \Lambda_{(k\times k)}$ is a matrix with elements Λ_{ij}: $\Lambda_{ij} \geq 0$, $i \neq j$, $\Lambda_{ii} = -\sum_{j\neq i} \Lambda_{ij}$).

Let $g = g(x)$ be a measurable finite function. Define the random process $Y^n = (Y_t^n)$ with

$$Y_t^n = \sqrt{n}\int_0^t [g(X_s^n) - Eg(X_s^n)]\, ds .$$

The aim of the present paper is to show that under some intrinsic conditions the sequence Y^n, $n \geq 1$, converges weakly (in Skorohod's topology of the space $D[0,\infty)$, [1]) to the random process bW, where $W = (W_t)$ is a Wiener process, and to calculate the constant b.

The necessity for results of this type is occasioned by investigations of some problems in queueing theory.

It should be noted that the process X^n coincides in the distribution sense with the process (X_{nt}), where $X = (X_t)$ is a

Markov process with finite state space J and intensity matrix of transition probabilities Λ, and hence the processes Y^n and $\tilde{Y}^n = (\tilde{Y}^n_t)$ with

$$\tilde{Y}^n_t = \frac{1}{\sqrt{n}} \int_0^{nt} [g(X_s) - Eg(X_s)]\, ds$$

coincide in the distribution sense. If X is a stationary ergodic process (in this case X^n is the same for each $n \geq 1$), the conditions of weak convergence of the sequence $\tilde{Y}^n$, $n \geq 1$, to bW (and hence Y^n, $n \geq 1$ to bW) are established in [2] - [4].

In the present paper, we give the condition of weak convergence of the sequence Y^n, $n \geq 1$, to bW without stationarity assumption on X^n, $n \geq 1$.

2

To formulate the main result we need the following auxiliary fact.

LEMMA 1. Let the equations

$$\sum_{i=1}^{k} \pi_i \Lambda_{ij} = 0, \quad j = 1,\dots,k, \qquad \sum_{i=1}^{k} \pi_i = 1 \tag{1}$$

have the unique positive solution $(\pi_i > 0,\ i = 1,\dots,k)$. Then the matrix $\tilde{\Lambda}_{(k-1)\times(k-1)}$ with elements

$$\tilde{\Lambda}_{ij} = \Lambda_{ij} - \Lambda_{kj}$$

is nonsingular.

Proof. Let us show that the equations (1) have a unique solution. Indeed, if $(x_1,\dots,x_k)$ is some solution of (1), then

$(y_1,\ldots,y_k)$ with $y_i = (1+c)^{-1}(c\pi_i + x_i)$, $i = 1,\ldots,k$ is the solution of (1) also for any $c > 0$. Evidently, it is possible to choose constant c such that $y_i > 0$, $i = 1,\ldots,k$. From this it follows that $y_i = \pi_i = x_i$, $i = 1,\ldots,k$.

The definition of $\tilde{\Lambda}$ and (1) imply that

$$\sum_{i=1}^{k-1} \pi_i \tilde{\Lambda}_{ij} = -\Lambda_{kj} , \qquad j = 1,\ldots,k-1 , \tag{2}$$

i.e., the equations (2) have a solution. Let $(\tilde{x}_1,\ldots,\tilde{x}_{k-1})$ be some solution of (2). Then $(x_1,\ldots,x_k)$ with $x_i = \tilde{x}_i$, $i \le k-1$, $x_k = 1 - \sum_{i \le k-1} \tilde{x}_i$ is a solution of (1) and by uniqueness of a solution of the equations (1) we have $x_i = \pi_i$, $i = 1,\ldots,k$. Hence $\tilde{x}_i = \pi_i$, $i = 1,\ldots,k-1$. It means that the equations (2) have a unique solution and consequently the matrix $\tilde{\Lambda}$ is nonsingular.

3

Let $\tilde{\Lambda}$ be a nonsingular matrix and let $(B_1,\ldots,B_{k-1})$ be a solution of the equations

$$\sum_{j=1}^{k-1} \tilde{\Lambda}_{ij} B_j = g(a_i) - g(a_k) , \qquad i = 1,\ldots,k-1 . \tag{3}$$

Now we formulate the main result.

THEOREM. Let the equation (1) have the unique positive solution $(\pi_1,\ldots,\pi_k)$.

Then the sequence Y^n, $n \ge 1$, converges weakly in topology of the space $D[0,\infty)$ to the random process bW; $W = (W_t)$ a Wiener process,

$$b^2 = \sum_{j=1}^{k-1} B_j^2 \sum_{i=1}^{k} |\Lambda_{ij}|\pi_i - \sum_{\substack{i,j=1 \\ i\neq j}}^{k-1} B_i B_j (\Lambda_{ij}\pi_i + \Lambda_{ji}\pi_j) \quad . \tag{4}$$

The proof of this Theorem makes use of some auxiliary results which we give below.

4

Let $X = (X_t)$ be a Markov process with trajectories in $D[0,\infty)$ and infinite state space $J = (\alpha,\beta,\gamma,\dots)$ and intensity matrix of transition probabilities $\|\lambda_{\alpha\beta}(t)\|$. The functions $\lambda_{\alpha\beta}(t)$, $\alpha,\beta \in J$ are continuous in t uniformly in $\alpha,\beta \in J$ and $|\lambda_{\alpha\beta}(t)| \le K$,

$$\lambda_{\alpha\beta}(t) \geq 0 \quad , \qquad \alpha \neq \beta$$

$$\lambda_{\alpha\alpha}(t) = -\sum_{\alpha\neq\beta} \lambda_{\alpha\beta}(t) \quad .$$

We shall assume that the process X is given on some probability space (Ω,F,P) and is adapted with respect to filtration $F = (F_t)$ of σ-algebras $F_t \subseteq F$, $t \ge 0$.

Denote $X_\beta(t) = I(X_t = \beta)$, $\beta \in J$. The process $X_\beta = (X_\beta(t))$ has a semimartingale representation

$$X_\beta(t) = X_\beta(0) + \int_0^t \sum_{\gamma\in J} X_\gamma(s)\, \lambda_{\gamma\beta}(s)\, ds + M_\beta(t) \tag{5}$$

with the square integrable purely discontinuous martingale

$M_\beta = (M_\beta(t))$ with respect to F (see Lemma 9.2 of [5]).

Denote by $\langle M_\beta \rangle$ and $\langle M_\alpha, M_\beta \rangle$ the quadratic characteristics of martingales M_β and M_α, M_β, respectively.

The following result is of independent interest.

LEMMA 2. The quadratic characteristics $\langle M_\beta \rangle$ and $\langle M_\alpha, M_\beta \rangle$ have the following representations:

$$\langle M_\beta \rangle_t = \int_0^t \sum_{\gamma \in J} X_\gamma(s) \, |\lambda_{\alpha\beta}(s)| \, ds \, , \qquad \beta \in J$$

$$\langle M_\beta, M_\beta \rangle_t = -\int_0^t [X_\alpha(s)\lambda_{\alpha\beta}(s) + X_\beta(s)\lambda_{\beta\alpha}(s)] \, ds \, , \qquad \alpha, \beta \in J \quad \alpha \neq \beta \, .$$

Proof. We use the fact that the quadratic variation $[M_\beta, M_\beta]$ of a square integrable martingale M_β has the Doob-Meyer decomposition $[M_\beta, M_\beta] = \langle M_\beta \rangle + N_\beta$ with a local martingale N_β ([6, p. 33, Section d and Lemma 2.26]), i.e., $[M_\beta, M_\beta]$ is a special semimartingale. On the other hand, for a purely discontinuous martingale we have

$$[M_\beta, M_\beta]_t = \sum_{s \le t} (M_\beta(s) - M_\beta(s-))^2 \, .$$

The representation (5) implies that

$$\begin{aligned} (M_\beta(s) - M_\beta(s-))^2 &= (X_\beta(s) - X_\beta(s-))^2 \\ &= X_\beta(s) - 2\, X_\beta(s-)\, X_\beta(s) + X_\beta(s-) \\ &= (1 - 2X_\beta(s-))(X_\beta(s) - X_\beta(s-)) \, . \end{aligned}$$

Using this fact, we have by virtue of (5)

$$[M_\beta, M_\beta]_t = A_\beta(t) + \tilde{N}_\beta(t) , \qquad (6)$$

where

$$A_\beta(t) = \int_0^t \sum_{\gamma \in J} X_\gamma(s) \lambda_{\gamma\beta}(s)[1 - 2X_\beta(s)] ds$$

$$= \int_0^t \sum_{\gamma \in J} X_\gamma(s) |\lambda_{\alpha\beta}(s)| ds ,$$

$$\tilde{N}_\beta(t) = \int_0^t [1 - 2X_\beta(s-)] dM_\beta(s) .$$

From the definition of $\tilde{N}_\beta(t)$ it follows that the process $\tilde{N}_\beta$ is a local martingale with respect to F. It means that the special semimartingale $[M_\beta, M_\beta]$ has the representation (6) with the predictable process $A_\beta = (A_\beta(t))$ and the local martingale $\tilde{N}_\beta$.

The processes A_β and $\langle M_\beta \rangle$ are indistinguishable by virtue of a uniqueness representation of the type (6) for special semimartingales ([6, Proposition 2.14]).

The second statement of our Lemma is proved in the analogous way. Indeed, on the one hand we have the representation

$$[M_\alpha, M_\beta] = \langle M_\alpha, M_\beta \rangle + N_{\alpha\beta}$$

with a local martingale $N_{\alpha\beta}$ and on the other hand by virtue of (5) for $\alpha \neq \beta$

$$[M_\alpha, M_\beta]_t = \sum_{s \le t} [X_\alpha(s) - X_\alpha(s-)][X_\beta(s) - X_\beta(s-)]$$

$$= - \sum_{s \le t} [X_\alpha(s-)(X_\beta(s) - X_\beta(s-)) + X_\beta(s-)(X_\alpha(s) - X_\alpha(s-))]$$

$$= A_{\alpha\beta}(t) + \tilde{N}_{\alpha\beta}(t) ,$$

where

$$A_{\alpha\beta}(t) = -\int_0^t [X_\alpha(s) \sum_{\gamma\in J} X_\gamma(s)\lambda_{\gamma\beta}(s) + X_\beta(s) \sum_{\gamma\in J} X_\gamma(s)\lambda_{\gamma\alpha}(s)]\, ds$$

$$= -\int_0^t [X_\alpha(s)\lambda_{\alpha\beta}(s) + X_\beta(s)\lambda_{\beta\alpha}(s)]\, ds$$

and

$$\tilde{N}_{\alpha\beta}(t) = -\int_0^t X_\alpha(s-)\, dM_\beta(s) - \int_0^t X_\beta(s-)\, dM_\alpha(s) .$$

5

Denote

$$p_j^n(t) = P(X_t^n = j) , \qquad x_j^n(t) = I(X_t^n = j) ,$$

$$j = 1,\ldots,k .$$

It is well known that $p_j^n(t)$, $j = 1,\ldots,k$ are defined by Kolmogorov's equations (see [7, Chapter VII, 2])

$$p_j^n(t) = p_j^n(0) + n\int_0^t \sum_{i=1}^k p_i^n(s)\, \Lambda_{ij}\, ds , \qquad j = 1,\ldots,k . \tag{7}$$

By the definition of the matrix $\tilde{\Lambda}$ it follows that

$$p_j^n(t) = p_j^n(0) + n\int_0^t \left[\sum_{i=1}^{k-1} p_i^n(s)\tilde{\Lambda}_{ij} + \Lambda_{kj}\right] ds , \qquad j = 1,\ldots,k-1$$

$$p_k^n(t) = 1 - \sum_{j=1}^{k-1} p_j^n(t) . \tag{8}$$

Without loss of generality, it is possible to assume that the Markov process X^n has the trajectories in $D[0,\infty)$, is defined on some probability space $(\Omega,\mathcal{F},P)$ and is adapted with respect to filtration $F = (F_t)$ of σ-algebras $F_t \subseteq F$, $t \geq 0$. Then by the representation (5) we have

$$X_j^n(t) = X_j^n(0) = n\int_0^t \sum_{i=1}^k X_i^n(s)\Lambda_{ij}\, ds + M_j^n(t), \qquad j = 1,\dots,k, \tag{9}$$

where $M_j^n = (M_j^n(t))$ is a square integrable martingale with respect to F with quadratic characteristics (see Lemma 2)

$$\langle M_j^n\rangle_t = n\int_0^t \sum_{i=1}^k X_i^n(s)|\Lambda_{ij}|\, ds, \qquad j = 1,\dots,k,$$

$$\langle M_i, M_j\rangle_t = -n\int_0^t [X_i^n(s)\Lambda_{ij} + X_j^n(s)\Lambda_{ji}]\, ds, \qquad i \neq j. \tag{10}$$

<u>LEMMA 3</u>. Let the equations (1) have the unique positive solution $(\pi_1,\dots,\pi_k)$. Then for each $t > 0$

$$P - \lim_n \int_0^t X_j^n(s)\, ds = \pi_j t, \qquad j = 1,\dots,k.$$

<u>Proof</u>. By the definition of the matrix $\tilde{\Lambda}$ it follows that (9) is equivalent to the representation

$$X_j^n(t) = X_j^n(0) + n\int_0^t \left[\sum_{i=1}^{k-1} X_i^n(s)\tilde{\Lambda}_{ij} + \Lambda_{jk}\right] ds + M_j^n(t), \qquad j = 1,\dots,k-1,$$

$$X_k^n(t) = 1 - \sum_{j=1}^{k-1} X_j^n(t). \tag{11}$$

Hence

$$\sum_{i=1}^{k-1} \frac{1}{t}\left(\int_0^t X_i^n(s)ds\right)\tilde{\Lambda}_{ij} + \Lambda_{jk} = \frac{1}{nt}[X_j^n(t) - X_j^n(0) - M_j^n(t)]$$

and the required statement will hold by virtue of Lemma 1 (see the equations (2)) if

$$\frac{1}{nt}[X_j^n(t) - X_j^n(0) - M_j^n(t)] \rightarrow 0$$

in probability as $n \rightarrow \infty$. To this end, it is sufficient to show that

$$\lim_n \frac{1}{(nt)^2} E(M_j^n(t))^2 = 0 \tag{12}$$

because $0 \leq X_j^n(t) \leq 1$, $t \geq 0$. To prove (12) we use the well-known property of square integrable martingales: $E(M_j^n(t))^2 = E\langle M_j^n\rangle_t$. Then by virtue of (10) we have

$$\frac{1}{(nt)^2} E(M_j^n(t))^2 = \frac{1}{nt^2} \int_0^t \sum_{i=1}^k p_j^n(s)|\Lambda_{ij}| \, ds \rightarrow 0 , \quad n \rightarrow \infty .$$

6

PROOF OF THE THEOREM

It should be noted that

$$\begin{aligned} g(x_s^n) - Eg(X_s^n) &= \sum_{i=1}^k [X_i^n(s) - p_i^n(s)]g(a_i) \\ &= \sum_{i=1}^{k-1} [X_i^n(s) - p_i^n(s)][g(a_i) - g(a_k)] . \end{aligned}$$

From that and (3) we obtain

$$g(X_s^n) - Eg(X_s^n) = \sum_{i,j=1}^{k-1} [X_i^n(s) - p_i^n(s)] \tilde{\Lambda}_{ij} B_j .$$

Therefore

$$Y_t^n = \sqrt{n} \int_0^t [g(X_s^n) - Eg(X_s^n)]\, ds$$

$$= \sqrt{n} \int_0^t \sum_{i,j=1}^{k-1} [X_i^n(s) - p_i^n(s)]\, \tilde{\Lambda}_{ij} B_j \, ds \quad . \tag{13}$$

Denote

$$C_j^n(t) = X_j^n(t) - X_j^n(0) - [p_j^n(t) - p_j^n(0)] \quad ,$$

$$A_t^n = \frac{1}{\sqrt{n}} \sum_{j=1}^{k-1} B_j \, C_j^n(t) \quad , \tag{14}$$

$$M_t^n = \frac{1}{\sqrt{n}} \sum_{j=1}^{k-1} B_j \, M_j^n(t) \quad .$$

From (11) and (8) it follows that

$$n \int_0^t \sum_{i=1}^{k-1} [X_i^n(s) - p_i^n(s)]\, \tilde{\Lambda}_{ij} \, ds = C_j^n(t) - M_j^n(t) \quad . \tag{15}$$

Consequently, (13) - (15) imply the following decomposition for Y_t^n:

$$Y_t^n = A_t^n - M_t^n \quad .$$

To prove the required statement, it is sufficient to show that the sequence $-M^n$, $n \geq 1$, converges weakly to bW because $\sup_{t \geq 0} |A_t^n| \leq n^{-\frac{1}{2}}$ const. (see Lemma 5 of [8]).

For each $n \geq 1$ the process M^n is a square integrable martingale with

$$\sup_{t>0} |M_t^n - M_{t-}^n| \leq n^{-\frac{1}{2}} \text{ const.} \tag{16}$$

and

$$\langle M^n\rangle_t = \frac{1}{n}\sum_{j=1}^{k-1} B_j \langle M_j^n\rangle_t + \frac{1}{n}\sum_{\substack{i,j=1\\ i\neq j}}^{k-1} B_i B_j \langle M_i^n, M_j^n\rangle_t$$

$$= \sum_{j=1}^{k-1} B_j \sum_{i=1}^{k} \int_0^t X_i^n(s)\, ds\, |\Lambda_{ij}| \tag{17}$$

$$- \sum_{\substack{i,j=1\\ i\neq j}}^{k-1} B_i B_j \left[\int_0^t X_i^n(s)\, ds\, \Lambda_{ij} + \int_0^t X_j^n(s)\, ds\, \Lambda_{ji}\right].$$

The inequality (16) implies the functional Lindeberg condition (see [8, Theorem 2, Corollary 2]) for the sequence M^n, $n \geq 1$. Therefore by Corollary 2 to Theorem 2 of [8] the convergence $\langle M^n\rangle \to b^2 t$ in probability for any $t > 0$ implies the weak convergence $-M^n$, $n \geq 1$, to bW. Here it should be noted that $P - \lim_n \langle M^n\rangle_t = b^2 t$ holds by Lemma 3 and by representation (17) for $\langle M^n\rangle_t$.

REFERENCES

[1] Lindvall, T. "Weak Convergence of Probability Measures and Random Functions in the Functional Space." *J. Appl. Probab.*, 10 (1983): 109-121.

[2] Gordin, M.I., and Lifšič, B.A. "The Central Limit Theorem for Stationary Markov Processes." *Soviet Math. Doklady*, vol.19, no.2 (1978): 392-394.

[3] Bhattacharya, R.N. "On the Functional Central Limit Theorem and the Law of the Iterated Logarithm for Markov Processes." *Z. Wahrscheinlichkeitstheorie und Verw. Gebiete*, vol.60 (1982): 185-201.

[4] Touati, A. "Thèorèmes de limite centrale functionneles pour les process de Markov." *Ann. Inst. H. Poincaré*, vol.19, no.1 (1983): 43-55.

[5] Liptser, R.Sh., and Shiryayev, A.N. *Statistics of Random Processes*. Vol.I. Berlin Heidelberg New York: Springer-Verlag Inc., 1977.

[6] Jacod, J. Calcul stochastique et problemès des martingales. *Lecture Notes in Math.*, vol.714. Berlin Heidelberg New York: Springer-Verlag Inc., 1979.

[7] Gikhman, I.I., and Skorokhod, A.V. *The Theory of Stochastic Processes*. Berlin Heidelberg New York: Springer-Verlag Inc., 1974-79.

[8] Liptser, R.Sh., and Shiryaev, A.N. "A Functional Central Limit Theorem for Semimartingales." *Theory Probab. Applications*, vol.25, no.4 (1980): 667-688.

V.K. MALINOVSKIJ

ON SOME ASYMPTOTIC RELATIONS AND IDENTITIES FOR HARRIS RECURRENT MARKOV CHAINS

Let $X_1, X_2, \ldots$ be a random regenerative sequence on a sample space Ω. As a popular example of such a sequence we can mention Markov chains with a recurrent state. Many problems in probability theory and mathematical statistics may be investigated by a regenerative technique, and we often need some relations which connect such characteristics of $X_1, X_2, \ldots$ as

$$D_\mu\left(\sum_{i=1}^{n} f(X_i)\right) ,$$

as $n \to \infty$, f is some measurable function on the state space, μ is some initial distribution, with the characteristics on the regeneration period.

In the present paper, some asymptotic relations of this type with the rate of convergence estimates are proved when $X_1, X_2, \ldots$ is a Harris recurrent Markov chain. The proof is based on the regeneration of such a chain as presented in [15] and on the limit theorems for sums of independent identically distributed m-lattice random vectors.

This method was successfully used for the construction of the asymptotic expansions in the central limit theorem for a

Harris recurrent Markov chain (see [11]). It was also used to prove the formula which expresses the deficiency of the asymptotically efficient test for a simple hypothesis versus a one-sided alternative based on Harris recurrent Markov observations (see [10]). This method is an extension of the one used in [13] and its demonstration is one of the main objectives of this paper.

We should like to point out that although we consider only Markov chains in this paper, the technique can be applied to any regenerative sequence (or a process in the continuous-time case) and any sequence which allows regenerative extension (see Section 2).

≪≫

1.

Let $X_1, X_2, \ldots$ be a homogeneous Markov chain with the state space $(X, \mathcal{B})$, $\mathcal{B}$ is separable.

DEFINITION 1. The chain $X_1, X_2, \ldots$ is called Harris recurrent if there exists a σ-finite measure ϕ on $(X, \mathcal{B})$, such that $\phi(X) > 0$, and if $A \in \mathcal{B}$, $\phi(A) > 0$, then $P_x\{\bigcup_{i=1}^{\infty}(X_i \in A)\} = 1$ for all x in X.

Let $p^k(x,y)$ stand for the transition k-step density with respect to some σ-finite measure ν.

DEFINITION 2. A measurable set A is called c-set with respect to the σ-finite measure ϕ on $(X, \mathcal{B})$ if $\phi(A) > 0$ and there exists an integer $k > 0$ such that $\inf\limits_{x,y \in A \times A} p^k(x,y) > 0$.

PROPOSITION. The chain $X_1, X_2, \ldots$ is Harris recurrent if and only if there exists an integer $k > 0$, a number $\lambda > 0$, a set

$A \in B$ and a probability measure ϕ_A concentrated on A, such that

$$P_x(X_i \in A \text{ for some } i \geq 1) = 1 \quad \text{for all } x \text{ in } X$$

and

$$P_x(X_k \in A) \geq \lambda \phi_A(A) \quad \text{for all } x \in A, \ A \subset A, \ A \in B.$$

The proof of this Proposition follows directly from Theorem 2.1 of [16] and is based on the existence of c-sets for the Harris recurrent Markov chain, so we can take any c-set as A and $\phi(\cdot \cap A)/\phi(A)$ as $\phi_A(\cdot)$. Such a constructive condition on the Markov chain contained in the Proposition was already investigated in [6]. When this condition is satisfied with $A = X$, $X_1, X_2, \ldots$ turns out to be Doeblin chain (see [7, V.3.3]) and vice versa for Doeblin ergodic chain.

Let us denote by μ the initial distribution of the Harris recurrent Markov chain $X_1, X_2, \ldots$, by π the invariant distribution. Let the measure ϕ_A, c-set A and $\lambda > 0$ be taken as in the Proposition and let us assume for simplicity that $k = 1$. This assumption can easily be removed, as we shall see from the proofs.

Let $B_{1,\lambda}$, $B_{2,\lambda}, \ldots$ stand for a sequence of mutually independent random variables taking the values 1 and 0 with probabilities λ and $1-\lambda$, respectively. We denote by P_α the distribution of $X_1, X_2, \ldots$ with the initial distribution α and by $P_{\alpha,\lambda}$ the joint distribution of this chain and the above mentioned sequence of random variables defined as in [12]. Let E_α and $E_{\alpha,\lambda}$ designate the corresponding expectations. We define

$$\tau = \min\{n > 0 : X_n \in A,\ B_{n,\lambda} = 1\} ,$$

$$\Sigma_{F,n} = \frac{1}{\sqrt{n}} \sum_{i=1}^{n} [f(X_i) - E_\pi f(X_i)] ,$$

$$F = \sum_{i=1}^{\tau} [f(X_i) - E_\pi f(X_i)] ,$$

f is a measurable function on $(X, \mathcal{B})$,

$$a = E_{\phi_A,\lambda}\, \tau , \qquad \sigma_F^2 = E_{\phi_A,\lambda}\, F^2 ,$$

$$\mu_3 = E_{\phi_A,\lambda}\, F^3 , \qquad \beta = E_{\phi_A,\lambda}\, (\tau - a)F ,$$

$$\eta = E_{\mu,\lambda} \sum_{i=1}^{\tau} [f(X_i) - E_\pi f(X_i)]$$

$$+ a^{-1} E_{\phi_A,\lambda} \sum_{i=1}^{\tau} (\tau - i)[f(X_i) - E_\pi f(X_i)] .$$

THEOREM 1. Let $X_1, X_2, \ldots$ be a Harris recurrent chain. If

- 1. $\sigma_F > 0$,
- 2. $\overline{\lim}_{|t|\to\infty} |E_{\phi_A,\lambda} \exp(itF)| < 1$,

there exists $0 < \delta \le 1$ such that

- 3. $E_{\mu,\lambda}\tau^2 < \infty$, $E_{\phi_A,\lambda}\tau^{3+\delta} < \infty$,
- 4. $E_{\mu,\lambda}\left(\sum_{i=1}^{\tau} |f(X_i)|\right)^2 < \infty$, $E_{\phi_A,\lambda}\left(\sum_{i=1}^{\infty} |f(X_i)|\right)^{3+\delta} < \infty$,

then as $n \to \infty$

$$E_\mu \Sigma_{F,n} - \frac{1}{\sqrt{n}}\left(\eta - \frac{\beta}{\alpha}\right) = O(n^{-(1+\delta)/2}) ,$$

$$E_\mu (\Sigma_{F,n})^2 - \frac{\sigma_F^2}{a} = O(n^{-(1+\delta)/2}) .$$

THEOREM 2. Let $X_1, X_2, \ldots$ be a Harris recurrent Markov chain. If conditions 1, 2 of Theorem 1 are satisfied and there exists $\frac{1}{2} < \delta \leq 1$ such that

♦3. $E_{\mu,\lambda}\tau < \infty, \quad E_{\phi_A,\lambda}\tau^{2+\delta} < \infty,$

♦4. $E_{\mu,\lambda}\left(\sum_{i=1}^{\tau} |f(X_i)|\right)^2 < \infty, \quad E_{\phi_A,\lambda}\left(\sum_{i=1}^{\tau} |f(X_i)|\right)^{2+\delta} < \infty,$

then as $n \to \infty$

$$E_\mu(\Sigma_{F,n})^2 - \frac{\sigma_F^2}{a} = O(n^{-(\delta-\frac{1}{2})}) \quad .$$

THEOREM 3. Let $X_1, X_2, \ldots$ be a Harris recurrent Markov chain. If conditions 1, 2 of Theorem 1 are satisfied and there exists $\frac{1}{2} < \delta \leq 1$ such that

♦3. $E_{\mu,\lambda}\tau^2 < \infty, \quad E_{\phi_A,\lambda}\tau^{3+\delta} < \infty,$

♦4. $E_{\mu,\lambda}\left(\sum_{i=1}^{\tau} |f(X_i)|\right)^3 < \infty, \quad E_{\phi_A,\lambda}\left(\sum_{i=1}^{\tau} |f(X_i)|\right)^{3+\delta} < \infty,$

then as $n \to \infty$

$$E_\mu(\Sigma_{F,n})^3 - \frac{1}{\sqrt{n}}\left(\frac{\mu_3}{a} + 3\frac{\sigma_F^2}{a}\eta - 6\frac{\beta\sigma_F^2}{a^2}\right) = O(n^{-\delta}) \quad .$$

REMARK 1. To check the "block"-moment conditions 4 of Theorems 1-3, the different inequalities decreasing the order of such conditions may be used (see, e.g., [13]). Relations between the conditions like condition 3 and the rate of strong mixing coefficients decreasing (for an aperiodic chain, certainly) were discussed in [14]. But in practice it looks easier to check directly these conditions, which means respectively accessibility and recurrence of the set A, with, e.g., testing functions methods (see [8]). For the chains with good ergodic properties these conditions may turn out to be much simpler than in the general situation, e.g., for Doeblin chain

$$\tau = \min \{n > 0: B_{n,\lambda} = 1\}$$

has a geometric distribution with a positive parameter and is independent of $X_1, X_2, \ldots$.

REMARK 2. Theorems 1-3 may be proved by the same method for every random moment, constructed by the regeneration moment of the extended chain (see Section 2), e.g., for

$$\nu = \min \{n > \tau: X_n \in A, B_{n,\lambda} = 1\} .$$

REMARK 3. Using Theorems 1-3 we can easily show that the results of [14], [10], [11] are independent from the method of proof. This result is suggestive but not obvious, because τ may be taken arbitrarily with the number $\lambda > 0$, c-set A and measure ϕ_A to satisfy the Proposition. For such different choices the characteristics a, σ_F^2, $\mu_3, \ldots$ are also different. But their combinations occurring in the theorems proved in [10], [11], [14] turn out to be independent of such choice.

REMARK 4. The uniformity of the used method of proof is one of its advantages. This method enables us to obtain a range of different results for random processes by the reduction to the well-developed limit theorems theory for independent random vectors.

2.

The typical example of regeneration time for a Markov chain is the time of the first passage to a fixed recurrent state (atom), if such a state does exist. The regeneration technique may also be used when there is no such state, but the regenerative extension of the chain needs to be constructed.

We shall say that $Y_1, Y_2, \ldots$ is a regenerative extension of $X_1, X_2, \ldots$ if it is a Markov chain with a distribution $\mathbb{P}_\mu$ defined on $(\Omega, \mathcal{F}^Y)$ $(\mathcal{F}^X \subset \mathcal{F}^Y,\ \mathcal{F}^X = \sigma\{X_i,\ i \geq 1\})$ such that

• 1. $\mathbb{P}_\mu(A) = P_\mu(A)$ for every $A \in \mathcal{F}^X$,

• 2. the chain $Y_1, Y_2, \ldots$ is regenerative (e.g., has a recurrent atom in the state space).

Such extension was constructed in [15] for the Harris recurrent Markov chain and is based on its ergodic properties. We shall briefly describe this construction. First the initial state space $(X, \mathcal{B})$ is extended:

$$X^* = \{x_i \overset{\text{def}}{=} (x,i) : x \in X,\ i = 0,1\} ,$$

$$\mathcal{B}^* = \{A_i \overset{\text{def}}{=} A \times \{i\} : A \in \mathcal{B},\ i = 0,1\} .$$

In what follows we identify any subset A of X with $A \times \{0,1\}$. Hence $X \subset X^*$ and $\mathcal{B} \subset \mathcal{B}^*$. Any $\mathcal{B}$-measurable function f on X is extended to the $\mathcal{B}$-measurable function f on X^* by defining for all $x \in X$, $f(x_0) = f(x_1) = f(x)$. Any nonnegative σ-finite measure μ on $(X, \mathcal{B})$ is extended to the measure μ on $(X^*, \mathcal{B}^*)$ by defining its values on the sets $A_i \in \mathcal{B}^*$:

$$\begin{aligned} \mu(A_0) &= \int_A (1 - \lambda 1_A(x))\, \mu(dx) , \\ \mu(A_1) &= \int_A \lambda 1_A(x)\, \mu(dx) . \end{aligned} \tag{1}$$

This extension is well defined because $A_0 \cup A_1 = A$ and $A_0 \cap A_1 = \emptyset$ (see also Proposition). The transition probability $P(x,A)$ is extended to the transition probability on $(X^*, \mathcal{B}^*)$ by applying (1) to transition measures

$$P(x_0, A) = \frac{P(x,A) - \lambda 1_A(x)\phi_A(A)}{1 - \lambda 1_A(x)} ,$$
$$P(x_1, A) = \phi_A(A) . \tag{2}$$

Thus we define the transition probability on the extended state space. With arbitrary initial distribution ν on $(X^*, \mathcal{B}^*)$ it determines a homogeneous Markov chain $Y_1, Y_2, \ldots$, $Y_i = (Y_i^1, Y_i^2)$, $Y_i^1 \in X$, $Y_i^2 \in \{0,1\}$ with distribution $\mathbb{P}_\nu$.

In [15] the following facts are proved:

•1. For every initial distribution μ on $(X, \mathcal{B})$ the $\mathbb{P}_\mu$-marginal distribution of the first coordinate of $Y_1, Y_2, \ldots$ and the P_μ-distribution of $X_1, X_2, \ldots$ are identical.

•2. Markov chain $Y_1, Y_2, \ldots$ has a recurrent atom $\Delta = X_1$ and hence is regenerative.

Property 1 and the fact that Δ is an atom of the extended chain follows directly from the definition (see (2)). The proof of recurrence of this atom is based on the checking of a Harris recurrence of the extended chain.

Let us note $\nu_0 = \min\{n > 0 : Y_n \in \Delta\}$,

$$\nu_i = \min\{n > \nu_{i-1} : Y_n \in \Delta\}, \qquad \tau_i = \nu_i - \nu_{i-1} ,$$

$$F_i = \sum_{j=\nu_{i-1}}^{\nu_i} (f(Y_j) - \mathbb{E}_\pi f(Y_j)) , \qquad i = 1,2,\ldots .$$

$\mathbb{P}_\Delta$ is a distribution of the extended chain with initial distribution on Δ, i.e., "after regeneration."

It turns out that the extended chain $Y_1, Y_2, \ldots$ may be described by the initial chain and a sequence $B_{1,\lambda}, B_{2,\lambda}, \ldots$ (see Section 1). The construction of this type is used in [12].

LEMMA. For every initial distribution μ on $(X, \mathcal{B})$

$$P_{\mu,\lambda}(\tau = k) = \mathbb{P}_\mu(\nu_0 = k) ,$$

$$P_{\phi_A,\lambda}(\tau = k) = \mathbb{P}_\Delta(\tau_i = k) , \qquad k,i = 1,2,\ldots .$$

The proof of this Lemma is practically contained in Section 3 of [15], and is based on regenerative extension construction. It is sufficient to note that

$$\mathbb{P}_\mu(\nu_0 = k) = \int \mu(dZ_1) \int [P(X_1, dZ_2) - \phi_A(dZ_2)\lambda 1_{A(Z_1)}] \cdots \int [P(Z_{k-1}, dZ_k) - \phi_A(dZ_k)\lambda 1_A(Z_{k-1})]\lambda 1_A(Z_k) ,$$

$$\mathbb{P}_\Delta(\tau_i = k) = \int \phi_A(dZ_1) \int [P(Z_1, dZ_2) - \phi_A(dZ_2)\lambda 1_A(Z_1)] \cdots \int [P(Z_{k-1}, dZ_k) - \phi_A(dZ_k)\lambda 1_A(Z_{k-1})]\lambda 1_A(Z_k) ,$$

$$k,i = 1,2,\ldots .$$

The proof of the relations

$$a = \mathbb{E}_\Delta \tau_1 , \qquad \sigma_F^2 = \mathbb{E}_\Delta(F_1)^2 , \qquad \mu_3 = \mathbb{E}_\Delta(F_1)^3 , \quad \text{etc.}$$

is analogous.

3. THE PROOF OF THEOREM 1

We shall consider the extended Markov chain described in the previous sections of this paper and use the notations introduced therein. We denote also

$$F^k = \frac{\sqrt{a}}{\sigma_F\sqrt{n}} \sum_{i=1}^{k} [f(Y_i) - \mathbb{E}_\pi f(Y_i)] \quad ,$$

covariance matrix

$$V = \begin{vmatrix} 1 & \rho \\ \rho & 1 \end{vmatrix} , \qquad \rho = \frac{\beta}{\sigma_F \sigma_T} , \qquad \lambda_{r,s,m} = \frac{n-r-s-am}{\sigma_T\sqrt{m}} \quad ,$$

and assume first that

$$\sigma_T^2 = \mathbb{E}_\Delta(\tau_1 - a)^2 > 0 \qquad \text{and} \qquad \det V > 0 \quad .$$

Using property 1 of the regenerative extension $Y_1, Y_2, \ldots$ of our Markov chain, we rewrite the formula under investigation in its terms. The system of $\mathcal{F}^Y$-measurable disjoint sets

$$U_{r,s,m} = \left\{ \nu_0 = r, \ \sum_{i=1}^{m} \tau_i = n-r-s, \ \tau_{m+1} > s \right\} , \qquad s,r,m = 0,1,\ldots,n,$$

form a division of Ω. So it is easy to check that

$$\frac{\sqrt{a}}{\sigma_F} E_\mu \Sigma_{F,n} = \sum_{m=1}^{n} \sum_{r=0}^{n} \sum_{s=0}^{n} \iint \mathbb{P}_\mu(F^r \in dv, \ \nu_0 = r) \ \mathbb{P}_\Delta(F^s \in du, \ \tau_{m+1} > s)$$

$$\times \int (y+u+v) \ \mathbb{P}_\Delta\left(\frac{\sqrt{a}}{\sigma_F\sqrt{n}} \sum_{i=1}^{m} F_i \in dy, \ \sum_{i=1}^{m} \tau_i = n-r-s \right)$$

$$+ \ \frac{\sqrt{a}}{\sigma_F} \mathbb{E}_\mu \left(1_{(\nu_1 > n)} \frac{1}{\sqrt{n}} \sum_{i=1}^{n} [f(Y_i) - \mathbb{E}_\mu f(Y_i)] \right) \qquad (3)$$

and $\left\{\frac{F_i}{\sigma_F}, \frac{\tau_i - a}{\sigma_T}\right\}$, $i = 1,2,\ldots$ is a sequence of independent identically distributed 1-lattice two-dimensional random vectors. It is easy to check also that those vectors have zero vector of expectations and covariance matrix V. The random variables $\frac{\tau_i - a}{\sigma_T}$ take values from the lattice $\left\{\frac{j-a}{\sigma_T}, \; j = 1,2,\ldots\right\}$ with the step σ_T^{-1}. We remark that in the general case of Harris chain with $k > 0$ (see Proposition and Assumption in Section 1) we can apply the usual cyclic chain arguments when constructing the extended chain and $\frac{\tau_i - a}{\sigma_T}$ will take values from the lattice with the step $k\sigma_T^{-1}$.

We apply now the Theorem of Section 4 with $d = 3$, $r = 1$. The conditions of this theorem are satisfied (see Section 2). We have

$$\left|\int y \, \mathbb{P}_\Delta\left(\frac{\sqrt{a}}{\sigma_F\sqrt{n}} \sum_{i=1}^m F_i \in dy, \; \sum_{i=1}^m \tau_i = n-r-s\right) - \sqrt{\frac{am}{n}} \; \frac{1}{\sigma_T\sqrt{m}} \left(\int y\left(1 + \frac{1}{\sqrt{m}} P_1(y, \lambda_{r,s,m})\right) \phi_V(y, \lambda_{r,s,m})dy\right)\right| \tag{4}$$
$$\leq \; cn^{-\frac{1}{2}} \, m^{-(1+\delta)/2} \left(1 + |\lambda_{r,s,m}|^{2+\delta}\right)^{-1}$$

and (see also [9], [3])

$$\left|\mathbb{P}_\Delta\left(\sum_{i=1}^m \tau_i = n-r-s\right) - \frac{1}{\sigma_T\sqrt{m}} \, \phi(\lambda_{r,s,m})\right| \tag{5}$$
$$\leq \; cm^{-(1+\delta)/2}\left(1 + |\lambda_{r,s,m}|^{2+\delta}\right)^{-1} ,$$

where ϕ_V stands for a density of two-dimensional normal distribution with zero expectations and covariance matrix V, ϕ is a density of a standard normal distribution and $P_1(\cdot,\cdot)$ is a polynomial of power 3 from the Cramer expansions of densities in the local limit theorem, see [3]. Here and in the sequel, $c, c_1, \ldots$ are positive constants which may be written down explicitly. We substitute now the estimates (4) and (5) in (3). Note at once that the limits of summation in the obtained expressions may be restricted preserving the required accuracy $O(n^{-(1+\delta)/2})$. To check it we use the decomposition

$$\sum_{m=1}^{n}\sum_{r=0}^{n}\sum_{s=0}^{n} = \sum_{m=1}^{n}\sum_{\eta=0}^{\sqrt{n}}\sum_{s=0}^{\sqrt{n}} + \sum_{m=1}^{n}\sum_{r=0}^{n}\sum_{s=\sqrt{n}+1}^{n} + \sum_{m=1}^{n}\sum_{r=\sqrt{n}+1}^{n}\sum_{s=0}^{n}$$

(the summation here is understood up to the integer part of these limits), the estimations following from the theorem moments conditions

$$\sum_{r=\sqrt{n}+1}^{n}\int |v|\, \mathbb{P}_\mu(F^r \in dv,\ \nu_0 = r) \le \frac{c}{\sqrt{n}}\,\mathbb{E}_\mu\left(\sum_{i=1}^{\nu_0} |f(Y_i)|\, 1_{(\nu_0>\sqrt{n})}\right)$$

$$= O(n^{-1}) \quad ,$$

$$\sum_{s=\sqrt{n}+1}^{n}\int |u|\, \mathbb{P}_\Delta(F^s \in du,\ \tau_1 > s) \le \frac{c}{\sqrt{n}}\sum_{s=\sqrt{n}+1}^{n}\mathbb{E}_\Delta\left(\sum_{i=1}^{\tau} |f(Y_i)|\, 1_{(\tau_1>s)}\right)$$

$$= O(n^{-1}) \quad ,$$

$$\sum_{s=\sqrt{n}+1}^{n}\mathbb{P}_\Delta(\tau_1 > s) \le c\sum_{s=\sqrt{n}+1}^{n} s^{-(\delta+3)} = O(n^{-1})$$

$$\sum_{r=\sqrt{n}+1}^{n}\mathbb{P}_\mu(\nu_0 = r) \le \mathbb{P}_\mu(\nu_0 > \sqrt{n}) = O(n^{-1})$$

and the relations (see $\lambda_{r,s,m}$ definition)

$$\sup_{0\le r+s\le n} \sum_{m=1}^{n} m^{-\frac{1}{2}}\phi(\lambda_{r,s,m}) = O(1) \quad ,$$

$$\sup_{0\le r+s\le n} n^{-\frac{1}{2}} \sum_{m=1}^{n} \int y(1+m^{-\frac{1}{2}}P_1(y,\lambda_{r,s,m}))\phi_V(y,\lambda_{r,s,m})dy = O(1) \quad , \quad n \to \infty \quad .$$

So $\frac{\sqrt{a}}{\sigma_F} E_\mu \Sigma_{F,n}$ is approximated by the sum of

$$A_1 = \sum_{m=1}^{n} \sum_{r=0}^{\sqrt{n}} \sum_{s=0}^{\sqrt{n}} \frac{1}{\sigma_T\sqrt{m}} \phi(\lambda_{r,s,m}) \times \left\{\int u\,\mathbb{P}_\Delta(F^s\in du,\tau_1>s)\,\mathbb{P}_\mu(\nu_0=r) + \int v\,\mathbb{P}_\mu(F^r\in dv,\nu_0=r)\,\mathbb{P}_\Delta(\tau_1>s)\right\},$$

$$A_2 = \sum_{m=1}^{n} \sum_{r=0}^{\sqrt{n}} \sum_{s=0}^{\sqrt{n}} \sqrt{\frac{am}{n}} \frac{1}{\sigma_T\sqrt{m}} \times \int y\left\{1+\frac{1}{\sqrt{m}} P_1(y,\lambda_{r,s,m})\right\}\phi_V(y,\lambda_{r,s,m})dy\,\mathbb{P}_\mu(\nu_0=r)\,\mathbb{P}_\Delta(\tau_1>s) \quad .$$

The remainder term of this approximation does not exceed the sum of

$$T_1 = \frac{c}{\sqrt{n}} \sum_{m=1}^{n} \sum_{r=0}^{\sqrt{n}} \sum_{s=0}^{\sqrt{n}} m^{-(1+\delta)/2}\left(1+|\lambda_{r,s,m}|^{2+\delta}\right)^{-1} \times \mathbb{P}_\mu(\nu_0=r)\,\mathbb{P}_\Delta(\tau_1>s) \quad ,$$

$$T_2 = c\sum_{m=1}^{n} \sum_{r=0}^{\sqrt{n}} \sum_{s=0}^{\sqrt{n}} m^{-(1+\delta)/2}\left(1+|\lambda_{r,s,m}|^{2+\delta}\right)^{-1} \times \left\{\mathbb{P}_\mu(\nu_0=r)\int |u|\,\mathbb{P}_\Delta(F^2\in du,\tau_1>s) + \int |v|\,\mathbb{P}_\mu(F^r\in dv,\nu_0=r)\,\mathbb{P}_\Delta(\tau_1>s)\right\} \quad ,$$

$$T_3 = \frac{c}{\sqrt{n}} \mathbb{E}_\mu \left\{ 1_{(\nu_0+\tau_1>n)} \sum_{i=1}^{n} |f(Y_i)| \right\} .$$

$T_1 - T_3$ are of the order $O(n^{-(1+\delta)/2})$ as $n \to \infty$. The proof of this fact is based on the theorem moment conditions and the relations

$$\sum_{r=0}^{\sqrt{n}} \int |v| \, \mathbb{P}_\mu(F^r \in dv, \ \nu_0 = r) = O(n^{-\frac{1}{2}}) \quad ,$$

$$\sum_{s=0}^{\sqrt{n}} \int |u| \, \mathbb{P}_\Delta(F^s \in du, \ \tau_1 > s) = O(n^{-\frac{1}{2}}) \quad , \tag{6}$$

$$\sup_{0 \le r, s \le \sqrt{n}} \sum_{m=1}^{n} m^{-(1+\delta)/2} \left(1 + |\lambda_{r,s,m}|^{2+\delta}\right)^{-1} = O(n^{-\delta/2}) \quad .$$

For checking (6) we divide the summation region on the intervals of length $O(\sqrt{n})$ and majorize (6) by the sum of the form

$$c\sqrt{n} \sum_{k=-c_1\sqrt{n}}^{c_2\sqrt{n}} (n + k\sqrt{n})^{-(1+\delta)/2} \left[\frac{\sqrt{1 + \frac{k}{\sqrt{n}}}}{k} \right]^{2+\delta} ,$$

as in [13]. Our next objective is to simplify A_1 and A_2, preserving the required accuracy of the approximation.

We shall prove first

$$\left| A_1 - \frac{\sqrt{a}}{\sigma_F} \frac{1}{\sqrt{n}} \eta \right| = O(n^{-(1+\delta)/2}) \quad . \tag{7}$$

Denote $\lambda_m = \dfrac{n-am}{\sigma_T\sqrt{m}}$ and note that

$$\lambda_{r,s,m} - \lambda_m = - \frac{r+s}{\sigma_T\sqrt{m}} \quad .$$

We apply the Taylor formula:

$$\left| A_1 - \sum_{m=1}^{n} \frac{1}{\sigma_T \sqrt{m}} \phi(\lambda_m) \left(\sum_{r=0}^{\sqrt{n}} \mathbb{P}_\mu(\nu_0 = r) \sum_{s=0}^{\sqrt{n}} \int u \, \mathbb{P}_\Delta(F^s \in du, \tau_1 > s) + \sum_{s=0}^{\sqrt{n}} \mathbb{P}_\Delta(\tau_1 > s) \sum_{r=0}^{\sqrt{n}} \int v \mathbb{P}_\mu(F^r \in dv, \nu_0 = r) \right) \right|$$

$$\leq c \sum_{m=1}^{n} \sum_{r=0}^{\sqrt{n}} \sum_{s=0}^{\sqrt{n}} \left\{ \frac{|r+s|}{m} \right\} \sup_{\lambda \in [\lambda_m, \lambda_{r,s,m}]} |\phi'(\lambda)| \tag{8}$$

$$\times \left\{ \int |u| \, \mathbb{P}_\Delta(F^s \in du, \tau_1 > s) \, \mathbb{P}_\mu(\nu_0 = r) + \mathbb{P}_\Delta(\tau_1 > s) \times \int |v| \, \mathbb{P}_\mu(F^r \in dv, \nu_0 = r) \right\} .$$

Reasoning as in the $T_1 - T_3$ estimation and using standard normal density properties we can show that the right-hand side of (8) is of order $O(n^{-(1+\delta)/2})$ as $n \to \infty$ and that

$$\left| a^{-1} \left\{ \sum_{r=0}^{\sqrt{n}} \mathbb{P}_\mu(\nu_0 = r) \sum_{s=0}^{\sqrt{n}} \int u \, \mathbb{P}_\Delta(F^s \in du, \tau_1 > s) + \sum_{s=0}^{\sqrt{n}} \mathbb{P}_\Delta(\tau_1 > s) \sum_{r=0}^{\sqrt{n}} \int v \, \mathbb{P}_\mu(F^r \in dv, \nu_0 = r) \right\} - \frac{\sqrt{a}}{\sigma_F} \eta \frac{1}{\sqrt{n}} \right|$$

$$= O(n^{-(1+\delta)/2}) \tag{9}$$

because

$$\left| \sum_{s=0}^{\sqrt{n}} \mathbb{P}_\Delta(\tau_1 > s) - a \right| = \sum_{s=\sqrt{n}+1}^{n} \mathbb{P}_\Delta(\tau_1 > s)$$

and

$$\left| \sum_{r=0}^{\sqrt{n}} \mathbb{P}_\mu(\nu_0 = r) - 1 \right| = \sum_{r=\sqrt{n}+1}^{\infty} \mathbb{P}_\mu(\nu_0 = r) = \mathbb{P}_\mu(\nu_0 > \sqrt{n}) .$$

The checking of (7) is reduced to the checking of

$$\left| \sum_{m=1}^{n} \frac{a}{\sigma_T \sqrt{m}} \phi(\lambda_m) - 1 \right| = O(n^{-(1+\delta)/2}) \quad . \tag{10}$$

We use the equality

$$\lambda_m - \lambda_{m+1} = \frac{a}{\sigma_T \sqrt{m}} + \lambda_{m+1}\left(\sqrt{1+\frac{1}{m}} - 1 \right)$$

and the Lemma of Section 4. Thus

$$\left| \sum_{m=1}^{n} \frac{a}{\sigma_T \sqrt{m}} \phi(\lambda_m) - \int_{\lambda_n}^{\lambda_1} \phi(z)dz + \tfrac{1}{2}\left(\frac{a^2}{\sigma_T} + \sigma_T \right) \sum_{m=1}^{n} \frac{\lambda_m \phi(\lambda_m)}{\sqrt{ma}} (\lambda_m - \lambda_{m+1}) \right|$$

$$\leq \sum_{m=1}^{n} m^{-3/2} \left| c_1 + c_2\lambda_m + c_3\lambda_m^2 \right| \exp\left(-c_u \lambda_m^2 \right) \tag{11}$$

and

$$\left| \sum_{m=1}^{n} \frac{\lambda_m \phi(\lambda_m)}{\sqrt{am}} (\lambda_m - \lambda_{m+1}) - \frac{1}{\sqrt{n}} \sum_{m=1}^{n} \lambda_m \phi(\lambda_m)(\lambda_m - \lambda_{m+1}) \right|$$

$$\leq c_1 \frac{1}{\sqrt{n}} \sum_{m=1}^{n} \left| \sqrt{\frac{n}{am}} - 1 \right| \frac{|\lambda_m|}{\sqrt{m}} \exp\left(-c_2 \lambda_m^2 \right) \quad . \tag{12}$$

Arguing as in checking (6) we can prove that

$$\sum_{m=1}^{n} m^{-3/2} \left| \lambda_m^2 + c_1\lambda_m + c_2 \right| \exp\left(-c_3 \lambda_m^2 \right) = O(n^{-1}) \tag{13}$$

and

$$\sum_{m=1}^{n} m^{-\frac{1}{2}} \left| \sqrt{\frac{n}{am}} - 1 \right| |\lambda_m| \exp\left(-c\lambda_m^2 \right) = O(n^{-\frac{1}{2}}) \quad . \tag{14}$$

We remark here that

$$\sqrt{\frac{n}{am}} - 1 = \frac{\sigma_T \lambda_m}{\sqrt{am}} \left(1 + \sqrt{\frac{am}{n}} \right)^{-1} \quad .$$

Arguing in the similar way and using the Lemma of Section 4, we can prove that

$$\left| \sum_{m=1}^{n} \lambda_m \phi(\lambda_m)(\lambda_m - \lambda_{m+1}) - \int_{\lambda_n}^{\lambda_1} z\phi(z)dz \right| = O(n^{-\frac{1}{2}}) , \tag{15}$$

and (10) follows from (11) - (15) because

$$\int z\ \phi(z)\ dz = 0 \quad \text{and} \quad \int \phi(z)\ dz = 1 .$$

The relation (7) is proved.

Analogously we can prove that

$$\left| A_2 + \frac{\sqrt{a}}{\sigma_F} \frac{1}{\sqrt{n}} \frac{\beta}{\alpha} \right| = O(n^{-(1+\delta)/2}) . \tag{16}$$

The statement of Theorem 1 follows from (7), (16) in the assumption $\sigma_T > 0$ and $\det v > 0$. But when $\sigma_T = 0$ or v is singular there are two possibilities: τ_i are degenerate random variables or τ_i and F_i are linearly connected. The proof of the Theorem in these cases is simpler and may be done by the technique discussed earlier.

The proof of the second statement of Theorem 1 and Theorems 2 and 3 goes along the same lines and we shall, therefore, omit the details.

4.

The proof of the main results of this paper is based on the appli- application of non-uniform estimations in the central limit theorem for the so-called m-lattice random vectors. The definition of such vectors and the investigation of their properties (necessary and sufficient conditions for convergence of the distribution of the normed sum of such vectors, the inversion formula and so on) seems to have appeared first in [1], [2].

Using the methods developed in [1] and [2], different limit theorems have been proved later, see, for instance, [4], [5]. The theorem formulated in this article can be proved using the methods developed in [1], [2], [4], and [5].

Let (X_i, T_i), $i = 1,2,\ldots$ be independent identically distributed random vectors, T_i takes values from the lattice

$$\{kh+a, \quad k = 0, \pm 1, \ldots, \quad a>0\}$$

with a maximal step $h > 0$. Let us assume for simplicity that these vectors have zero vector of expectations and identical covariance matrix I (if not, we use diagonal norming procedure).

<u>THEOREM</u>. Let for the described earlier random vectors (X_i, T_i), $i = 1,2,\ldots$

- 1. $\lim\limits_{|t|\to\infty} E\left|e^{itX_1}\right| < 1$,
- 2. for some $0 < \delta \leq 1$ and integer $d \geq 2$

$$E|X_1|^{d+\delta} < \infty, \qquad E|T_1|^{d+\delta} < \infty .$$

Then for every integer k, $0 < r \leq d$

$$\left|\frac{\sqrt{n}}{h}\int z^r P\left(\frac{1}{\sqrt{n}}\sum_{i=1}^n X_i \in dz, \ \sum_{i=1}^n T_i = kh+an\right) - \int z^r \phi_{d,n}\left(z, \frac{kh+an}{\sqrt{n}}\right) dz\right|$$

$$\leq cn^{-(d-2+\delta)/2}\left(1 + \left|\frac{kh+an}{\sqrt{n}}\right|^{d-r+\delta}\right)^{-1},$$

$$\phi_{d,n}(x,y) = \phi(x,y)\left[1 + \sum_{i=1}^{d-2} n^{-i/2} P_i(x,y)\right],$$

the expression arising in Cramer expansions, see [3].

LEMMA. Let $f(x)$ have k derivatives,

$$f^{(i)}(x) = \frac{d^i}{dx^i} f(x), \qquad i = 1,2,\dots,k, \qquad a, \sigma_T > 0,$$

$$\lambda_m = \frac{n-am}{\sigma_T\sqrt{m}}, \qquad 1 \le m \le n.$$

Then

$$\left| \int_{\lambda_n}^{\lambda_1} f(z)dz - \sum_{i=1}^{k-1} \sum_{m=1}^{n-1} (-1)^i \frac{f^{(i)}(\lambda_m)}{(i+1)!} (\lambda_m - \lambda_{m+1})^{i+1} \right|$$

$$\le \sum_{m=1}^{n} \frac{1}{(k+1)!} |\lambda_m - \lambda_{m+1}|^{k+1} \sup_{\lambda \in [\lambda_{m+1}, \lambda_m]} |f^{(k)}(\lambda)| .$$

This lemma is easy to prove using the Taylor formula.

The author wishes to thank D.M. Chibisov and A.M. Zubkov for their useful comments.

REFERENCES

[1] Bikelis, A. "On the Central Limit Theorem in R^k." I,II. *Litovskij matematicheskij Sbornik*, vol.11, no.1 (1971): 27-58; vol.12, no.3 (1972): 19-35.

[2] Bikelis, A. "Zum zentralen Grenwertzatz in R^k." *Chalmers Instit. of Tech. and the University of Geterborg*, no.9 (1972): 14-21.

[3] Bhattacharya, R.N., and Rao, R.R. *Normal Approximation and Asymptotic Expansions*. New York: Wiley & Sons, 1976.

[4] Dubinskaité, J. "Asymptotic Expansions in Local Limit Theorem for Sums of Independent m-Latticed Random Vectors." I.II. *Lithuanian Math. Journal*, vol.18, no.4 (1978): 464-471; vol. 19, no.1 (1979): 40-47.

[5] Dubinskaitė, J. "Limit Theorems in R^k." I. *Lithuanian Math. Journal*, vol.22, no.2 (1982): 129-140.

[6] Nagaev, S.V. "Some Questions on the Theory of Homogeneous Markov Process with Discrete Time." *Soviet Math. Doklady*, vol.2, no.4 (1961): 867-869.

[7] Neveu, J. *Mathematical Foundations of the Calculus of Probability*. San Francisco: Holden-Day, 1965.

[8] Kalashnikov, V.V. *Kachestvennyj analiz povedeniya slozhnykh sistem po metodu probnykh funktsij* (Qualitative Analysis of the Behavior of Complex Systems by the Method of Test Functions). Moskva: Nauka, 1978.

[9] Petrov, V.V. *Sums of Independent Random Variables*. Berlin Heidelberg New York: Springer-Verlag, 1975.

[10] Malinovsky, V.K. "On Computation of the Defect of an Asymptotically Efficient Test in the Case of Markov Observations." *Soviet Math. Doklady*, vol.26, no.3 (1982): 736-740.

[11] Malinovskii, V.K. "On Asymptotic Expansions in the Central Limit Theorem for Harris Recurrent Markov Chains." *Soviet Math. Doklady*, vol.29, no.3 (1984).

[12] Athreya, K.B., and Ney, P. "A New Approach to the Limit Theory of Recurrent Markov Chains." *Trans. Amer. Math. Soc.*, vol.245 (1978): 493-501.

[13] Bolthausen, E. "The Berry-Esseen Theorem for Functionals of Discrete Markov Chains." *Z. Wahrscheinlichkeitstheorie und Verw. Gebiete*, vol.54 (1980): 59-73.

[14] Bolthausen, E. "The Berry-Esseen Theorem for Strongly Mixing Harris Recurrent Markov Chains." *Z. Wahrscheinlichkeitstheorie und Verw. Gebiete*, vol.60 (1982): 283-289.

[15] Nummelin, E. "A Splitting Technique for Harris Recurrent Markov Chains." *Z. Wahrscheinlichkeitstheorie und Verw. Gebiete*, vol.43 (1978): 309-319.

[19] Orey, S. *Limit Theorems for Markov Chains of Transition Probabilities*. London: Van Nostrand, 1971.

A.V. MEL'NIKOV and D.I. HADJIEV

BOUNDARY VALUE PROBLEMS FOR GAUSSIAN MARTINGALES

Let $(\Omega, \mathcal{F}, P)$ be a complete probability space equipped with a non-decreasing family of σ-algebras $(\mathcal{F}_t)$ satisfying the usual conditions. We suppose that there exists a cadlag Gaussian martingale (see e.g., [1] for notation) $m = (m_t, \mathcal{F}_t)_{t\geq 0}$, $m_0 = 0$, with quadratic characteristic $\langle m \rangle_t = Em_t^2$. Also, we consider the "boundary" G_t, $t \geq 0$.

The asymptotics of the following type of probabilities (1) and (2) will be studied in this paper:

$$P(m_t \geq G_t,\ t \leq T), \quad P(|m_t| \leq G_t,\ t \leq T) \quad \text{as} \quad T \uparrow \infty \tag{1}$$

and

$$P(|X_t| \leq \varepsilon G_t,\ t \leq T) \quad \text{as} \quad \varepsilon \downarrow 0, \tag{2}$$

where $X = (X_t, \mathcal{F}_t)_{t\leq T}$ is a suitable semimartingale with a Gaussian martingale part m.

The problem of deriving asymptotical formulas for these probabilities was treated in several papers by Novikov (see [2] - [3] for the case of a Wiener process or of a Gaussian random walk, and also [4]). Novikov developed an effective method for obtaining such asymptotical formulas by means of an absolutely continuous change of the initial probability measure and by reducing the problem to the corresponding one with a constant boundary.

Our purpose is to get the asymptotics for probabilities (1) and (2) by the same method applied to the processes considered here. Some results of the paper were announced in [5] - [6].

1. ASYMPTOTICS FOR PROBABILITIES (1)

Let $g = g(s)$ be a nonnegative nonincreasing function such that

$$G_t = G_0 + \int_0^t g(s)\, d\langle m\rangle_s ,$$

$$\int_0^T g^2(s)d\langle m\rangle_s < \infty , \qquad G_0 < 0 .$$

Let Y be a Gaussian random variable with $EY = 0$, $EY^2 > 0$ not depending on the Gaussian martingale m. We denote

$$m_t^Y = Y + m_t , \qquad \tau = \inf\{t\colon m_t^Y < G_t\} .$$

The statement of the following theorem can be regarded as a unified form of Novikov's results concerning both a Wiener process and a Gaussian random walk (see [2] - [3]).

Denote $\phi(x) = o(\psi(x))$ and $\phi(x) \sim \psi(x)$ as $x \to x_o$, where ϕ and ψ are nonnegative functions, if $\lim_{x\to x_o} \phi(x)/\psi(x) = 0$ and $\lim_{x\to x_o} \phi(x)/\psi(x) = 1$, respectively.

<u>THEOREM 1</u>. Suppose we have a Gaussian martingale $m = (m_t, \mathcal{F}_t)$ and a function g such that

$$\ln \langle m^Y \rangle_T = o\left[\int_0^T g^2(s)\, d\langle m\rangle_s \right], \qquad T \uparrow \infty . \tag{3}$$

Then as $T \uparrow \infty$

$$\ln P(m_t^Y \geq G(t),\ t \leq T) = -\tfrac{1}{2} \int_0^T g^2(s)\, d\langle m\rangle_s (1 + o(1)) .$$

Proof. It is sufficient to show that

$$\varliminf_{T\to\infty} \frac{\ln P(\tau > T)}{\int_0^T g^2(s) d\langle m\rangle_s} \geq -\tfrac{1}{2} , \qquad \varlimsup_{T\to\infty} \frac{\ln P(\tau > T)}{\int_0^T g^2(s) d\langle m\rangle_s} \leq -\tfrac{1}{2} . \tag{4}$$

Setting

$$Z_T = \exp\left\{ \int_0^T g(s)\, dm_s - \tfrac{1}{2} \int_0^T g^2(s)\, d\langle m\rangle_s \right\} ,$$

$$\tilde{P}_T(A) = EI_A Z_T , \qquad A \in F_T ,$$

and applying the Hölder inequality with exponent $p > 1$, $q = \frac{p}{1-p}$, we have

$$\tilde{P}T(\tau > T) = EI_{\{\tau > T\}} Z_T \leq (P(\tau > T))^{1/q} (EZ_T^p)^{1/p} . \tag{5}$$

Using the fact that the martingale m is a Gaussian process, we get

$$(EZ_T^p)^{1/p} = \exp\left\{ \frac{p-1}{2} \int_0^T g^2(s)\, d\langle m\rangle_s \right\} . \tag{6}$$

Now we formulate two auxiliary results.

LEMMA 1. (Girsanov's theorem on Gaussian martingales [7] - [9]). The process $M_t = m_t - \int_0^t g_s d\langle m\rangle_s$ is a Gaussian martingale with the quadratic characteristics $\langle M\rangle_t = \langle m\rangle_t$ with respect to $\tilde{P}_T$.

LEMMA 2. (See [2], [8] for the case of discrete time). Let

$M = (M_t, \mathcal{F}_t)_{0 \le t \le T}$ be a cadlag square integrable martingale such that $EM_t = 0$, $P(M_o < -\lambda) > 0$, $\lambda > 0$, and $\sigma = \inf\{t: M_t \le -\lambda\}$. Then $P\{\sigma > T\} \ge c/EM_T^2$, where c is a constant.

Now we estimate the probability $\tilde{P}(\tau > T)$. It is clear that

$$\tilde{P}_T(\tau > T) = \tilde{P}_T(m_t^Y \ge G(t),\ t \le T)$$
$$= \tilde{P}_T\left\{m_t^Y - \int_0^T g(s)\ d\langle m\rangle_s \ge G(0),\quad t \le T\right\} .$$

By Lemma 1, $M_t = m_t - \int_0^t g(s)\ d\langle m\rangle_s$ is a Gaussian martingale with respect to $\tilde{P}_T$. Applying Lemma 2 to the square integrable martingale $M^Y = Y + M$, we have

$$\tilde{P}_T\{\tau > T\} \ge \frac{c}{E(M_Y^Y)^2} = \frac{c}{\langle m^T\rangle_T} . \tag{7}$$

The inequalities (5) - (7) give for every $p > 1$

$$P\{\tau > T\} \ge c_p \exp\left\{-\frac{p}{2}\int_0^T g^2(s)\ d\langle m\rangle_s - \frac{p}{p-1}\ln\langle m^Y\rangle_T\right\} . \tag{8}$$

Therefore, by (8) and (3) in view of the arbitrariness of $p > 1$ we obtain the first correlation of (4).

In order to obtain the second inequality of (4), we note that

$$P\{\tau > T\} = \tilde{E}_T\ I_{\{\tau > T\}}\ Z_T^{-1} . \tag{9}$$

Integrating by parts we have

$$\int_0^T g(s)\ dm_s = g(T)m_T - \int_0^T m_{s-}\ dg(s) .$$

Therefore

$$\int_0^T g(s)\, dm_s \;\geq\; g(T)G(T) - Yg(0) - \int_0^T G(s-)\, dg(s)$$

on the set $\{m_t \geq G(t)-Y,\ \ t \leq T\}$.

Using the formula of integrating by parts, we get (in view of nonincreasing g)

$$\int_0^T g(s)\, dm_s \;\geq\; g(T)G(T) - Yg(0) - g(T)G(T) + \int_0^T g(s-)g(s)\, d\langle m\rangle_s$$

$$\geq\; -g(0)Y + \int_0^T g^2(s)\, d\langle m\rangle_s$$

on the set $\{m_t \geq G(t)-Y,\ \ t \leq T\}$. This and (9) imply the following estimate:

$$P(\tau>T) \;=\; \tilde{E}_T\, I_{\{\tau>T\}}\, Z_T^{-1}$$

$$\leq\; \exp\left\{-\tfrac{1}{2}\int_0^T g^2(s)d\langle m\rangle_s\right\} \tilde{E}_T \exp\{Yg(0)\}\, I_{\{\tau>T\}}\ .$$

Since

$$\tilde{E}_T\, I_{\{\tau>T\}} \exp\{Yg(0)\} \;\leq\; E\, Z_T \exp\{Yg(0)\}$$

$$=\; E \exp\{Yg(0)\} \;<\; \infty\ ,$$

we obtain

$$P\{\tau>T\} \;\leq\; C \exp\left\{-\tfrac{1}{2}\int_0^T g^2(s)\, d\langle m\rangle_s\right\}. \qquad ///$$

Next we study the asymptotics of the second probability in (1). Let m be a continuous Gaussian martingale, $G(t) = G(0) + \int_0^T g(s)d\langle m\rangle_s$, where $G(0) > 0$, the function g is nonnegative and increasing.

THEOREM 2. Let $G(T) \to 0$, $\int_0^T g^2(s)d\langle m\rangle_s = o(\ln G(T))$ as $T \uparrow \infty$. Then as $T \uparrow \infty$

$$\ln P\{|m_t| \le G(t),\ t \le T\} = \tfrac{1}{2}\ln G(T)(1+o(1)) + \ln P\{|m_t^*| \le 1,\ t \le T\},$$

where m^* is a Gaussian martingale with the quadratic characteristic $\langle m^*\rangle_t = \int_0^T G^{-2}(s)\, d\langle m\rangle_s$.

Proof. Consider the process N defined by the relation

$$N_t = G(t)\int_0^t G^{-1}(s)\, dm_s .$$

By Itô's formula [9] we have

$$N_t = \int_0^t g(s)\, G^{-1}(s)\, N_s\, d\langle m\rangle_s + m_t .$$

It is well known [7] that the measures generated by processes N and m are equivalent and the corresponding density is

$$Z_T(m) = \exp\left\{\int_0^T g(s)G^{-1}(s)m_s\, dm_s - \tfrac{1}{2}\int_0^T g^2(s)G^{-2}(s)m_s^2\, d\langle m\rangle_s\right\} .$$

Therefore

$$P\{|N_t| \le G(t),\ t \le T\} = E\, I_{\{|m_t| \le G(t), t \le T\}}\, Z_T \tag{10}$$

and

$$P\{|N_t| \le G(t),\ t \le T\} = P\{|m_t^*| \le 1,\ t \le T\}, \tag{11}$$

where $m_t^* = \int_0^t G^{-1}(s)\, dm_s$ is a Gaussian martiangale with the quadratic characteristic $\langle m^*\rangle_t = \int_0^t G^{-2}(s)\, d\langle m\rangle_s$.

Let us estimate $\ln Z_T$. Integrating by parts, we obtain

$$\ln Z_T = \int_0^T g(s)G^{-1}(s)m_s\, dm_s - \tfrac{1}{2}\int_0^T g^2(s)G^{-2}(s)m_s^2\, d\langle m\rangle_s \tag{12}$$

$$= \tfrac{1}{2}\, g(T)\, G^{-1}(T)m_T - \tfrac{1}{2}\, g(T)\, G(T)\langle m\rangle_T - \tfrac{1}{2}\int_0^T m_s^2\, d(g(s)G^{-1}(s))$$

$$+ \tfrac{1}{2}\int_0^T \langle m\rangle_s\, d(g(s)G^{-1}(s)) - \tfrac{1}{2}\int_0^T g^2(s)G^{-2}(s)m_s^2\, d\langle m\rangle_s \; .$$

Now integrating by parts and using the fact that g is non-increasing and positive, we have on the set $\{|m_t| \le G(t),\ t \le T\}$

$$\ln Z_T \le -\tfrac{1}{2}\, g(T)\, G^{-1}(T)\langle m\rangle_T - \tfrac{1}{2}\int_0^T m_s^2\, d(g(s)G^{-1}(s))$$

$$+ \tfrac{1}{2}\int_0^T \langle m\rangle_s\, d(g(s)G^{-1}(s))$$

$$= -\tfrac{1}{2}\, g(T)\, G^{-1}(T)\langle m\rangle_T - \tfrac{1}{2}\int_0^T m_s^2\, d(g(s)G^{-1}(s))$$

$$+ \tfrac{1}{2}\langle m\rangle_T g(T)G^{-1}(T) - \tfrac{1}{2}\int_0^T g(s)G^{-1}(s)\, d\langle m\rangle_s$$

$$\le -\tfrac{1}{2}\int_0^T m_s^2 G^{-1}(s)\, dg(s) + \tfrac{1}{2}\int_0^T m_s^2 g^2(s)\, d\langle m\rangle_s\, G^{-2}(s)$$

$$- \tfrac{1}{2}\int_0^T g(s)G^{-1}(s)\, d\langle m\rangle_s$$

$$\le \tfrac{1}{2}\int_0^T g^2(s)\, d\langle m\rangle_s - \tfrac{1}{2}\ln \frac{G(T)}{G(0)} \; .$$

As in the previous case, we find from (12) that on the same set

$$\ln Z_T \geq \tfrac{1}{2} g(T)\, G^{-1}(T) m_T^2 - \tfrac{1}{2} g(T)\, G^{-1}(T) \langle m \rangle_T$$

$$- \tfrac{1}{2} G(T)(g(T) - g(0)) + \tfrac{1}{2} g(T)\, G^{-1}(T) \langle m \rangle_T$$

$$- \tfrac{1}{2} \int_0^T g(s) G^{-1}(s)\, d\langle m \rangle_s - \tfrac{1}{2} \int_0^T g^2(s) G^{-2}(s) m_s^2\, d\langle m \rangle_s$$

$$\geq \tfrac{1}{2} g(T)\, G(T) - \tfrac{1}{2} \ln \frac{G(T)}{G(0)} - \tfrac{1}{2} \int_0^T g^2(s)\, d\langle m \rangle_s .$$

These estimates and the conditions of Theorem 2 imply that

$$\ln Z_T = -\tfrac{1}{2} \ln G(T)(1 + o(1)) , \qquad T \uparrow \infty .$$

Using the equalities (10) and (11), we get the statement of Theorem 2.

<u>REMARK</u>. The asymptotic of $\ln P(|m_t^*| \leq a,\ t \leq T)$, where $a > 0$, m^* is a continuous Gaussian martingale, can be calculated by using the time change $\tau_t = \inf(s: \langle m^* \rangle_s > t)$. It is known that the process $m^*_{\tau_t}$ is standard Wiener. Then (see [3])

$$\ln P(|m_t^*| \leq a,\ t \leq T) = \ln P\left(|w_t| \leq 1,\ t \leq \langle m^* \rangle_{Ta^{-2}}\right) \sim \frac{\pi^2}{8} \langle m^* \rangle_{Ta^{-2}} ,$$

$$T \uparrow \infty .$$

Here w_t is a standard Wiener process.

2. ASYMPTOTICS FOR PROBABILITY (2)

Consider the d-dimensional semimartingale $X = A + m$ with a Gaussian martingale part $m = (m^1, \ldots, m^d)^*$. Denote by $\langle\langle m \rangle\rangle_t = (\langle m^i, m^j \rangle)$ a covariance $(d \times d)$-matrix of m, $\langle m \rangle_t = \mathrm{tr}\, \langle\langle m \rangle\rangle_t$, $Q_t = d\langle\langle m \rangle\rangle_t \,/\, d\langle m \rangle_t$.

$$G_t = G(0) + \int_0^t g(s)\, d\langle m\rangle_s \ , \qquad G(0) > 0 \ ,$$

and

$$A_t = \int_0^t Q_s\, a_s\, d\langle m\rangle_s \ .$$

THEOREM 3. Let m be a continuous Gaussian martingale starting from 0. Suppose non-random functions $g(s)$ and $a(s) = (a_s^1,\dots,a_s^d)^*$ are such that $g(s)$ is continuous and

$$\int_0^T [(a_s, Q_s a_s) + g^2(s)]\, d\langle m\rangle_s < \infty \ .$$

Then as $\varepsilon \downarrow 0$

$$P\{|X_t| \le \varepsilon G(t),\ t \le T\} \sim \exp\left\{-\tfrac{1}{2}\int_0^T \left[(a_s, Q_s a_s) - \frac{g_s}{G_s}\right] d\langle m\rangle_s\right\} \times P\{|m_t^*| \le \varepsilon,\ t \le T\} \ ,$$

where m^* is a d-dimensional Gaussian martingale with a diagonal covariance matrix $\langle\langle m^*\rangle\rangle_t = (\langle m^{*i}, m^{*j}\rangle_t)$ and

$$\langle m^{*i}, m^{*i}\rangle_t = \int_0^t G^{-2}(s)\, d\langle m^i, m^i\rangle_s \ , \qquad i = 1,\dots,d \ .$$

Proof. Using the theorem on absolute continuity of distributions of multi-dimensional semimartingales with Gaussian martingale parts (see [9] - [10]), we have

$$P\{|X_t| \le \varepsilon G(t),\ t \le T\}$$

$$= E\, I_{\{|m_t|\le\varepsilon G(t),t\le T\}} \exp\left\{\int_0^T (a_s,\ dm_s) - \tfrac{1}{2}\int_0^T (a_s, Q_s a_s)\ d\langle m\rangle_s\right\}$$

$$= \exp\left\{-\tfrac{1}{2}\int_0^T (a_s, Q_s a_s)\ d\langle m\rangle_s\right\}$$

$$\times \left[P\{|m_t|\le\varepsilon G(t),\ t\le T\} + E\, I_{\{|m_t|\le\varepsilon G(t),t\le T\}} \left(\exp\left\{\int_0^T (a_s, dm_s)\right\} - 1\right)\right].$$

The lower bound of this probability immediately follows from

$$E\, I_{\{|m_t|\le\varepsilon G(t),t\le T\}} \left[\exp\left\{\int_0^T (a_s,\ dm_s)\right\} - 1\right]$$

$$\ge\ E\, I_{\{|m_t|\le\varepsilon G(t),t\le T\}} \int_0^T (a_s,\ dm_s) = 0 .$$

In order to obtain the upper estimate, we approximate the function $a(s)$ in the space $L^2_{[0,T]}(d\langle\langle m\rangle\rangle)$ by a sequence of bounded variation functions $a_n(s)$.

Applying the Novikov method ([4]), we get

$$\overline{\lim_{\varepsilon\to 0}}\ P^{-1}\{|m_t|\le\varepsilon G(t),\ t\le T\}\, E\, I_{\{|m_t|\le\varepsilon G(t),t\le T\}} \left[\exp\left\{\int_0^T (a_s, dm_s)\right\} - 1\right]$$

$$\le \exp\left\{\tfrac{1}{2}\int_0^T (a_s - a_s^n,\ Q_s(a_s - a_s^n))\ d\langle m\rangle_s\right\} - 1 .$$

Therefore as $\varepsilon \downarrow 0$

$$P\{|X_t|\le\varepsilon G(t),\ t\le T\} \sim \exp\left\{-\tfrac{1}{2}\int_0^T (a_s, Q_s a_s)\ d\langle m\rangle_s\right\} P\{|m_t|\le\varepsilon G(t),\ t\le T\}. \tag{13}$$

Now we find the asymptotic of the probability

$P\{|m_t| \le \varepsilon G_t,\ t \le T\}$ as $\varepsilon \downarrow 0$. Denote by $\bar{G}(s)$ a diagonal $(d\times d)$-matrix with diagonal elements $G^{-1}(s)$. Consider the process

$$N_t = G(t) \int_0^t \bar{G}(s)\, dm_s$$

with the following integral representation:

$$N_t = \int_0^t g(s)\, G^{-1}(s)\, N_s\, d\langle m\rangle_s + m_t .$$

Another application of the theorem on absolute continuity ([9] - [10]) gives

$$P\{|N_t| \le \varepsilon G(t),\ t \le T\}$$

$$= E\, I_{\{|m_t| \le \varepsilon G(t), t \le T\}} \exp\left\{ \int_0^T g(s)\, G^{-1}(s)\, (m_s, dm_s) - \tfrac{1}{2}\int_0^T |g(s)\, G^{-1}(s)\, m_s|^2\, d\langle m\rangle_s \right\} .$$

From this and Itô's formula we can deduce that

$$\int_0^T g(s)\, G^{-1}(s)\, (m_s, dm_s)$$

$$= \tfrac{1}{2}\int_0^T g(s)\, G^{-1}(s)\, d\left(|m|_s^2 - \langle m\rangle_s\right)$$

$$= \tfrac{1}{2} g(T)\, G^{-1}(s)\, |m_T|^2 - \tfrac{1}{2}\int_0^T |m_s|^2\, d(g(s)G^{-1}(s)) - \tfrac{1}{2}\int_0^T g(s)G^{-1}(s)\, d\langle m\rangle_s .$$

Then it follows from (14) that as $\varepsilon \downarrow 0$

$$P\{|m_t| \le \varepsilon G(t),\ t \le T\} \sim \exp\left\{\tfrac{1}{2}\int_0^T g(s)G^{-1}(s)d\langle m\rangle_s\right\} P\{|m_t^*| \le \varepsilon,\ t \le T\} , \tag{14}$$

where m^* is a Gaussian martingale with the covariance matrix $\langle\langle m^* \rangle\rangle$. Now we get the statement of Theorem 3 from (13) by taking into account (15).

REFERENCES

[1] Elliot, R.J. *Stochastic Calculus and Applications.* Berlin Heidelberg New York: Springer-Verlag Inc., 1982.

[2] Novikov, A.A. "On Estimates and the Asymptotic Behavior of the Probability of Nonintersection of Moving Boundaries by Sums of Independent Random Variables." *Math. USSR, Izvestiya,* vol.17, no.1 (1981): 129-145.

[3] Novikov, A.A. "On Estimates and the Asymptotic Behavior of Nonexit Probabilities of a Wiener Process to a Moving Boundary." *Math. USSR, Sbornik*, vol.38, no.4 (1981): 495-505.

[4] Novikov, A.A. "On Small Deviations of Gaussian Processes." *Mathematical Notes*, vol.29, no.2 (1981): 291-302.

[5] Mel'nikov, A.V., and Hadjiev, D.I. "Asymptotics of Small Deviations of Probability for Gaussian Martingales." *Comptes rendus Acad. Bulg. Sci.*, vol.34, no.11 (1981): 1485-1486.

[6] Mel'nikov, A.V. "On the Boundary Value Problem for Gaussian Processes with Independent Increments." *The Fourteenth USSR School-Colloquium on Probability Theory and Mathem. Statist., Abstracts*, pp. 151-152. Tbilisi, 1979.

[7] Kabanov, Yu.M., Liptser, R.S., and Shiryayev, A.N. "On Semimartingales with Gaussian Martingale Part." *Internat. Sympos. on Stochastic Differential Equations, Abstracts,* pp. 57-59. Vilnius, 1978.

[8] Shiryaev, A.N. *Probability.* Graduate Texts in Math., vol.95. Berlin Heidelberg New York: Springer-Verlag Inc., 1984.

[9] Gal'chuk, L.I. "Generalization of a Change Measure Theorem of Girsanov." *Theory Probab. Applications*, vol.22, no.2 (1977): 271-285.

[10] Skorokhod, A.V. *Random Processes with Independent Increments.* The English translation is published by Wright Patterson Air Force Base, USA, 1966.

R. MIKULEVICHYUS

NECESSARY AND SUFFICIENT CONDITIONS FOR CONVERGENCE TO SINGULAR PROCESSES

In [7], [9], sufficient conditions were obtained for weak convergence to processes with singularities. In the present article we investigate the necessity of these conditions. It turns out that they actually guarantee convergence stronger than the weak, namely, the extended weak convergence (see [3], [8], cf. [1]), which was applied in [8] to derive the conditions of another type. In a somewhat different way necessary and sufficient conditions were investigated in [3] for partial one-dimensional cases. We remark that in the case of functional central limit theorem, necessary and sufficient conditions for weak convergence can be obtained under some a priori assumptions ([5]).

1. DEFINITIONS, NOTATION AND MAIN RESULTS

Let E_i, $i = 1,\dots,m$ be a locally compact separable space such that $E_i \subset E_{i+1}$, E_i is closed in E_{i+1}, and topology E_i coincides with induced one. Let L be some countable set of bounded continuous functions on E_m, which generates E_m topology and ρ is the corresponding metric on E_m. We consider on a probability space (Ω, F, p) a sequence of Skorokhod processes

$(X^n)_{n\geq 0}$ (i.e., right-continuous and left-hand limited processes with values in E_m which are $\mathbb{F}^n$-adapted ($\mathbb{F}^n = (F^n_t)$ are right-continuous filtering families of σ-algebras). Assume that for each $f \in L$

$$M^t_n(f) = f(X^n_t) - A^n_t(f)$$

is a $(p, \mathbb{F}^n)$-local martingale, where $A^n_\cdot f$ is an $\mathbb{F}^n$-adapted Skorokhod process with p-a.s. locally bounded variation.

Let $G \subset \bigcap_n F^n_t$ be a separable σ-algebra such that (Ω, F, p) is a complete probability space. Let δ be a positive $G \otimes B(E_m)$-measurable bounded function on $\Omega \times E_m$ (by $B(\cdot)$ we denote Borel σ-algebras). Let $A = (A_\omega)$ be the family of mappings from L onto the set of functions on E_m, such that for each $f \in L$, $(\omega, x) \to A_\omega f(x)$ is a $G \otimes B(E_m)$-measurable bounded function. Assume that

$$A^0_t f = \int_0^t Af(X^0_s)\, d\phi^0_s ,$$

where ϕ^0 is an $\mathbb{F}^0$-adapted increasing process with p-a.s. continuous trajectories, such that $\delta(X^0_s)\, d\phi^0_s = ds$.

If E is a Polish space, we denote by M(E) the set of probability measures on $B(E)$ with topology of weak convergence Let $\mu_c(\Omega \times E)$ be a set of $G \otimes B(E_m)$-measurable bounded functions f such that $f(\omega, \cdot)$ is continuous for each ω. We denote by $M(\Omega \times E)$ the set of probability measures on $G \otimes B(E)$ with topology generated by mappings

$$\mu \to \mu(f) = \int f\, d\mu \qquad (f \in \mu_c(\Omega \times E)) .$$

The space $M(\Omega \times E)$ is metrizable (see [4]). We denote by $D^s = D^s(E_m)$ the space of all right-continuous and left-hand limited mappings from $[s,\infty)$ to E_m endowed with Skorokhod's J_1-topology. Let

$$\mathcal{D}_t^s = \omega(X_u, \ s \le u \le t) , \qquad X_u = X_u(w) = w(u) , \qquad w \in D^s ;$$

$$\mathcal{D}^s = (\mathcal{D}_{t+}^s)_{t \ge s} , \qquad D = D^0 , \quad \mathcal{D}_t = \mathcal{D}_t^0 , \quad \mathcal{D}^0 = \mathcal{D} .$$

Let $P(s,X,A_\omega,\delta_\omega)$ be the set of probability measures μ on $(D^s, \mathcal{B}(D^s))$, such that for each $f \in L$

$$f(X_t) - \int_s^t Af(X_u)\, d\phi_u \quad \text{is a} \quad (\mu,\mathcal{D}^s)\text{-local martingale,}$$

where ϕ is a $(\mathcal{D}^s)$-adapted increasing process with μ-a.s. continuous trajectories and

$$\int_s^t \delta(x_u)\, d\phi_u = t - s , \qquad \mu(X_s = X) = 1 .$$

Since D is a Polish space, there are $\mathbb{F}^n$-adapted $M(D)$-valued Skorokhod processes Z^n such that

$$p(X^n \in dX \mid F_t^n) = Z_t^n (dx) \qquad p\text{-a.s.}$$

We write $(X^n,\mathbb{F}^n) \to (X^0,\mathbb{F}^0)$ if for each $p \ge 1$

$$f \in \mu_c(\Omega \times M(D)^p) , \qquad (t_1,\dots,t_p) \in [0,\infty)^p$$

$$Ef(Z_{t_1}^n, \dots, Z_{t_p}^n) \qquad Ef(Z_{t_1}^0, \dots, Z_{t_p}^0) .$$

For every $T > 0$ we denote by $S(T)$ the set of sequences

(s_n) $(s_n \in \mathcal{T}(\mathbb{F}^n)$, $\mathcal{T}(\mathbb{F}^n)$ is the set of $\mathbb{F}^n$ stopping times) such that $s_n \le T$ and $p(s_n < T) \to 0$ as $n \to \infty$.

Further we shall always assume that

• A) there are $\Phi \in \alpha$ and $\mathbb{F}^n$-adapted increasing processes H^n such that for every $T > 0$ and for some $(s_n) \in S(T)$, $H^n_T - T \overset{p}{\to} 0$, $(H^n_{s_n})$ are uniformly bounded, $\inf_{X,\omega} (\delta + A\Phi) > 0$, $d\phi^n dp$ are positive measures on $\mathcal{B}([0,\infty)) \otimes \left(\bigcup_t \mathcal{F}^n_t\right)$, where

$$d\phi^n_t = \left(\delta(X^n_t) + A\Phi(X^n_t)\right)^{-1}\left(dH^n_t + dA^n_t(\Phi)\right) \quad ;$$

• B) for each $f \in L$, $\omega \in \Omega$ the restrictions $A_\omega f\big|_{E_{i+1}\setminus E_i}$ are continuous $(i = 0,\dots,m-1,\ E_0 = \emptyset)$.

We denote by $\rho(\mu)$ the image of a measure μ under a mapping ρ. Let K be the set of compact subsets of E_m and let

$$I^n_t : \omega \to \left(\omega, X^n_t(\omega) \in \Omega \times E_m\right), \qquad t \ge 0 \quad .$$

The following two statements are true.

THEOREM 1. Let

♦ 1) $(X^n, \mathbb{F}^n) \to (X^0, \mathbb{F}^0)$;

♦ 2) for all $\varepsilon > 0$, $k \in K$, $i = 1,\dots,m-1$, $t > 0$

$$\lim_{c\to 0} \overline{\lim_n}\, P\left\{\int_0^t 1_{k\cap\{\rho(E_i,\cdot)<c\}}(X^n_s)\, d\phi^n_s > \varepsilon\right\} = 0 \quad ;$$

♦ 3) for each $f \in L$ the set of distributions of $\{A^n_\cdot f,\ n = 1,\dots\}$ is relatively compact and all limiting points are concentrated on the set of continuous functions of locally bounded variation; moreover, for each $T > 0$ there is $s_n \in S(T)$ such that $\left(\sup_{s\le s_n} |A^n_s f|\right)$ is uniformly integrable.

Then for all $t > 0$, $\varepsilon > 0$

$$\lim_n p\left\{\sup_{s\le t} \left| \int_0^s Af(X_u^n)\, d\phi_u^n - A_s^n f \right| > \varepsilon\right\} = 0 .$$

THEOREM 2. Let

♦1') $I_0^n(p) \to I_0^0(p)$ in $M(\Omega \times E_m)$;

♦2') for each $t > 0$, $I_t^0(p)$-a.s. $P(t,X,A_\omega,\delta_\omega)$ has only one element;

♦3') there is an increasing sequence (u_j) of open relatively compact subsets of E_m $(\bar{u}_j \subset u_{j+1},\ \bigcup_j u_j = E_m)$, a sequence of functions (f_j) $(f_j \in L)$, a sequence of numbers $c_j \downarrow 0$ and $c > 0$ such that $0 \le f_j \le 1$

$$f_j\big|_{E_m \setminus u_{2j+2}} = 1 ,$$

$$f_j\big|_{u_{2j+1}} = 0 ,$$

$$Af_j\big|_{u_{2j+2}} \le c_j + c\, 1_{u_{2j+2}\setminus u_{2j}} ;$$

♦4') for each $k \in K$, $\varepsilon > 0$ there is $\Phi_\varepsilon \in \alpha$, $c_\varepsilon > 0$, $c_\varepsilon > 0$ such that $0 \le \Phi_\varepsilon \le \varepsilon$, $c_\varepsilon\delta + A\Phi_\varepsilon \ge c_\varepsilon$ on k and $\varepsilon/c_\varepsilon \to 0$ as $\varepsilon \to 0$.

♦5') for each $k \in K$, $0 \le i < m$, there exists a sequence (V_j^i) of neighborhoods of $k \cap E_i$ $\left(V_j^i \subset V_j^{i+1},\ \bigcap_j V_j^i = k \cap E_i\right)$, a sequence of functions (g_j) $(g_j \in L)$, a sequence of constants $c_j \downarrow 0$, $c > 0$, $c > 0$ such that on k

$$Ag_j\Big|_{V_j^i\setminus E_i} \geq c\, 1_{V_j^i\setminus E_i} - c_j - c\sum_{k=1}^{m-i-1} 1_{V_j^{i+k}\setminus E_{i+k}}$$

and $\sup_X |g_j(X)| \to 0$;

♦6') for each $T > 0$, $f \in L$ there is $(s_n) \in S(T)$ such that for all $\tau_n \in \mathcal{T}(\mathbb{F}^n)$ $(\tau_n \leq s_n)$, $(\Delta A^n_{\tau_n} f)$ is uniformly integrable;

♦7') for each $T > 0$, $\varepsilon > 0$, $f \in \alpha$, $k \in K$ there is $(s_n) \in S(T)$ such that $(\phi^n_{s_n})$ is uniformly integrable, $\lim_n p\left\{\sup_{S\leq T} |\Delta\phi^n_S| > \varepsilon\right\} = 0$ and

$$\lim_n p\left\{\sup_{u\leq T\wedge T^n_k} \left|A^n_u f - \int_0^u Af(X^n_S)\, d\phi^n_S\right| > \varepsilon\right\} = 0 ,$$

where $T^n_k = \inf\,(t\colon X^n_t \notin k)$.

Then conditions 1, 2, 3 of Theorem 1 are satisfied.

2. PROOF OF THEOREM 1

In this section we assume that conditions 1, 2 and 3 are satisfied. At first we prove several auxiliary statements.

LEMMA 1. (Lenglart's inequality). Let $\mathbb{H} = (H_t)_{t\geq 0}$ be a filtering family of σ-algebras in (Ω, F, p) satisfying usual conditions. Let B be an increasing $\mathbb{H}$-adapted process and let Y be a positive measurable process. Suppose that for every bounded $T \in \mathcal{T}(\mathbb{H})$

$$EY_T \leq EB_T .$$

Then for $\lambda > 0$ and bounded $T \in \mathcal{T}(\mathbb{H})$ we have

$$EY_T \wedge \lambda \leq E \sup_{S \leq T} \Delta B_S 1_{\{B_T > \lambda\}} + \lambda P(B_T > \lambda) + EB_T \wedge \lambda \quad .$$

Proof. We shall follow [2]. Let $s = \inf(s: B_s \geq \lambda) \wedge T$. It is clear that $s \in \mathcal{T}(\mathbb{H})$ and $Y_T \wedge \lambda \leq Y_s + \lambda 1_{\{s<T\}}$. Hence

$$\begin{aligned} EY_T \wedge \lambda &\leq EB_s + \lambda P(s<T) \\ &\leq EB_s 1_{\{s=T\}} + \lambda P(s<T) + EB_{s-} 1_{\{s<T\}} + E\Delta B_s 1_{\{s<T\}} \\ &\leq EB_T \wedge \lambda + E \sup_{s \leq T} \Delta B_s 1_{\{B_T > \lambda\}} + \lambda P(B_T > \lambda) \quad . \end{aligned}$$

Let $f \in L$, $T > 0$ and let $(t_j^k)_{1 \leq j \leq p(k)}$ be the sequence of partitions of the interval $[0,T]$ such that

$$0 = t_0^k < \cdots < t_{p(k)}^k = T \quad ,$$

$$\max_j |t_{j+1}^k - t_j^k| \to 0 \quad ,$$

$$\{t_j^k, \ j = 1,\ldots,p(k)\} \subset \{t_j^{k+1}, \ j = 1,\ldots,p(k+1)\} \quad .$$

From assumption A and assumption 3 we have that for some $(s_n) \in S(T)$

$$\left(\phi_{s_n}^n + \sup_{S \leq s_n} |A_S^n f| \right)$$

is uniformly integrable and $(H_{s_n}^n)$ is uniformly bounded. We define

$$\bar{X}_t^n = X_{t \wedge s_n}^n \quad , \qquad \bar{\phi}_t^n = \phi_{t \wedge s_n}^n \quad , \qquad \bar{H}_t^n = H_{t \wedge s_n}^n \quad ,$$

$$y_t^{n-1} = A_{t \wedge s_n}^n f \quad , \qquad y^{n,2} = \int_0^t Af(X_S^n) \, d\bar{\phi}_S^n \quad ,$$

$$Y_j^{n,k,i} = Y^{n,i}(t_{j+1}^k) - Y^{n,i}(t_j^k) ,$$

$$\bar{Y}_t^{n,k,i} = \sum_{t_{j+1}^k \le t} \left\{ Y_j^{n,k,i} - E\left(Y_j^{n,k,i} \mid \mathcal{F}_{t_j^k}^n \right) \right\} , \quad i = 1,2 .$$

LEMMA 2. For each $\varepsilon > 0$, $i = 1,2$, $t \in \bigcup_{k,j} \{t_j^k\}$

$$\lim_k \overline{\lim_n}\, p(|Y_t^{n,i} - \bar{Y}_t^{n,k,i}| > \varepsilon) = 0 .$$

Proof. It is easy to see that

$$E\left(Y_j^{n,k,1} \mid \mathcal{F}_{t_j^k}^n \right) = E\left[\{ f(\bar{X}^n(t_{j+1}^k)) - f(\bar{X}^n(t_j^k)) \} \mid \mathcal{F}_{t_j^k}^n \right] ,$$

$$E\left(Y_j^{n,k,2} \mid \mathcal{F}_{t_j^k}^n \right) \le \text{const } E\left[(\bar{\phi}^n(t_{j+1}^k) - \bar{\phi}^n(t_j^k)) \mid \mathcal{F}_{t_j^k}^n \right]$$

$$\le \text{const } E\Big[\Phi(\bar{X}^n(t_{j+1}^k))$$

$$- \Phi(\bar{X}^n(t_j^k) + \bar{H}^n(t_{j+1}^k) - \bar{H}^n(t_j^k)) \mid \mathcal{F}_{t_j^k}^n \Big] .$$

We define $\mathcal{F}_t^{n,k} = \mathcal{F}_{t_j^k}^n$ if $t_j^k \le t \le t_{j+1}^k$, $\mathbb{F}^{n,k} = (\mathcal{F}_t^{n,k})$. Obviously, $\bar{Y}^{n,k,i}$ are $(\mathbb{F}^{n,k}, p)$-martingales. Therefore there is $c > 0$ such that for all bounded $\tau \in \tau(\mathbb{F}^{n,k})$

$$E \sup_s |\bar{Y}_{S\wedge\tau}^{n,k,i}| \le c\, E\left[\left(\sum_{S \le \tau} (\Delta \bar{Y}_S^{n,k,i})^{\frac{1}{2}} \right) \right] .$$

From assumption A and assumption 3 we have for each $\varepsilon > 0$

$$\lim_k \overline{\lim_n}\, p\left(\sum_j (\bar{\phi}^n(t_{j+1}^k) - \bar{\phi}^n(t_j^k))^2 > \varepsilon \right) = 0 ,$$

$$\lim_k \overline{\lim_n}\, p\left(\sum_j (Y_j^{n,k,1})^2 > \varepsilon\right) = 0 . \tag{1}$$

Let us denote $\delta_j = E\left[(\phi^0(t_{j+1}^k) - \phi^0(t_j^k)) \mid F^0_{t_j^k}\right]$,

$$\delta_j^\varepsilon = E\left[(\phi^0(t_{j+1}^k) - \phi^0(t_j^k))\, 1_{\{\max_\ell(\phi^0(t_{\ell+1}^k)-\phi^0(t_\ell^k))>\varepsilon\}} \mid F^0_{t_j^k}\right] ,$$

$$\varepsilon > 0 .$$

Since $(X^n, \mathbb{F}^n) \to (X^0, \mathbb{F}^0)$,

$$\overline{\lim_n}\, E\sum_j \left(E\left[Y_j^{n,k,1} \mid F^n_{t_j^k}\right]\right)^2 = E\sum_j \left(E\left[f(X^0(t_{j+1}^k)) - f(X^0(t_j^k)) \mid F^0_{t_j^k}\right]\right)^2$$

$$= E\sum_j \left(E\left[\int_{t_j^k}^{t_{j+1}^k} Af(X_s^0)\, d\phi_s^0 \mid F^0_{t_j^k}\right]\right)^2$$

$$\leq \text{const } E\sum_j \left(E\left[(\phi^0(t_{j+1}^k) - \phi^0(t_j^k)) \mid F^0_{t_j^k}\right]\right)^2$$

$$= \text{const } E\sum_j \delta_j(\phi^0(t_{j+1}^k) - \phi^0(t_j^k))$$

$$\leq \text{const } (\varepsilon\, E\,\phi_T^0 + E \max_j \delta_j^\varepsilon \phi_T) ,$$

and

$$E \max_j \delta_j^\varepsilon \leq E\sum_j \delta_j^\varepsilon = E\,\phi_T^0\, 1_{\{\max_j(\phi^0(t_{j+1}^k)-\phi^0(t_j^k))>\varepsilon\}} ,$$

$(\delta_j^\varepsilon \leq \delta_j \leq \text{const})$. Hence

$$\lim_k \overline{\lim_n}\, E\sum_j \left(E\left|Y_j^{n,k,1} \mid F^n_{t_j^k}\right|\right)^2 = 0$$

and similarly

$$\lim_k \overline{\lim_n} E \sum_j \left\{ E\left[Y_j^{n,k,2} \mid \mathcal{F}^n_{t^k_j} \right] \right\}^2 = 0 .$$

From the last two equalities, Lemma 1 and (1) we obtain the statement of Lemma 2.

<u>LEMMA 3</u>. For each $S \le t \le T$, $\varepsilon > 0$

$$\lim_n p\left\{ \left| E\left[\int_{S\wedge s_n}^{t\wedge s_n} dA^n_u f - \int_S^t Af(X^n_u)\, d\bar{\phi}^n_u \mid \mathcal{F}^n_S \right] \right| > \varepsilon \right\}$$

$$= \lim_n p\left\{ \left| E\left[f(\bar{X}^n_t) - f(\bar{X}^n_S) - \int_S^t Af(X^n_u) d\bar{\phi}^n_u \mid \mathcal{F}^n_S \right] \right| > \varepsilon \right\}$$

$$= 0 .$$

<u>Proof</u>. We define $\mathbb{F}^n$-adapted Skorokhod processes $\bar{Z}^n$ with values in $M([0,T]\times D)$ such that for each bounded measurable function g on $[0,T]\times D$

$$\bar{Z}^n_t(g) = E\left[\int_0^T g(S, X^n)\, d\tilde{\phi}^n_S \mid \mathcal{F}^n_t \right] , \qquad p-a.s.$$

(here $d\tilde{\phi}^n_t = (\delta(X^n_t) + A\Phi(X^n_t))d\bar{\phi}^n_t$). If $K \subset D$ is compact, then

$$p(Z^n_S(D\backslash K) > \varepsilon) + p(\bar{Z}^n_S((D\backslash K)\times[0,T]) > \varepsilon)$$

$$\le \frac{1}{\varepsilon}\,(p(X^n \notin K) + E\bar{\phi}^n_T 1_{\{X^n \notin K\}} . \tag{2}$$

Thus the set of distributions of $(Z^n_S, \bar{Z}^n_S)$ is relatively compact. Let Q be a measure on

$$(\Omega \times M(D) \times M(D_T),\quad \mathcal{G}\otimes\mathcal{B}(M(D)\times M(D_T)))$$

($D_T = [0,T] \times D$) such that for some subsequence n_k for each $F \in M_c(\Omega \times M(D) \times M(D_T))$, $EF(Z_S^{nk}, \bar{Z}_S^{nk}) \to QF$ as $k \to \infty$. For each $u \geq S$, $m > 0$

$$g \in M_c(\Omega \times D \times M(D) \times M(D_T))$$

$$E \int_S^u \int_D g(X, Z_S^n, \bar{Z}_S^n) \bar{Z}_S^n(dv, dX) \leq |g|_\infty E \tilde{\phi}_T^n 1_{\{\tilde{\phi}_T^n > m\}} + \tag{3}$$

$$+ mE \int_D g(X, Z_S^n, \bar{Z}_S^n) Z_S^n(d \;)$$

(here $|g|_\infty$ is supremum of $|g|$). Therefore

$$\int \int_S^u \int_D g(w, X, Z, \bar{Z}) \; \bar{Z}(dv, dX) \; Q(dw, dZ, d\bar{Z})$$

$$\leq |g|_\infty c(m) + m \int \int_D g(w, X, Z, \bar{Z}) \; Z(dX) \; Q(dw, dZ, d\bar{Z}) \; ,$$

where $c(m) \to 0$ as $m \to \infty$.

From Theorem 65 in [2] we have that there is an increasing process ϕ on $\Omega \times D \times M(D) \times M(D_T)$ such that for every $g \in M_c(\Omega \times [0,T] \times D \times M(D) \times M(D_T))$

$$\int \int_0^T \int_D g(w, v, X, Z, \bar{Z}) \; Z(dv, dX) \; Q(dw, dZ, d\bar{Z})$$

$$= \int \int_0^T \int_D g(w, v, X, Z, \bar{Z}) \; d\phi_v(w, X, Z, \bar{Z}) \; Z(dX) \; Q(dw, dZ, d\bar{Z}) \; .$$

Since g is arbitrary. $d\bar{Z} = d\phi dZ$ on $[t,T] \times D$ Q - a.s.

Let $F \in M_c(\Omega \times M(D) \times M(D_T))$ and let h be a D_u-measurable bounded a.s. continuous function on D with respect to the distribution of X^0, $S \leq u \leq v \leq T$. Then

$$E[F(Z_S^n, \bar{Z}_S^n)\ \bar{Z}_S^n(1_{(u,v]} \otimes h)]$$

$$= E[F(Z_S^n, \bar{Z}_S^n)]\ E[(\bar{H}_v^n - \bar{H}_u^n + \Phi_u^n(\bar{X}_v^n) - \Phi(\bar{X}_u^n))\ h(X^n)\ |\ F_S^n] \ .$$

Therefore

$$\int F(w, Z, \bar{Z})\ \bar{Z}(1_{(u,v]} \otimes h)\ Q(dw, dZ, d\bar{Z})$$

$$= \int \int_D F(w, Z, \bar{Z}) h(X)(v - u + \Phi(X_v) - \Phi(X_u)) Z(dX) Q(dw, dZ, d\bar{Z}) \ .$$

Since F is arbitrary, Q - a.s.

$$\bar{Z}(1_{(u,v]} \otimes h) = \int h(x)(\Phi(X_v) - \Phi(X_u) + v - u) Z(dX)$$

$$= Z(h(\phi_v - \phi_u)) \ . \qquad (4)$$

Let us denote $\bar{F}_t^0 = \bigcap_{u>t} G \otimes \sigma(X_v^0,\ v \le u)$, $\bar{\mathbb{F}}^0 = (\bar{F}_t^0)$. It is easy to see that there exists an $\bar{\mathbb{F}}^0$-adapted increasing process $\hat{\phi}$ on $\Omega \times D$ such that $\hat{\phi}(X^0)$ is the $(\bar{\mathbb{F}}^0, p)$-dual predictable projection of $\int_0^t (\delta(X_S^0) + A\Phi(X_S^0)) d\phi_S^0$. Thus from (3) we obtain that Q - a.s.

$$Z(h(\phi_v - \phi_u)) = Z(h(\hat{\phi}_v - \hat{\phi}_u)) \ . \qquad (5)$$

Assumptions 2 and 4, 5 ensure that

$$\lim_n E\left|E\left[\left(f(\bar{X}_t^n) - f(\bar{X}_S^n) - \int_S^t Af(X_u^n)\ d\bar{\phi}_u^n\right)\ |\ F_S^n\right]\right|$$

$$= \int \left|\int_D f(X_t) - f(X_S) - \int_S^t Af(X_u)(\delta_\omega(X_u) + A_\omega \Phi(X_u))^{-1} \bar{Z}(du, dX)\right|$$

$$\times\ Q(d\omega, dZ, d\bar{Z})$$

$$= \int \left|\int_D f(X_t) - f(X_S) - \int_S^t Af(X_u)(\delta_\omega(X_u) + A_\omega \Phi(X_u))^{-1} d\hat{\phi}_u\right|$$

$$\times\ Q(d\omega, dZ, d\bar{Z}) \ .$$

Now the proof of Theorem 1 easily follows from Lemmas 2 and 3. (The set of distributions of

$$A^n_{\bullet} f - \int_0^{\bullet} Af(X^n_S)\, d\phi^n_S$$

is relatively compact and from those Lemmas we have that for all $t \in \bigcup_{k,j} \{t^j_k\}$

$$A^n_t f - \int_0^t Af(X^n_S)\, d\phi^n_S \xrightarrow{p} 0 \quad . \;)$$

3. PROOF OF THEOREM 2

In this Section we assume that conditions 1' - 7' are satisfied.

LEMMA 4. The sequence of distributions of X^n is relatively compact.

Proof. For each $T > 0$ there is $(s_n) \in S(T)$ such that $(\phi^n_{s_n})$ is uniformly integrable and $H^n_{s_n}$ is uniformly bounded. We define

$$\bar{\phi}^n_t = \phi_{t \wedge s_n} , \qquad B^n_t = t + \bar{\phi}^n_t ,$$

$$T^j_n = \inf\,(t\colon X^n_t \notin U_{2j+2},\ X^n_0 \in U_{2j}) \wedge T ,$$

$$s^j_n = \inf \left(t\colon \sup_{S \le t} |A^n_S f_j| > 1 + |Af_j|_\infty \phi^n_t\right) \wedge s_n \wedge \tilde{s}^j_n$$

(here $(\tilde{s}^j_n) \in 1(T)$ is a sequence from assumption 6' corresponding to the function f_j). If $R_n \in T(\mathbb{F}^n)$ and $R_n \le T$, then

$$p(T^j_n < R_n) \le p(f_j(X^n(R_n \wedge T^j_n \wedge s^j_n)) \ge 1) + p(s^j_n < T^j_n) .$$

In view of assumptions 3', 6' and 7'

$$\overline{\lim_n}\ p(T_n^j < R_n) \le \overline{\lim_n} \int_0^{R_n \wedge T_n^j \wedge S_n^j} Af_j(X_S^n)\, d\phi_S^n$$

$$\le c_j\ \overline{\lim_n}\ E\, B_T^n + c\ \overline{\lim_n}\ E \int_0^{R_n} 1_{\{T_n^{j-1}<t\}}\, dB_t^n \ .$$

We define the stopping times

$$\tau_n^0 = 0 \ ,$$

$$\tau_n^{\ell+1} = \inf\left\{t > \tau_n^\ell :\ B_t^n - B_{\tau_n^\ell}^n \ge \tfrac{1}{4}c\right\} \wedge T\ , \qquad \ell = 0,1,\ldots \ .$$

If $R_n = \tau_n^{\ell+1}$, then

$$\overline{\lim_j}\ \overline{\lim_n}\ p(T_n^j < \tau_n^{\ell+1})$$

$$\le c\ \overline{\lim_j}\ \overline{\lim_n}\ E \int_0^{\tau_n^\ell} 1_{\{T_n^{j-1}<t\}}\, dB_t^n + c\ \overline{\lim_j}\ \overline{\lim_r}\ E \int_{\tau_n^\ell}^{\tau_n^{\ell+1}} 1_{\{T_n^{j-1}<t\}}\, dB_t^n \ ,$$

$$c\, E \int_{\tau_n^\ell}^{\tau_n^{\ell+1}} 1_{\{T_n^{j-1}<t\}}\, dB_t^n \le \tfrac{1}{2} p(T_n^{j-1} < \tau_n^{\ell+1}) + CE\Delta B_{\tau_n}^n \ .$$

Therefore

$$\overline{\lim_j}\ \overline{\lim_n}\ p(T_n^j < \tau_n^{\ell+1}) \le 2c\ \overline{\lim_j}\ \overline{\lim_n}\ E \int_0^{\tau_n^\ell} 1_{\{T_n^{j-1}<t\}}\, dB_t^n \ .$$

Starting with $\ell = 0$, we have by induction that

$$\overline{\lim_j}\ \overline{\lim_n}\ p(T_n^j < \tau_n^{\ell+1}) = 0 \ , \qquad \ell = 0,1,2,\ldots$$

Since $p(\tau_n^\ell < T) \le p(B_T^n \ge \ell/4c) \le (4c/\ell)\, EB_t^n$, we have

$$\overline{\lim_j}\ \overline{\lim_n}\ p(T_n^j < T) = 0 .$$

Conditions 4' and 7' imply that for each $\varepsilon > 0$, $T > 0$, $\tau_n \in \mathcal{T}(\mathbb{F}^n)$, $\tau_n \le 7$ and $\delta_n \downarrow 0$

$$\lim_n p\left(\bar{\phi}_{\tau_n+\delta_n} - \bar{\phi}_{\tau_n} > \varepsilon\right) = 0 . \tag{6}$$

Hence for each $\varepsilon > 0$, $f \in L$

$$\lim_n p\left\{\left|f\left(X^n_{\tau_n+\delta_n}\right) - f(X^n_{\tau_n})\right| > \varepsilon\right\} = 0 .$$

The statement follows now from the fact that L generates the topology of E_m.

In view of the equality (6) and assumptions 5', 6', 7', in the same manner as in [6] one can easily prove the following lemma.

LEMMA 5. Conditions 2 and 3 of Theorem 1 are satisfied.

From the inequality (2) in the proof of Lemma 3 we have the following corollary.

COROLLARY 1. For each $p \ge 1$, $(t_1,\dots,t_p) \in [0,\infty)^p$ the set of distributions of $(X^n, Z^n_{t_1},\dots,Z^n_{t_p})$ is relatively compact.

REMARK 1. Let us define $p_t\colon D \to D^t$ by $p_t(X) = X|_{[t,\infty)}$, $t \ge 0$. It is easy to see that p-a.s. $p_t(Z_t) \in P(t, X^0_t, A, \delta)$.

LEMMA 6. If $Q \in M(\Omega \times D \times M(D))$, $t \ge 0$, are such that for some subsequence n_k and for all $f \in M_c(\Omega \times D \times M(D))$, $Ef(X^{n_k}, Z^{n_k}_t) \to Q(f)$, then

$$Q(f) = Ef(X^0, Z^0_t) .$$

Proof. There is $(s_n) \in S(T)$ such that $(\phi^n_{s_n})$ is uniformly integrable. Let $k \in K$, $T > 0$, $\varepsilon > 0$, $\bar{\phi}^n_t = \phi^n_{t \wedge s_n}$. We define $\mathbb{F}^n$-adapted $M([0,T] \times D)$-valued Skorokhod processes $\bar{Z}^n$, $\bar{Z}^{n,k}$, $\bar{Z}^{n,k,\varepsilon}$ such that for each bounded continuous function f on $[0,T] \times D$ p-a.s.

$$
\begin{aligned}
\bar{Z}^n_t(f) &= E\left[\int_0^T f(S, X^n)\, d\bar{\phi}^n_S \mid F^n_t\right] , \\
\bar{Z}^{n,k}_t(f) &= E\left[\int_0^T f(S, X^n)\, d\bar{\phi}^n_{S \wedge T^n_k} \mid F^n_t\right] , \qquad (7) \\
\bar{Z}^{n,k,\varepsilon}_t(f) &= E\left[\int_0^T f(S, X^n)(c_\varepsilon \delta(X^n_S) + A\Phi_\varepsilon(X^n_S))d\bar{\phi}^n_{S \wedge T^n_k} \mid F^n_t\right]
\end{aligned}
$$

(here $T^n_k = \inf\,(t\colon X^n_t \notin k)$; Φ_ε, c_ε satisfy assumption 4' for $k \in K$ and $\varepsilon > 0$). Let

$$
\begin{aligned}
X &= D \times M(D) \times M([0,T] \times D) , \\
Y^n &= (X^n, Z^n_t, \bar{Z}^n_t) , \\
\bar{X} &= \quad \times M([0,T] \times D)^2 , \\
Y^{n,k,\varepsilon} &= (X^n, Z^n_t, \bar{Z}^n_t, \bar{Z}^{n,k}_t, \bar{Z}^{n,k,\varepsilon}_t) \qquad t \le T .
\end{aligned}
$$

We define the measure M_n on $G \otimes B(\bar{X})$ by

$$
M_n(d\omega, dX) = p(d\omega)\, \varepsilon_{Y^n}(dx) ,
$$

and the measure $M^{k,\varepsilon}_n$ on $G \otimes B(\bar{X})$ by

$$
M^{k,\varepsilon}_n(d\omega, d\bar{x}) = p(d\omega)\, \varepsilon_{Y^{n,k,\varepsilon}}(d\bar{x}) ,
$$

where ε_a is the Dirac measure. Let M be a limiting point of the sequence (M_n) in $M(\Omega \times X)$. We choose a limiting point $M^{k,\varepsilon}$ of $(M_n^{k,\varepsilon})$ such that the image of $M^{k,\varepsilon}$ under canonical projection of $\Omega \times \bar{X}$ onto $\Omega \times X$ equals M (this is possible because of the relative compactness). Let us denote the canonical projections of $\Omega \times X$ and $\Omega \times \bar{X}$ by $(\hat{\omega}', X', Z', \bar{Z}')$ and $(\hat{\omega}, X, Z, \bar{Z}, \bar{Z}^k, \bar{Z}^{k,\varepsilon})$, respectively. We define the measure $\tilde{M}^{k,\varepsilon}$ on $\Omega \times D \times X$ by

$$\tilde{M}^{k,\varepsilon}(f) = \iint f(x,a) Z(a)(dx)\, M^{k,\varepsilon}(da) , \quad f \in M_c(\Omega \times D \times \bar{X}).$$

Similarly as in the proof of Lemma 3 (see the inequality (3) and subsequent arguments) we obtain that on the probability space $(\Omega \times D \times \bar{X},\ G \otimes B(D \times \bar{X}))$ there exist increasing processes ϕ, ϕ^k, $\phi^{k,\varepsilon}$ such that $M^{k,\varepsilon}$-a.s.

$$d\bar{Z} = d\phi\, dZ , \qquad d\bar{Z}^k = d\phi^k dZ ,$$

$$d\bar{Z}^{k,\varepsilon} = d\phi^{k,\varepsilon}\, dZ \qquad \text{on} \quad [t,T] \times D .$$

In the measurable space $(\Omega \times D \times \bar{X},\ G \otimes B(D \times \bar{X}))$ we consider the right-continuous filtering family of σ-algebras

$$\hat{D} = (\hat{D}_v)_{v \geq t} ,$$

$$\hat{D}_v = \bigcap_{u > v} \sigma(X_S,\ S \leq u,\ Z,\ \bar{Z},\ \hat{\omega}) .$$

Denote by $\bar{\Phi}$, $\bar{\Phi}^\varepsilon$, $\bar{\Phi}^{k,\varepsilon}$ the $(\hat{D}, M^{k,\varepsilon})$-dual predictable projections of increasing processes ϕ, ϕ^k, $\phi^{k,\varepsilon}$. Passing to the limit, we can see that $\bar{\phi}_v^{k,\varepsilon} - c_\varepsilon v - \Phi_\varepsilon(X_v)$ is a $(\hat{D}, \hat{M}^{k,\varepsilon})$ martingale; hence for every bounded predictable $\tau \in T(\hat{D})$

$$\tilde{M}^{k,\varepsilon}(\Delta\bar{\phi}_\tau^{k,\varepsilon}) \le 2\varepsilon .$$

From (7) it is clear that

$$\tilde{M}^{k,\varepsilon}(\Delta\bar{\phi}_\tau^{k}) \le \frac{2\varepsilon}{c_\varepsilon}, \qquad \tilde{M}^{k,\varepsilon}(\Delta\bar{\phi}_\tau) \le \frac{2\varepsilon}{c_\varepsilon} + c(k)$$

and $c(k) \to 0$ as $k \uparrow E_m$. Since k and ε are arbitrary, there exists an increasing $\tilde{D}$-adapted

$$(\tilde{\mathcal{D}} = (\tilde{D}_v)_{v\ge t}, \quad \tilde{D}_v = \bigcap_{u>v} \sigma(X_s,\ s\le u;\ Z',\ \bar{Z}'\ \hat{\omega}')\ )$$

process $\tilde{\phi}$ on $\Omega \times D \times X$ with M-a.s. continuous trajectories such that $d\bar{Z}' = d\phi dZ'$ on $([t,T] \times D, P)$ (here P is a D^t-predictable σ-algebra). Conditions 2, 3 of Theorem 1 are satisfied and passing to the limit we have that M-a.s. for all $f \in L$

$$f(X_v) - \int_t^v Af(X_s)\, d\tilde{\phi}_s$$

and

$$v - t = \int_t^v \delta(X_s)\, d\tilde{\phi}_s$$

is the (D^t, Z)-martingale. For every bounded continuous h on R and bounded continuous g on E_m

$$M\left\{h\left(\int_D g(y_t)\, Z'(dy) - g(X_T)\right)\right\} = h(0) .$$

Therefore

$$Z'(X_t \in dy) = \varepsilon_{X_t}(dy) \qquad M\text{-a.s.} \tag{8}$$

If $t = 0$ it follows from assumptions 1', 2' that M-a.s. $Z' \in P(0,X_0,A,\delta)$, i.e., distributions of Z' and of Z_0^0 coincide. Hence

$$p(d\omega)\varepsilon_{X^n}(dy) \to p(d\omega)\varepsilon_{X^0}(dy) \qquad \text{in} \quad M(\Omega \times D)$$

and

$$I_t^n(p) \to I_t^0(p) \qquad \text{in} \quad M(\Omega \times E_m) \quad .$$

In view of (8) and condition 2, for each $t \geq 0$

$p_t(Z') \in P(t, X_t, A, \delta)$ a.s. and

$$p_t(Z') = p_t(M(\cdot \mid G \otimes B(X_S, S \leq t)) \quad . \quad ///$$

The statement of Theorem 2 easily follows from Corollary 1, Remark 1 and Lemmas 5 and 6.

4. EXAMPLE

Let $A^n = (A_\omega^n)$, $n \geq 0$, be the sequence of families of mappings from L onto the set of functions on E_m, such that for each $f \in L$, $n \geq 0$, $(\omega, X) \to A_\omega^n f(X)$ is $G \otimes B(E_m)$ measurable, and $\sup_n |A_\omega^n f(X)|$ is bounded. Let δ^n be the sequence of positive $G \otimes B(E_m)$-measurable functions such that $\sup_n \delta^n$ is bounded.

Assume that

- a. $A_t^n f = \int_0^t A^n f(X_S^n) d\phi_S^n$, $t = \int_0^t {}^n(X_S^n)\, d\phi_S^n$

for some $\mathbb{F}^n$-adapted increasing p-a.s. continuous process ϕ^n;

- b. (A^n) satisfies uniformly the assumptions 3', 4', 5', i.e., the corresponding constants and functions are the same for all n;
- c. for every (ω, X), $P(0, X, A_\omega^0, \delta_\omega^0)$ has only one element;
- d. if $f \in L$ then $f^2 \in L$;
- e. A^0 satisfies condition B;
- f. there is $\Phi \in L$ such that $\sup_{n,x,\omega} |A_\omega^n \Phi(X)| > 0.$

Under assumptions a - f the following statement holds.

PROPOSITION 1. $(X^n, \mathbb{F}^n) \to (X^0, \mathbb{F}^0)$ if and only if

$$I_n^0(p) \to I_0^0(p) \qquad \text{in } M(\Omega \times E_m) ,$$

$$\lim_n p\left\{\sup_{S\le t} \left|A_S^n f - \int_0^S A^0 f(X_S^n)\, d\phi_S^n\right| > \varepsilon\right\} = 0$$

for each $\varepsilon > 0$, $t > 0$, $f \in L$.

Proof. It suffices to note that

$$\langle M(f)\rangle_t = \int_0^t (A_S^n f^2(X_S^n) - 2f(X_S^n)A^n f(X_S^n))\, d\phi_S^n .$$

From assumption f we have that $\sup_n E\phi_t^n < \infty$, hence

$$\sup_n E\langle M^n(f)\rangle_t < \infty , \qquad \sup_n E(\phi_t^n)^2 < \infty .$$

Now it is clear that the conditions of Theorem 1 and 2 are satisfied and Proposition 1 is true.

REFERENCES

[1] Aldous, D. *Weak Convergence of Stochastic Processes for Processes Viewed in the Strasbourg manner*. Priprint, 1978.

[2] Dellacherie, C., and Meyer, P.A. *Probabilities and Potential*. Paris, Hermann; Amsterdam, New York: North-Holland, 1978-82.

[3] Helland, I.S. *Convergence to Diffusions with Regular Boundaries*. Preprint, 1978.

[4] Jacod, J., and Mémin, J. "Sur un type de convergence intermediaire entre la convergence en loi et la convergence en probabilite." *Lecture Notes in Math.*, pp. 529-560, vol.850. Berlin Heidelberg New York: Springer-Verlag Inc., 1981.

[5] Liptser, R., and Shiryaev, A.N. "On a Problem of Necessary and Sufficient Conditions in Functional Central Limit Theorem." *Z. Wahrscheinlichkeitstheorie und Verw. Gebiete*, vol.59 (1980): 311-318.

[6] Grigelionis, B., and Mikulevicius, R. "On Weak Convergence to Random Processes with Boundary Conditions." *Lecture Notes in Math.*, pp. 260-275. Vol. 972. Berlin Heidelberg New York: Springer-Verlag Inc., 1982.

[7] Grigelionis, B, and Mikulevichyus, R. *On the Diffusion Approximations in Queueing Theory*. Preprint. Vilnius, 1978.

[8] Kubilis, K., and Mikulevichyus, R. "Necessary and Sufficient Conditions of Convergence of Semimartingales and Point Processes." I,II. *Litovskij matematich. sbornik*, vol.24, no.1 (1984).

[9] Mikulyavichyus, R. "On the Martingale Problem." *Russian Math. Surveys*, vol.37, no.6 (1982): 137-150.

T.P. MIROSHNICHENKO

A TEST FOR MINIMIZING THE MAXIMUM EXPECTATION OF OBSERVATIONS WITH DELAY

Let non-decreasing families of σ-algebras (F_t), $t \geq 0$, and a Wiener process $W = (W_t)$ be specified in a probability space (Ω, F, P). We are observing a process $\xi = (\xi_t)$ with differential

$$d\xi_t = \theta \, dt + dW_t , \qquad \xi_0 = 0 ,$$

where θ is an unknown non-random parameter taking on two values $\theta = 0$ (hypothesis H_0) and $\theta = 1$ (hypothesis H_1). Let $C([0,\infty))$ be a space of continuous functions $x = (x_t)$ and B_∞ be the σ-algebra of its cylindrical subsets. We use the notation

$$B_t = \sigma\{(x\colon x_\zeta < a),\ \zeta \leq t,\ a \in (-\infty,\infty)\} .$$

Let P_i be a distribution of probabilities in B_∞ when the process under consideration satisfied the hypothesis H_i $(i = 0,1)$. M_i will stand for the corresponding averaging. We shall consider decision rules (tests) $\delta = (\tau, d)$ specified by a stopping time $\tau = \tau(x)$ and a final decision function $d = d(x)$. $\tau(x)$ is supposed to be a Markov time with respect to the family (B_t); $d = d(x)$ is $B_{\tau+m}$-measurable and takes on only two values 0 and 1; m is a known positive constant. The decision $d(x) = 0$ will imply an adoption of the hypothesis H_0; if $d(x) = 1$ the hypothesis H_1 will be adopted. By $M = \{\tau(x)\}$ we denote the class of Markov times $\tau = \tau(x)$ with respect to the family (B_t) in the space of continuous functions such that $M_0\tau < \infty$, $M_1\tau < \infty$.

We shall relate each rule $\delta = (\tau, d)$ to the quantities

$$\alpha(\delta) = P_0(d = 1) , \qquad \beta(\delta) = P_1(d = 0)$$

(the probabilities of errors of the first and second types).

Suppose $\Delta(\alpha,\beta)$ is a class of the decision rules $\delta = (\tau, d)$ for which $\tau \in M$ and d is a $B_{\tau+m}$-measurable function, $\alpha(\delta) \leq \alpha$, $\beta(\delta) \leq \beta$, α and β are specified positive constants, $\alpha + \beta < 1$.

Observing the process ξ we have to find a rule $\delta = (\tau, d) \in \Delta(\alpha,\beta)$ minimizing the functional

$$\max\ [M_0\tau,\ M_1\tau] \quad .$$

This problem arises naturally from [2] proving that there is no rule to minimize both the expectations $M_0\tau$ and $M_1\tau$. It seems natural to find the rule minimizing the maximal expectation. To make the problem non-trivial we suppose that $0 < m < t(\alpha,\beta)$, where $t(\alpha,\beta)$ is the least non-random duration of observations to provide errors of the first and second types no greater than α, β respectively.

In Theorem 1 [2] for any $\lambda \in [0,1]$ a rule $\delta_\lambda = (\tau_\lambda, d_\lambda) \in \Delta(\alpha,\beta)$ was found which minimizes $\lambda M_0\tau + (1-\lambda)M_1\tau$. We use the properties of this rule to solve the problem. So, let us describe δ_λ in detail. Consider a system of equations (see system 3 in [2])

$$\alpha = X\left(\frac{a}{bA}\right)\frac{B-1}{B-A} + X\left(\frac{a}{bB}\right)\frac{1-A}{B-A}, \tag{1}$$

$$\beta = Y\left(\frac{a}{bA}\right)\frac{A(B-1)}{B-A} + Y\left(\frac{a}{bB}\right)\frac{B(1-A)}{B-A}, \tag{2}$$

$$2\lambda K_1 - 2(1-\lambda)K_2 B = a\left\{X\left(\frac{a}{bB}\right) - X\left(\frac{a}{bA}\right)\right\} + bB\left\{Y\left(\frac{a}{bB}\right) - Y\left(\frac{a}{bA}\right)\right\}, \tag{3}$$

$$2\lambda K_2 - 2(1-\lambda)K_1 A = a\left\{X\left(\frac{a}{bB}\right) - X\left(\frac{a}{bA}\right)\right\} + bA\left\{Y\left(\frac{a}{bB}\right) - Y\left(\frac{a}{bA}\right)\right\}, \tag{4}$$

where

$$X(\zeta) = P\left\{W_m > \ln\zeta + \frac{m}{2}\right\},$$

$$Y(\zeta) = P\left\{W_m > \ln\zeta - \frac{m}{2}\right\},$$

$$K_1 = \frac{-(B-A)}{A} - \ln AB^{-1},$$

$$K_2 = \frac{-(B-A)}{B} - \ln AB^{-1}.$$

It is proved in [2] that this system has a unique solution $(A_\lambda, B_\lambda, a_\lambda, b_\lambda)$ such that $A_\lambda < 1 < B_\lambda$ and τ_λ is the first exit time of $\phi = \exp\left\{\xi_t - \frac{t}{2}\right\}$ from (A_λ, B_λ).

For the rule $\delta_\lambda = (\tau_\lambda, d_\lambda)$ of Theorem 1 [2] it is easy to calculate the expectation $M_0\tau_\lambda$ and $M_1\tau_\lambda$ (see [1, Chapter IV, Section 2, Lemma 5])

$$M_1\tau_\lambda = \frac{(B_\lambda - A_\lambda B_\lambda)\ln\left(\frac{B_\lambda}{A_\lambda}\right)}{(B_\lambda - A_\lambda)} + \ln A_\lambda,$$

$$M_0\tau_\lambda = \frac{(B_\lambda - 1)\ln\left(\frac{B_\lambda}{A_\lambda}\right)}{(B_\lambda - A_\lambda)} - \ln B_\lambda.$$

We shall prove an auxiliary lemma.

LEMMA 1. If $M_0\tau_\lambda = M_1\tau_\lambda$ then $A_\lambda B_\lambda = 1$; if $M_0\tau_\lambda < M_1\tau_\lambda$ then $A_\lambda B_\lambda > 1$; if $M_0\tau_\lambda > M_1\tau_\lambda$ then $A_\lambda B_\lambda < 1$.

Proof. We have

$$M_0\tau_\lambda - M_1\tau_\lambda = \frac{(A_\lambda B_\lambda+1)\ln\left(\frac{B_\lambda}{A_\lambda}\right) + (B_\lambda - A_\lambda)\ln B_\lambda A_\lambda}{(B_\lambda - A_\lambda)},$$

But $B_\lambda - A_\lambda > 0$ always. The sign of the difference $M_0\tau_\lambda - M_1\tau_\lambda$ depends on the sign of the function

$$f(A_\lambda, B_\lambda) = (A_\lambda B_\lambda - 1)\ln\left(\frac{B_\lambda}{A_\lambda}\right) + (B_\lambda - A_\lambda)\ln A_\lambda B_\lambda .$$

We substitute variables $A_\lambda B_\lambda = t$, $\frac{B_\lambda}{A_\lambda} = u > 1$. Then $f(A_\lambda, B_\lambda)$ becomes the function $F(t,u)$

$$F(t,u) = (t-1)\ln u - \left(\sqrt{tu} - \frac{\sqrt{t}}{\sqrt{u}}\right)\ln t .$$

Let us investigate the function $F(t,u)$. Note that $F(1,u) = 0$. Suppose $t > 1$. We investigate the case $F(t,u) < 0$, i.e., $(t-1)\ln u < \left(\sqrt{u} - \frac{1}{\sqrt{u}}\right)\sqrt{t}\ln t$ or $\frac{\sqrt{u}\ln u}{u-1} < \frac{\sqrt{t}\ln t}{t-1}$. Since $\frac{\sqrt{u}\ln u}{u-1}$ decreases with respect to u for $u > 1$ and remains unchanged when u is replaced by u^{-1}, then $f(A_\lambda, B_\lambda) < 0$ for $t > 1$, $u > t$, i.e., for $A_\lambda B_\lambda > 1$, $A_\lambda < 1$ or for $t < 1$, $ut < 1$, i.e., for $A_\lambda B_\lambda < 1$ and $B_\lambda < 1$. The latter is wrong while A_λ is always smaller than 1. Thus Lemma 1 is proved.///

LEMMA 2. If for some λ, $0 \le \lambda \le 1$, A_λ and B_λ defined by the system 3 [2] are such that $A_\lambda B_\lambda < 1$ and for the other λ, $0 \le \lambda \le 1$, $A_\lambda B_\lambda > 1$ then there exists λ, $0 \le \lambda \le 1$ such that

$$A_\lambda B_\lambda = 1 \tag{5}$$

Proof. We shall show that the solution of the system 3 [2] (that is, the system (1) - (4)) depends continuously on the parameter λ. We take a sequence $\lambda_n \to \lambda$ such that $0 \le \lambda_n \le 1$; due to Theorem 1 [2] there is a unique solution of the system 3 [2] $(A_{\lambda_n}, B_{\lambda_n}, a_{\lambda_n}, b_{\lambda_n})$ for any λ_n. These sequences are bounded (boundedness of A_{λ_n} and B_{λ_n} follows from boundedness of the expectations). In any bounded sequence it is possible to single out a convergent subsequence. Let these subsequences converge to (A,B,a,b) respectively. We substitute A_{λ_n}, B_{λ_n}, a_{λ_n}, b_{λ_n} into the system 3 [2] and pass to the limit on the right and on the left. The equations of the system depend continuously on A_{λ_n}, B_{λ_n}, a_{λ_n}, b_{λ_n}. Since the solution of the system 3 [2] is unique, it depends continuously on the parameter λ. Since for some λ, $A_\lambda B_\lambda > 1$ and for the others $A_\lambda B_\lambda < 1$, then due to continuous dependence, we can find λ such that $A_\lambda B_\lambda = 1$. Lemma 2 is proved. ///

REMARK. It follows from this Lemma that under its conditions the solution of the extended system (1) - (5) exists and $\lambda \in [0,1]$.

We shall use D_1 to denote a class of tests $\delta = (\tau, d) \in \Delta(\alpha,\beta)$ for which $M_0\tau \ge M_1\tau$ and D_2 to denote a class of tests from $\Delta(\alpha,\beta)$ for which $M_1\tau \ge M_0\tau$. For the rule δ_1 and δ_0 defined in Section 2 of [2] (δ_λ is described above) there are the following possibilities:

- a) $\delta_1 \in D_2$ and $\delta_0 \in D_2$;
- b) $\delta_0 \in D_1$ and $\delta_1 \in D_1$;

•c) $\delta_1 \in D_1$ and $\delta_0 \in D_2$.

Note that if $\delta_1 \in D_2$ then $M_0\tau_0 \le M_0\tau_1 < M_1\tau_1 \le M_1\tau_0$, consequently $\delta_0 \in D_2$; so the situation $\delta_0 \in D_1$ and $\delta_1 \in D_2$ is impossible.

In the case a), $\max(M_1\tau_1, M_0\tau_1) = M_1\tau_1$. For any $\delta \in \Delta(\alpha,\beta)$, $M_1\tau \ge M_1\tau_1$. It implies that for the case a), the optimal test is δ_1.

The case b) is analogous to the case a), here the optimal test is δ_0.

The case c): According to the Remark, the system (1) - (5) has the solution $(\hat{A},\hat{B},\hat{a},\hat{b})$. The corresponding test will be denoted by $\hat{\delta} = (\hat{\tau},\hat{d})$. According to Lemma 1 $M_0\hat{\tau} = M_1\hat{\tau}$.

Now let me remind you that in the proof of Theorem 1 [2] the Lagrange multiplier method was employed and it was shown that

$$\hat{\lambda}M_0\hat{\tau} + (1-\hat{\lambda})M_1\tau + \hat{a}\alpha(\delta) + \hat{b}\beta(\delta)$$

$$\ge \hat{\lambda}M_0\hat{\tau} + (1-\hat{\lambda})M_1\hat{\tau} + \hat{a}\alpha(\hat{\delta}) + \hat{b}\beta(\hat{\delta})$$

for any test δ. Besides $\alpha(\hat{\delta}) = \alpha, \beta(\hat{\delta}) = \beta$ whence for $\delta \in \Delta(\alpha,\beta)(M_0\hat{\tau} = M_1\hat{\tau})$,

$$\max(M_0\tau, M_1\tau)$$

$$\ge \hat{\lambda}M_0\tau + (1-\hat{\lambda})M_1\tau + \hat{a}\alpha(\delta) + \hat{b}\beta(\delta) - \hat{a}\alpha - \hat{b}\beta$$

$$\ge \max(M_0\hat{\tau}, M_1\hat{\tau}) \quad .$$

Thus, we have the following theorem.

<u>THEOREM</u>. The optimal rule for finding $\inf\limits_{\delta\in\Delta(\alpha,\beta)} \max[M_0\hat{\tau}, M_1\hat{\tau}]$

exists and is of the form:

• 1) If $A_0B_0 < 1$ and $A_1B_1 < 1$ the optimal test will be $\delta_0 = (\tau_0, d_0)$, where

$$\tau_0 = \inf\{t \geq 0: \phi_t \notin (A_0, B_0)\} ,$$

$$d_0 = \left\{1 \text{ if } \phi_{\tau_0+m} \geq \frac{a_0}{b_0};\ 0 \text{ if } \phi_{\tau_0+m} < \frac{a_0}{b_0}\right\} .$$

• 2) If $A_0B_0 > 1$ and $A_1B_1 > 1$ the optimal test will be $\delta_1 = (\tau_1, d_1)$, where

$$\tau_1 = \inf\{t \geq 0: \phi_t \notin (A_1, B_1)\} ,$$

$$d_1 = \left\{1 \text{ if } \phi_{\tau_1+m} \geq \frac{a_1}{b_1};\ 0 \text{ if } \phi_{\tau_1+m} < \frac{a_1}{b_1}\right\} .$$

• 3) If $A_1B_1 \leq 1$ and $A_0B_0 \geq 1$ the optimal test will be $\hat{\delta} = (\hat{\tau}, \hat{d})$, where

$$\hat{\tau} = \inf\{t \geq 0: \phi_t \notin (\hat{A}, \hat{B})\} ,$$

$$\hat{d} = \left\{1 \text{ if } \phi_{\hat{\tau}+m} \geq \frac{\hat{a}}{\hat{b}};\ 0 \text{ if } \phi_{\hat{\tau}+m} < \frac{\hat{a}}{\hat{b}}\right\} ,$$

and $\hat{A}$, $\hat{B}$, $\hat{a}$, $\hat{b}$ are the solution of the system (1) - (5).

The problem of sequential analysis with delayed observations was suggested to the author by A.N. Shiryaev.

REFERENCES

[1] Shiryaev, A.N. *Statistical Sequential Analysis*. Providence, R.I.: American Math. Soc., 1973.

[2] Miroshnichenko, T.P. "Testing of Two Simple Hypotheses in the Presence of Delayed Observations." *Theory Probab. Applications*, vol.24, no.3 (1979): 467-479.

M. NIKUNEN and E. VALKEILA

METRIC DISTANCES BETWEEN COUNTING PROCESSES

1. INTRODUCTION, DEFINITIONS AND RESULTS

1

Martingale methods are often used in problems of weak convergence of stochastic processes. When we restrict ourselves to counting processes, then sufficient conditions for the weak convergence of a sequence of counting processes to a limiting counting process can be given in terms of the compensators of the processes. Weak convergence can be metrized by the Prokhorov distance. We shall give upper bounds for the Prokhorov distance between a Poisson process and a counting process with a continuous compensator. These bounds depend only on the compensators of the processes. Moreover, we get an upper bound also for the corresponding Wasserstein distance and for the total variation distance between one-dimensional distributions.

We start by giving definitions of the spaces and metrics involved. Then we recall the definition of a counting process. We end this section by formulating the results. Section 2 contains auxiliary results and Section 3 contains the proofs.

2

Denote by $\mathbb{D}[0,T]$ the space of right continuous real functions x on $[0,T]$ with left-hand limits. Let Λ_T be the space of continuous strictly increasing real functions λ on $[0,T]$ with $\lambda(0) = 0$ and $\lambda(T) = T$. For two functions $x, y \in \mathbb{D}[0,T]$ define the Skorokhod distance $d_T(x,y)$ by

$$d_T(x,y) = \inf_{\lambda \in \Lambda_T} \left\{ \sup_{0 \le t \le T} \left(|x(t) - y(\lambda(t))| + |\lambda(t) - t| \right\} . \quad (1)$$

Then the space $(\mathbb{D}[0,T], d_T)$ is a separable metric space (see, for instance, Gikhman and Skorokhod [4]. The uniform distance between two functions $x, y \in \mathbb{D}[0,T]$ is defined by the formula

$$m_T(x,y) = \sup_{0 \le t \le T} |x(t) - y(t)| . \quad (2)$$

3

For a nonnegative random variable X defined on a probability space $(\Omega, \mathcal{F}, P)$ we put

$$\nu(X) = \inf \{\varepsilon \ge 0 : P\{X \ge \varepsilon\} \le \varepsilon\} .$$

It is easy to see that $\nu(X)^2 \le E(X)$ and for any $\varepsilon > 0$ we have $\nu(X) \le \varepsilon + P\{X \ge \varepsilon\}$. Moreover, if X takes only nonnegative integer values, then $\nu(X) = P\{X \ge 1\}$. Suppose that Y is another nonnegative random variable on $(\Omega, \mathcal{F}, P)$. Then it is not difficult to show that $\nu(X+Y) \le \nu(X) + \nu(Y)$, and if $X \le Y$ (P-a.s.), then $\nu(X) \le \nu(Y)$. If X is a constant K (P-a.s.), then we have $\nu(X) = K \wedge 1$.

Let (S,d) be a separable metric space and let X, Y be random elements from a probability space $(\Omega, \mathcal{F}, P)$ into S. We define the Ky-Fan metric $\alpha(X,Y; d)$ by

$$\alpha(X,Y; d) = \nu(d(X,Y)) . \tag{4}$$

We recall that this metric metrizes convergence in probability.

Denote by $\mathcal{B}(S)$ the σ-algebra of Borel sets of S and for $C \in \mathcal{B}(S)$, $\varepsilon > 0$ define

$$C^{\varepsilon} = \{x \in S: d(x,C) < \varepsilon\} .$$

Let R and Q be two probability measures defined on $(S, \mathcal{B}(S))$. We define the Prokhorov distance $\rho(R,Q; d)$ by

$$\rho(R,Q; d) = \inf \{\varepsilon > 0: R(F) \le \varepsilon + Q(F^{\varepsilon}) \text{ for all closed } F\} . \tag{5}$$

If $Q, Q_1, Q_2, \ldots$ are probability measures on $(S, \mathcal{B}(S))$ such that (Q_n) converges weakly to Q, then $\rho(Q_n,Q; d) \to 0$ as $n \to \infty$. The investigation of the rate of convergence of $\rho(Q_n,Q; d)$ is often based on the following result. Let X and Y be S-valued random elements defined on a common probability space $(\Omega, \mathcal{F}, P)$ such that X induces the measure R on S and Y induces the measure Q on S. The pair (X,Y) is called a coupling of the measures R and Q. Let $C(R,Q)$ be the set of all possible couplings of R and Q (where also the space $(\Omega, \mathcal{F}, P)$ may vary). Strassen and Dudley (see Dudley [3]) proved that

$$\rho(R,Q; d) = \inf \{\alpha(X,Y; d): (X,Y) \in C(R,Q)\} . \tag{6}$$

Hence $\rho(R,Q; d) \le \alpha(X,Y; d)$ for any coupling (X,Y) of R and Q.

If R and Q are as above, we define the Wasserstein distance $W(R,Q; d)$ between R and Q by

$$W(R,Q; d) = \inf \{E(d(X,Y)) : (X,Y) \in C(R,Q)\} .$$

If X and Y are real random variables defined on a common probability space $(\Omega, \mathcal{F}, P)$, then the total variation distance between their distributions will be denoted by $\pi(X,Y)$, where

$$\pi(X,Y) = \sup \left\{ |P\{X \in G\} - P\{Y \in G\}| : G \in \mathcal{F} \right\} .$$

4

Let $(\Omega, \mathcal{F}, P)$ be a probability space with a filtration $\mathbb{F} = (\mathcal{F}_t)_{t \ge 0}$. We suppose that $\mathbb{F}$ is right continuous and $\mathcal{F}_0$ contains all P-null sets. A counting process $(N,\mathbb{F})$ is a process having increasing right-continuous piecewise constant paths with unit jumps and $N_0 = 0$. The notation $(N,\mathbb{F})$ means that N_t is $\mathcal{F}_t$-measurable for each $t \ge 0$. We suppose that $P\{N_t < \infty\} = 1$ for each $t \ge 0$. The compensator $(A,\mathbb{F})$ of a counting process $(N,\mathbb{F})$ is a predictable increasing process such that $N - A$ is a square-integrable local $\mathbb{F}$-martingale. We say that a counting process $(N,\mathbb{F})$ with compensator $(A,\mathbb{F})$ is a Poisson process if A is deterministic and continuous. If $A_t = \mu t$ for all $t \ge 0$, where μ is a positive constant, then N is called a Poisson process with intensity μ. For more details on counting processes see Liptser and Shiryayev [8].

5

Now we shall formulate the results. Let $T > 0$ be fixed. If $X = (X_t)_{t \geq 0}$ and $Y = (Y_t)_{t \geq 0}$ are stochastic processes with paths that are right continuous and have left-hand limits, then their restrictions to the interval $[0,T]$ induce two probability measures R and Q on $(\mathbb{D}[0,T], d_T)$. We shall write $\rho_T(X,Y)$ instead of $\rho(R,Q; d_T)$ and $W_T(X,Y)$ instead of $W(R,Q; d_T)$. If X and Y have continuous paths, then we shall write $\alpha_T(X,Y)$ for $\alpha(X,Y: m_T)$. We denote by I the identity function, $I(t) = t$ for all $t \geq 0$.

THEOREM 1. Let $(M,\mathbb{H})$ and $(N,\mathbb{F})$ be two counting processes defined on the same probability space $(\Omega,\mathcal{F},P)$. Suppose that M has a continuous compensator A and N is a Poisson process with a strictly increasing compensator B. Then

$$\rho_T(M,N) \leq \alpha_T(B^{-1} \circ A, I) + E(\varepsilon \wedge |A_T - B_T|) + P\{|A_T - B_T| \geq \varepsilon\} \qquad (8)$$

for any $\varepsilon > 0$ and

$$W_T(M,N) \leq E\left(\sup_{0 \leq t \leq T} |(B^{-1} \circ A)_t - t|\right) + E(|A_T - B_T|) \quad . \qquad (9)$$

Moreover,

$$\pi(M_T, B_T) \leq E(\varepsilon \wedge |A_T - B_T|) + P\{|A_T - B_T| \geq \varepsilon\} \qquad (10)$$

for any $\varepsilon > 0$.

COROLLARY 1. If A is also deterministic, then

$$\rho_T(M,N) \leq \sup_{0 \leq t \leq T} |(B^{-1} \circ A)_t - t| + |A_T - B_T| \quad . \qquad (11)$$

REMARK 1. We note that on the right-hand side of (8) the term $\alpha_T(B^{-1} \circ A, I)$ is bounded from above by

$$\delta + P\left\{\sup_{0 \le t \le T} |(B^{-1} \circ A)_t - t| \ge \delta\right\}$$

for any $\delta > 0$ or by

$$\left[E\left(\sup_{0 \le t \le T} |(B^{-1} \circ A)_t - t|\right)\right]^{\frac{1}{2}} .$$

This follows from the definition of Ky-Fan metric and the above-mentioned properties of ν. Note also that

$$\inf_{\varepsilon > 0} \left\{E(\varepsilon \wedge |A_T - B_T|) + P\{|A_T - B_T| \ge \varepsilon\}\right\} \le E|A_T - B_T| .$$

REMARK 2. Let $(N^n, \mathbb{F}^n)$ be a sequence of counting processes on a probability space $(\Omega, \mathcal{F}, P)$ with compensators A^n. Suppose that $(N, \mathbb{F})$ is a Poisson process on $(\Omega, \mathcal{F}, P)$ with a strictly increasing compensator B. If for each $t > 0$ we have

$$A_t^n \overset{P}{\to} B_t , \tag{12}$$

then the sequence (N^n) converges weakly to N (here $\overset{P}{\to}$ means convergence in probability as $n \to \infty$). This is a special case of the results of Brown [1] and Kabanov, Liptser, and Shiryayev [6]. Because A^n and B are increasing and B is continuous, it is easy to show that for any $T > 0$

$$\sup_{0 \le t \le T} |(B^{-1} \circ A)_t - t| \overset{P}{\to} 0 . \tag{13}$$

Hence this result follows directly from our Theorem 1. Moreover, we get an estimate for the rate of convergence to a Poisson process.

REMARK 3. It is shown in [6] that the condition (12) is not necessary for the weak convergence to a Poisson process. Hence it is not always possible to get good rates of convergence in terms of the compensators.

REMARK 4. The inequality (10) is a slight improvement of the corresponding results of Brown [2] and Kabanov, Liptser, and Shiryayev [6]. They have also considered the case when A is not necessarily continuous. In this case we can get the following bound:

$$\pi(M_T, N_T) \leq E(\varepsilon \wedge |A_T - B_T|) + P\{|A_T - B_T| \geq \varepsilon\} + E\left\{\sum_{t \leq T} (\Delta A_t)^2\right\}$$

for any $\varepsilon > 0$. To prove this we use the fact that there exists a counting process $(M', \mathbb{F}')$ with a continuous compensator A' such that $A'_T = A_T$ and $\pi(M_T, M'_T) \leq E\left\{\sum_{t \leq T} (\Delta A_t)^2\right\}$ (see [6]). The rest of the proof follows now easily from (10) and the triangle inequality.

For another kind of bound where N is a counting process with a deterministic (not necessarily continuous) compensator B we refer to Nikunen and Valkeila [9].

We note, however, that Brown [2] and also Kabanov, Liptser, and Shiryayev [6] consider the more general case of distances between all finite-dimensional distributions and also the total variation distance in function space.

2. AUXILIARY RESULTS

We start with a technical lemma that will be needed in the sequel.

LEMMA 1. Let f be a right-continuous increasing real function and g a continuous strictly increasing real function with $g(0) = 0$. Then for any $T > 0$ we have

$$d_T(f,\ f\circ g) \le \sup_{0\le t\le T}|h(t)-t| + |f(g(T))-f(T)| + \Delta f(T) + \Delta f(g(T)), \tag{14}$$

where h is the inverse function of g and $\Delta f(T) = f(T)-f(T-)$.

Proof. Let $T > 0$ be fixed. Suppose first that $h(T) = T$. Then $g \in \Lambda_T$ and

$$d_T(f,\ f\circ g) \le m_T(f\circ g,\ f\circ g) + m_T(g,I) = m_T(h,I) .$$

If $h(T) < T$, then there exists a $T_0 < T$ such that for all $t \in [T_0,T]$ we have $h(t) \le t$. Now we choose a $\delta \in (0, T-T_0)$, put $K = T/\delta$ and define

$$\lambda(t) = \begin{cases} h(t)\ , & t \in [0,T-\delta]\ , \\ [h(t)+K(t-T+\delta)]\wedge t\ , & t \in [T-\delta,T]\ . \end{cases} \tag{15}$$

It is clear that $\lambda \in \Lambda_T$, $\lambda \ge h$ on $[0,T]$ and $m_T(\lambda,I) \le m_T(h,I)$. Moreover,

$$\begin{aligned} d_T(f,\ f\circ g) &\le m_T(f,\ f\circ g\circ\lambda) + m_T(\lambda,I) \\ &\le f(g(T)) - f(T-\delta) + m_T(h,I) \ . \end{aligned}$$

Letting $\delta \to 0$, we see that (14) holds.

Let us now consider the case $h(T) > T$. Define a function

λ by the formula (15) with h replaced by g. Then $\lambda \in \Lambda_T$, $\lambda \geq g$ on $[0,T]$ and

$$d_T(f, f \circ g) \leq m_T(f \circ \lambda, f \circ g) + m_T(\lambda, I)$$

$$\leq f(T) - f(g(T-\delta)) + m_T(g, I) \quad .$$

Letting $\delta \to 0$, we get $d_T(f, f \circ g) \leq f(T) - (f \circ g)(T-) + m_T(g,I)$. It remains to note that in this case

$$m_T(g, I) = \sup_{0 \leq t \leq T} |g(t) - t| = \sup_{0 \leq h(t) \leq T} |g(h(t)) - h(t)|$$

$$= \sup_{0 \leq t \leq g(T)} |t - h(t)| \leq m_T(h, I) \; . \; ///$$

The following result is easy to deduce from Lenglart's inequality (see Lenglart [7]).

LEMMA 2. Let $(M,\mathbb{F})$ be a counting process with a continuous compensator A and let σ, τ be finite $\mathbb{F}$-stopping times such that $\tau \leq \sigma$. Then for any $\varepsilon > 0$

$$P\{M_\sigma - M_\tau \geq 1\} \leq E(\varepsilon \wedge (A_\sigma - A_\tau)) + P\{A_\sigma - A_\tau \geq \varepsilon\} \quad .$$

Let $(M,\mathbb{H})$ be a counting process on $(\Omega, \mathcal{F}, P)$ with compensator A and suppose that A is continuous, strictly increasing and $A_\infty = \infty$ $(P-a.s.)$. Let B be a continuous increasing real function with $B_0 = 0$. Define a time change σ by

$$\sigma(t) = \inf\{s: A_s > B_t\} \;, \qquad t \geq 0 \quad . \tag{16}$$

Note that $(\sigma(t))_{t \geq 0}$ is an increasing family of predictable $\mathbb{H}$-stopping times and $\sigma = A^{-1} \circ B$. Define a new process $(M^\sigma, \mathbb{H}^\sigma)$

by

$$M_t^\sigma = M_{\sigma(t)} \quad \text{and} \quad \mathbb{H}^\sigma = (H_{\sigma(t)})_{t\geq 0} \; . \tag{17}$$

Then the process $(M^\sigma, \mathbb{H}^\sigma)$ is a Poisson process with compensator B. For the proof see Liptser and Shiryayev [8].

3. THE PROOFS OF THE RESULTS

Proof of Theorem 1. Suppose first that the compensator A is strictly increasing and $A_\infty = \infty$. Define a new Poisson process M^σ by (17). By taking $f = M$ and $g = \sigma$ in Lemma 1 we see that

$$d_T(M, M^\sigma) \leq \sup_{0\leq t\leq T} |(B^{-1} \circ A)_t - t| + |M_{\sigma(T)} - M_T| + \Delta M_T + \Delta M_T^\sigma \; . \tag{18}$$

It follows from the continuity of A that the process M is quasi-left-continuous (Liptser and Shiryayev [8, Lemma 18.3]. Hence

$$d_T(M, M^\sigma) \leq \sup_{0\leq t\leq T} |(B^{-1} \circ A)_t - t| + |M_{\sigma(T)} - M_T| \quad , \quad P\text{-a.s.} \tag{19}$$

Now we use (6), (19) and the properties of ν listed above to get

$$\rho_T(M,N) = \rho_T(M, M^\sigma) \leq \alpha(M, M^\sigma; d_T) = \nu(d_T(M, M^\sigma))$$

$$\leq \alpha_T(B^{-1} \circ A, I) + P\{|M_{\sigma(T)} - M_T| \geq 1\} \; .$$

Take in Lemma 2 $\sigma = \sigma(T) \vee T$, $\tau = \sigma(T) \wedge T$. Because M and A are increasing processes we have $M_\sigma - M_\tau = |M_{\sigma(T)} - M_T|$ and $A_\sigma - A_\tau = |A_{\sigma(T)} - A_T| = |B_T - A_T|$. This gives us

$$P\left\{|M_{\sigma(T)} - M_T| \geq 1\right\} \leq E(\varepsilon \wedge |A_T - B_T|) + P\left\{|A_T - B_T| \geq \varepsilon\right\} . \tag{20}$$

This proves the inequality (8). For σ and τ as above we get $E\{|M_{\sigma(T)} - M_T| = E(M_\sigma - M_\tau) = E(A_\sigma - A_\tau) = E(|A_T - B_T|)$. This together with the inequality (19) proves (9). To prove (10) we recall that

$$\pi(M_T, N_T) = \pi(M_T, M_{\sigma(T)}) \leq P\{M_T \neq M_{\sigma(T)}\}$$
$$\leq P\{|M_{\sigma(T)} - M_T| \geq 1\} .$$

Now use (20). Theorem 1 is proved in the case when A is strictly increasing and $A_\infty = \infty$. The general case follows easily by adding to M an independent Poisson process with intensity μ and then letting $\mu \to 0$. ///

In conclusion, the authors want to point out that this problem comes from the "problem jar" of Professor A.N. Shiryaev.

REFERENCES

[1] Brown, T.C. "A Martingale Approach to the Poisson Convergence of Simple Point Processes." *Ann. Probab.*, 6 (1978): 615-628.

[2] Brown, T.C. "Some Poisson Approximations Using Compensators." *Ann. Probab.*, 11 (1983): 726-744.

[3] Dudley, R.M. "Distances of Probability Measures and Random Variables." *Ann. Math. Statist.*, 39 (1968): 1563-1572.

[4] Gikhman, I.I., and Skorokhod, A.V. *The Theory of Stochastic Processes*. Vol.1. Berlin Heidelberg New York: Springer-Verlag Inc., 1974.

[5] Kabanov, Yu.M., Liptser, R.S., and Shiryayev, A.N. "Some Limit Theorems for Simple Point Processes (A Martingale Approach)." *Stochastics*, 3 (1980): 203-216.

[6] Kabanov, Yu.M., Liptser, R.Sh., and Shiryaev, A.N. "Weak and Strong Convergence of the Distributions of Counting Processes." *Theory Probab. Applications*, vol.28, no.2 (1983): 303-335.

[7] Lenglart, E. "Relations de domination entre deux processus." *Ann. Inst. H. Poincaré*, 13 (1977): 171-179.

[8] Liptser, R.Sh., and Siryayev, A.N. *Statistics of Random Processes*. Vol.II. Berlin Heidelberg New York: Springer-Verlag Inc., 1978.

[9] Nikunen, M., and Valkeila, E. "A Note on One-Dimensional Distances Between Two Counting Processes." *Teoriya veroyatnostei i ee primeneniya*, vol.29, no.3 (1984): 558-561.

A.A. NOVIKOV

CONSISTENCY OF LEAST SQUARES ESTIMATES IN REGRESSION MODELS WITH MARTINGALE ERRORS

1. INTRODUCTION

In the classic model, for estimating parameters by the least squares method (LSM) it is supposed that the observed vector process $y_t = (y_{1t},\dots,y_{nt})'$ satisfies the following equation:

$$y_t = f_t\theta + v_t, \qquad t \in Z^+ = (0,1,\dots), \tag{1}$$

where $\theta = (\theta_1,\dots,\theta_k)'$ are unknown parameters, f_t is an observed matrix, and v_t are unobserved random errors ([1], [2], [3]). In this model of the least squares estimate for θ for each t has the form:

$$\theta_t = F_t^{-1} \sum_{s=0}^{t} f_s y_s \qquad \text{with} \quad F_t = \sum_{s=0}^{t} f_s f_s', \qquad t \in Z^+ \tag{2}$$

(here and below the negative power means the pseudoinverse).

In recent years, there has been considerable interest in the question of consistency of the least squares estimate θ_t, $t \to \infty$, in the case where v_t is a martingale-difference sequence, replacing the classical assumption of the independence of the v_t

([4], [5], [6], [7]). One may naturally interpret such results as a robust property of least-squares estimates. For the continuous-time case some related results have been obtained in the papers [8]-[11], and in many other works.

In the present paper, we consider a general model which allows us to consider the cases of continuous and discrete time simultaneously. In the framework of this model we establish general conditions of consistency of LSM-estimate in terms of eigenvalues of the matrix F_t, which are somewhat an improvement of the results obtained by Lai and Wei [6] (in the case of square integrable errors). The method used is similar to the method of [6], but here we consider the case of multi-dimensional observation process.

In Section 2 we give notations, a description of the model for the observation process and some auxiliary results. The strong consistency of the LMS-estimate in the model with square integrable errors is discussed in Section 3.

In Section 4 we discuss the case of non-square integrable errors and give the conditions for the weak consistency of the LMS-estimate.

2. NOTATION AND A MODEL FOR THE OBSERVED PROCESS

We shall use the terminology of the general theory of random processes (see [12] - [15]; following this theory it is possible to consider the case of continuous and discrete time simultaneously).

Let (Ω, G_t, P) be a probability space with standard requirement for right continuity of the filtration (G_t), $t \in R^+ = [0,\infty)$, and of completeness with respect to the measure P. Let:

M_{loc} (M^2_{loc}, M^c_{loc} and M^d_{loc}) be a class of local martingales with respect to the measure P and to the filtration (G_t);

$\underline{\underline{P}}$ be a class of predictable processes with respect to the measure P and to the filtration (G_t);

A^+ be a class of G_t-adapted nonnegative nondecreasing random processes.

By ξ we denote positive and finite (with probability one) random variables whose values are not important. In the proofs C denotes a positive constant whose value may change. All the inequalities and equalities are assumed to hold almost surely with respect to the measure P. The symbol lim, $O(\cdot)$ and $o(\cdot)$ denote limit relations (with probability one) as $t \to \infty$ (the symbol "$t \to \infty$" will be omitted).

In Section 4 we shall consider a convergence in probability which will be denoted by P-lim.

A MODEL FOR THE OBSERVED PROCESS. We suppose that the observed process $Y_t = (Y_{1t},\dots,Y_{\ell t})'$ satisfies the equation

$$Y_t = \int_0^t f'_s \, da_s \, \theta + m_t , \qquad t \in R^+, \tag{3}$$

where $m_t \in M_{loc}$, $a_t \in A^+ \cap \underline{\underline{P}}$, $f_t \in \underline{\underline{P}}$, $\int_0^t ||f_s|| \, da_s < \infty$ and

$$\int_0^t \|f_s\|^2 \, da_s \qquad t \in R^+ . \tag{4}$$

Here $\|A\|^2 = \mathrm{Sp}(AA')$ and the integrals with respect to a_s are usual Lebesgue-Stieltjes integrals. The condition (4) together with some other conditions will guarantee the existence of the stochastic integral $\int_0^t f_s \, dm_s$.

The discrete-time model (1) is imbedded into the model (3) in the following way. Let v_t be a martingale difference, that is,

$$E(v_t \mid G_{t-1}) = 0 , \qquad t \in Z^+.$$

Put $m_t = v_t$, $m_0 = 0$, $Y_t = y_t$ and complete these definitions for $t \in (n,n+1)$ in a piecewise manner. Now take

$$a_s = \text{integer part } s.$$

Then it is easy to see that the equation (3) is satisfied and the LSM-estimate (2) can be rewritten:

$$\theta_t = F_t^{-1} \int_0^t f_s \, dY_s \qquad \text{with} \qquad F_t = \int_0^t f_s f_s' \, da_s, \qquad t \in R^+. \tag{5}$$

By analogy with this case we take the same estimate for θ in the general model (3). Below we shall impose some conditions which will guarantee the existence of a stochastic integral from (5). Then by the property of stochastic integrals

$$\int_0^t f_s \, dY_s = \int_0^t f_s f_s' \, da_s \, \theta + \int_0^t f_s \, dm_s . \tag{6}$$

Note that for $m_t \in M^c_{loc}$ we may assume that f_t is only G_t-adapted (instead of the condition $f_t \in \underline{\underline{P}}$).

The proof of a strong consistency θ_t will be based on the following lemma, which is a generalization of the well-known Kronecker lemma.

LEMMA 1. (Liptser [16]). Let m_t and K_t be scalar processes, $m_t \in M_{loc}$, $1 \le K_t \in A^+ \cap \underline{\underline{P}}$ and $\lim K_t = \infty$. Denote

$$h_t = \int_0^t K_s^{-1}\, dm_s , \qquad B_t = \sum_{s \le t} |\Delta h_s|^2 (1 + |\Delta h_s|)^{-1} .$$

Then

$$\int_0^\infty K_t^{-2}\, d\langle m^c\rangle_s + \lim \tilde{B}_t < \infty \Rightarrow \lim \frac{m_t}{K_t} = 0 .$$

In this lemma and below we use the following notation:

$\Delta m_t = m_t - m_{t-}$;

m^c_t is a *continuous* part of m_t;

$\tilde{B}_t$ is a *compensator* of B_t, that is, $\tilde{B}_t \in \underline{\underline{P}}$ and $B_t - \tilde{B}_t \in M_{loc}$;

$$\langle m\rangle_t \stackrel{d}{=} \widetilde{\|m_t\|^2} \in A^+ \cap \underline{\underline{P}} .$$

In what follows we shall use also the following notation:

$$[m,m]_t \stackrel{d}{=} \langle m^c, m^c\rangle_t + \sum_{s \le t} \Delta m_s\, \Delta m_s'$$

is a matrix of quadratic variations of $m_t \in M_{loc}$;

$$\langle m,m\rangle_t \;=\; \widetilde{[m,m]_t}$$

is a matrix of quadratic characteristics of $m_t \in M^2_{loc}$. (Note $Sp(\langle m,m\rangle_t) = \langle m\rangle_t$.)

Lemma 2 (see [16]) will be used in the proof of Theorem 2.

LEMMA 2. Let $B_t \in A^+$ and $E \sup_{t\geq 0} \Delta B_t < \infty$. Then

$$B_t \sim \tilde{B}_t \qquad \text{on the set} \quad (\lim B_t = \infty).$$

In the proof of Theorems 1, 2 and 3 we shall use the notion of a *compensator* (dual predictable projection) $dq = q(dt,dx)$ of counting measure $dp = p(dt,dx)$ generated by a martingale m_t ([12] - [13]). In terms of these notions we have the following representations:

$$m_t \;=\; m_t^c \;+\; \int_0^t\!\!\int x\, d(p-q) \quad ,$$

$$[m,m]_t \;=\; \langle m^c,m^c\rangle_t \;+\; \int_0^t\!\!\int x\,x'\, dp \quad .$$

Here $x = (x_1,\ldots,x_k)'$ and the internal integral is over $R^n\backslash(0)$. For $m_t \in M^2_{loc}$ we have also

$$\widetilde{[m,m]}_t \;=\; \langle m,m\rangle_t \;=\; \langle m^c,m^c\rangle_t \;+\; \int_0^t\!\!\int x\,x'\, dq \quad ,$$

$$\langle m\rangle_t \;=\; \langle m^c\rangle_t \;+\; \int_0^t\!\!\int ||x||^2\, dq \quad .$$

3. CONSISTENCY OF THE LSM-ESTIMATE IN THE CASE OF SQUARE INTEGRABLE ERRORS

Let $\lambda_i(A)$ denote eigenvalues of the matrix $A = A'$ (size $k \times k$). Then $\lambda_1(A) = \min \lambda_i(A) \le \lambda_k = \max \lambda_i(A)$. Introduce a class of the functions

$$\underline{\underline{W}} = \left(w:\ w(u) \ge 1, \quad w(u)\uparrow, \quad \int_0^\infty \frac{1}{w(u)}\, du < \infty \right) .$$

<u>THEOREM 1</u>. Let $w \in \underline{\underline{W}}$, $\lim \lambda_1(F_t) = \infty$ and

$$\frac{d\langle m\rangle_t}{da_t} < \xi . \tag{7}$$

Then

$$\lim \frac{w(\log \lambda_k(F_t))}{\lambda_1(F_t)} = 0 \Rightarrow \lim ||\theta_t - \theta|| = 0 . \tag{8}$$

<u>Proof</u>. Introduce the stopping time $\tau = \inf(t > 0:\ \lambda_1(F_t) > 1)$, assuming as usual $\inf(\emptyset) = \infty$. From (3) and (6) it follows that

$$\theta_t - \theta = F_t^{-1} q_t \qquad \text{for any } t > \tau$$

and

$$||\theta_t - \theta||^2 \le \frac{Q_t}{\lambda_1(F_t)} , \tag{9}$$

where

$$q_t = \int_0^t f_s\, dm_s , \qquad Q_t = q_t' F_t^{-1} q_t .$$

Applying Itô's formula ([12]) we have for $t \ge s > \tau$

$$Q_t - Q_s = -\alpha_t + 2\beta_t + \gamma_t , \tag{10}$$

where

$$\alpha_t = \int_s^t q'_{u-} F_{u-}^{-1} (dF_u) F_u^{-1} q_{u-} ,$$

$$\beta_t = \int_s^t q'_{u-} F_{u-}^{-1} f_u \, dm_u ,$$

$$\gamma_t = \mathrm{Sp}\left(\int_s^t F_u^{-1} f_u \, d[m,m]_u \, f'_u\right) .$$

Note that $\alpha_t \in A^+ \cap \underline{\underline{P}}$, $\beta_t \in M^2_{loc}$. Since

$$\frac{d\langle m,m\rangle_t}{d\langle m\rangle_t} \leq E_k , \qquad E_k \text{ is an identity matrix,}$$

then

$$\langle \beta\rangle_t = \int_s^t q'_u F_{u-}^{-1} f_u \, d\langle m,m\rangle_u \, f'_u F_{u-}^{-1} q_{u-}$$

$$\leq \int_s^t q'_u F_{u-}^{-1} f_u \, d\langle m\rangle_u \, f'_u F_{u-}^{-1} q_{u-} \leq \xi\alpha_t , \tag{11}$$

where the last inequality holds by (7).

It is well known that for any $\beta \in M^2_{loc}$

$$\beta_t = o(\langle\beta\rangle_t) \qquad \text{on the set } (\lim \langle\beta\rangle_t = \infty),$$

$$|\lim \beta_t| < \infty \qquad \text{on the set } (\lim \langle\beta\rangle_t < \infty).$$

Hence by (11) we have

$$-\alpha_t + 2\beta_t < \xi . \tag{12}$$

In order to prove Theorem 1, it suffices to show that

$$\gamma_t = o(w(\log \lambda_k(F_t))) \tag{13}$$

for the function $w \in \underline{\underline{W}}$. To verify this relation, let us consider the process

$$K_t = Sp\left(\int_s^t F_u^{-1} f_u f_u' \, da_u\right)$$

and note that similar to the proof of the inequality (11) one can easily obtain that

$$\tilde{\gamma}_t = Sp\left(\int_s^t F_u^{-1} f_u \, d\langle m,m\rangle_u \, f_u'\right) \leq \xi K_t \ .$$

Since $t = o(w(t))$, then

$$\tilde{\gamma}_t = o(w(K_t)) \qquad \text{on the set} \quad (\lim K_t = \infty). \quad (14)$$

To apply Lemma 1 to the process

$$\gamma_t - \tilde{\gamma}_t = Sp\left(\int_s^t F_u^{-1} f_u \, d([m,m]_u - \langle m,m\rangle_u) \, f_u'\right) \in M_{loc}^d \ ,$$

let us consider the local martingale

$$h_t = \int_s^t w^{-1}(K_u) \, d(\gamma_u - \tilde{\gamma}_u)$$

$$= Sp\left(\int_s^t \int w^{-1}(K_u) F_u^{-1} f_u x x' f_u' \, d(p-q)\right)$$

By the inequality $||x-y|| \leq ||x|| + ||y||$ and by (7) we have

$$\sum_{u=s}^{t} |\Delta h_u| \leq 2\int_0^\infty \int w^{-1}(K_u)\ Sp(F_u^{-1} f_u x x' f_u')\ dq$$

$$\leq 2\int_0^\infty w^{-1}(K_u)\ Sp(F_u^{-1} f_u f_u') \int ||x||^2\ dq$$

$$\leq \xi \int_0^\infty w^{-1}(K_u)\ Sp(F_u^{1} f_u f_u')\ da_u \leq \xi \ .$$

Hence by Lemma 1 $\gamma_t - \tilde{\gamma}_t = o(w(K_t))$ on the set $(\lim K_t = \infty)$. Therefore by (14) we have proved that

$$\gamma_t = o(w(K_t)) \qquad \text{on the set } (\lim K_t = \infty).$$

To prove (13) we need to show only that

$$K_t \leq C \log \lambda_k(F_t) \ . \tag{15}$$

First note that for any $\delta \geq 0$ and $u \geq \delta + \tau$

$$\det (F_{t-\delta}) = \det (F_t - \Delta_\sigma F_t) = \det (F_t) \prod_{i=1}^{k} (1 - \lambda_i(F_t^{-1} \Delta_\delta F_t)) ,$$

where $\Delta_\delta F_t = F_t - F_{t-\delta}$. Since $0 \leq \lambda_i(F_t^{-1}\Delta_\delta F_t) \leq 1$, then by the inequality $1 - x \leq \exp(-x)$, $x \geq 0$,

$$Sp(F_t^{-1}\Delta_\delta F_t) = \sum_{i=1}^{k} \lambda_i(F_t^{-1}\Delta_\delta F_t)$$

$$\leq \log \left[\frac{\det (F_t)}{\det (F_{t-})}\right] \tag{16}$$

It is easy to see that

$$K_n = \lim_{n\to\infty} Sp\left(\int_s^t F_{n,u}^{-1}\ dF_u\right) ,$$

where $F_{n,u}^{-1}$ is a step function equal to F_u^{-1} at the right end-points (to construct such a step function, one may separate the discontinuous and continuous parts of F_u^{-1} and use for the continuous part the usual uniform approximation). By (16) we have now

$$K_t \leq \log \left[\frac{\det (F_t)}{\det (F_s)}\right] \leq \log \det (F_t) \quad ,$$

and hence the estimate (15) holds. ///

<u>COROLLARY 1</u>. In the model (1) let $\lim \lambda_1(F_t) = \infty$ and v_t be a martingale difference such that

$$E(\|v_t\|^2 \mid G_{t-1}) < \infty \quad , \qquad t \in Z^+.$$

Then

$$\lim \frac{w(\log \lambda_k(G_t))}{\lambda_1(F_t)} = 0 \Rightarrow \lim \|\theta_t - \theta\| = 0$$

for any function $w \in \underline{\underline{W}}$.

For the case $w(x) = x^p$, $p > 1$, and a one-dimensional v_t this result has been obtained in Lai and Wei [6].

In Theorem 2 below we show that the implication (8) holds also for $w(x) = x$, $x > 1$, but under a slightly stronger condition for moments of v_t. To formulate this theorem, we need some additional notations. Let:

$$m_t^1 = m_t^c + \int_0^t\!\!\int I(\|x\| \leq 1)\ x\ d(p-q);$$

$I(A)$ be an indicator function of the set A;

a_t^d be a *discontinuous* part of the process a_t;

<u>THEOREM 2</u>. Let $w \in \underline{\underline{W}}$, $\lim \lambda_1(F_t) = \infty$, $(\Delta a_t^d)^{-1} \leq c$ and

$$\frac{d\langle m^1\rangle_t}{da_t} + \frac{d}{da_t}\int_0^t\int I(||x|| > 1)\; w(||x||^2)\, dq \;\leq\; \xi \;. \tag{17}$$

Then

$$\lim \frac{\log \lambda_k(F_t)}{\lambda_1(F_t)} = 0 \;\Rightarrow\; \lim ||\theta_t - \theta|| = 0 \;. \tag{18}$$

<u>Proof</u>. Looking through the proof of Theorem 1 it is easy to see that one needs only to verify that

$$\gamma_t = O(\log \lambda_k(F_t)) \;. \tag{19}$$

To prove this relation, we note first that

$$\gamma_t = \phi_t + \psi_t \;,$$

where

$$\phi_t = Sp\left[\int_s^t F_u^{-1} f_u\; d\langle m^c, m^c\rangle_u f_u'\right] + Sp\left[\int_0^t \int I(||x|| \leq 1) F_u^{-1} f_u xx' f_u'\; dp\right] \tag{20}$$

and

$$\psi_t = Sp\left(\int_s^t I(||x|| > 1)\; F_u^{-1} f_u xx' f_u'\; dp\right) \;.$$

The process ϕ_t has bounded jumps since in (20) the first term is continuous and jumps of the second one do not exceed the value

$$\sup_{t \geq s} Sp(F_s^{-1} f_s f_s') \;\leq\; c \sup_{t \geq s} Sp(F_s^{-1} \Delta_\delta F_s) \;\leq\; c \;.$$

Hence by Lemma 2 $\phi_t \sim \tilde{\phi}_t$. On the other hand,

$$\tilde{\phi}_t = Sp\left\{\int_s^t F_u^{-1} f_u \, d\langle m^c, m^c\rangle_u \, f_u'\right\} + Sp\left\{\int_s^t I(\|x\| \le 1) F_u^{-1} f_u xx' f_u' \, dg\right\}$$

$$\le Sp\left\{\int_s^t F_u^{-1} f_u \, d\langle m^1\rangle_u \, f_u'\right\} .$$

By (17) and (15) we have $\phi_t = O(\log \lambda_k(F_t))$.

To estimate ψ_t, let us apply Lemma 1 to the local martingale

$$h_t = h_t' + h_t''$$

with

$$h_t' = \int_s^t (K_u + 1)^{-1} I(1 < \|x\| < K_u + 1) \, Sp(F_u^{-1} f_u xx' f_u') \, d(p-q) ,$$

$$h_t'' = \int_s^t (K_u + 1)^{-1} I(\|x\| \ge K_u + 1) \, Sp(F_u^{-1} f_u xx' f_u') \, d(p-q) .$$

To apply Lemma 1, we need to verify that

$$\lim \widetilde{\sum_{s \le t} |\Delta h_s'|} < \infty , \qquad \lim \langle h''\rangle_t < \infty . \tag{21}$$

By (17) we have

$$\widetilde{\sum_{s \le t} |\Delta h_s'|} \le 2\int_s^t (K_u + 1)^{-1} Sp(F_u^{-1} f_u f_u') \int I(1 < \|x\| < K_u + 1) \, \|x\|^2 \, dq$$

$$\le 2\int_s^t [w(K_u + 1)]^{-1} \, Sp(F_u^{-1} f_u f_u') \, da_u \le \xi .$$

Since by (17)

$$\langle h''\rangle_t = \int_s^t [(K_u+1)^{-1} Sp(F_u^{-1}f_uf_u')]^2 \int I(\|x\| \geq K_u+1)\ \|x\|^2\ dq$$

$$\leq \xi\int_s^t [w(K_u+1)]^{-1} Sp(F_u^{-1}f_uf_u')\ da_u \leq \xi ,$$

then both conditions (21) are satisfied. Hence by Lemma 1 $\psi_t - \tilde{\psi}_t = o(K_t)$. It is easy to see that $\psi_t = O(\log \lambda_k(F_t))$ and hence $\psi_t = O(\log \lambda_k(f_t))$. Summarizing we have (19). ///

<u>EXAMPLE</u>. This example is a continuous-time version of the example given by Lai and Wei in [6] who showed that the condition of the implication (18) cannot be weakened. We consider here only the case of continuous martingale errors, but the same technique may be used for constructing a similar example with discontinuous martingale errors. Let m_t be a scalar process, $m_t \in M^c_{loc}$ and

$$Y_t = \int_0^t (\theta_1 + \theta_2 x_s)\ d\langle m\rangle_s + m_t , \qquad t \in R^+.$$

Using the notation $\bar{x}_t = \langle m\rangle_t^{-1}\int_0^t x_s\ d\langle m\rangle_s$, we can write

$$\det (F_t) = \langle m\rangle_t\left[\int_0^t x_s^2\ d\langle m\rangle_s - \left(\int_0^t x_s\ d\langle m\rangle_s\right)^2\right]$$

$$= \langle m\rangle_t\left[\int_0^t (x_s - \bar{x}_s)^2\ d\langle m\rangle_s\right] .$$

Let $\lim \langle m\rangle_t = \infty$ and $x_t = \int_0^t (1 + \langle m\rangle_s)^{-1} dm_s$. Then $\lim \langle x\rangle_t < \infty$ and hence $\lim x_t$ is finite random. Denote $\lim x_t = \eta$. Using Itô's formula and Lemma 1, one can easily verify that

$$\int_0^t (x_s - \bar{x}_s)^2 \, d\langle m\rangle_s \sim \log \langle m\rangle_t ,$$

$$\lim \frac{\int_0^t x_s \, d\langle m\rangle_s}{\langle m\rangle_t} = \eta ,$$

$$\lim \frac{\int_0^t x_s^2 \, d\langle m\rangle_s}{\langle m\rangle_t} = \eta^2 .$$

The simple calculations show that

$$\lambda_2(F_t) \sim \langle m\rangle_t \zeta$$

with

$$\zeta = \frac{1 + \eta^2 + \sqrt{(1-\eta)^2 + 4\eta^2}}{2} > 0$$

and

$$\lambda_1(F_t) = \frac{\det (F_t)}{\lambda_2(F_t)} \sim \frac{\log \langle m\rangle_t}{\zeta}$$

Hence $\lim \dfrac{\log \lambda_2(F_t)}{\lambda_1(F_t)} > 0$ and therefore the condition of the implication (18) does not hold. On the other hand,

$$\theta_{2t} - \theta_2 = \left[\int_0^t (x_s - \bar{x}_s)^2 \, d\langle m\rangle_s\right]^{-1} \left[\int_0^t x_s \, dm_s - m_t \bar{x}_t\right]$$

Using again Itô's formula and Lemma 1, one can easily show that

$$\left[m_t x_t - \int_0^t x_s \, dm_s\right] \sim \log \langle m\rangle_t .$$

Hence $\lim \theta_{2t} - \theta_2 = -1$ and therefore in this model the LSM-estimate θ_t is not consistent.

COROLLARY 2. In the model (1) let $\lim \lambda_1(F_t) = \infty$ and let v_t be a martingale difference such that

$$E(w(\|v_t\|) \mid G_{t-1}) \leq \xi , \qquad t \in Z^+,$$

for a function $w \in \underline{\underline{W}}$. Then

$$\lim \frac{\log \lambda_k(F_t)}{\lambda_1(F_t)} = 0 \Rightarrow \lim \|\theta_t - \theta\| = 0 .$$

4. WEAK CONSISTENCY OF THE LSM-ESTIMATE IN THE CASE OF NON-SQUARE INTEGRABLE MARTINGALE ERRORS

In this Section we consider the general model (3) under an additional assumption that F_t is a deterministic matrix, but we do not suppose now that m_t is a square integrable martingale.

THEOREM 3. Let $\lim \lambda_1(F_t) = \infty$, $m_t \in M_{loc}$, F_t be a deterministic matrix and let

$$\frac{d\langle m^1\rangle_t}{da_t} + \frac{d}{da_t}\int_0^t \int I(\|x\| > 1)\, \|x\|^\alpha\, dq < \xi$$

for some $\alpha \in [1,2]$.

Then

$$\lim \frac{(a_t)^{2/\alpha - 1}}{\lambda_1(F_t)} = 0 \Rightarrow P\text{-}\lim \|\theta_t - \theta\| = 0 .$$

Proof. Introduce the stopping time

$$\tau_N = \inf\left(t > 0 : \frac{d\langle m^1\rangle_t}{da_t} > N \quad \text{or} \quad \frac{d}{da_t}\int_0^t \int I(\|x\| > 1)\|x\|^\alpha dq > N\right) .$$

Then

$$P(\tau_N < t) \le P(\xi > N) \to 0 \qquad \text{as } N \to \infty$$

and by Chebyshev's inequality

$$P(\|\theta_t - \theta\| > \varepsilon)\varepsilon^{\alpha} \le P(\xi > N)\varepsilon^{\alpha} + EI(\tau_N > t)\|\theta_t - \theta\|^{\alpha} . \quad (22)$$

Note that τ_N is a predictable stopping time and by representations (5) and (6) we have

$$I(\tau_N > t)\|\theta_t - \theta\|^{\alpha} \le C\left\|\int_0^t I(\tau_N > s)F_t^{-1}f_s \, dm_s^1\right\|^{\alpha} \quad (23)$$

$$+ C\left\|\int_0^t I(\tau_N > s)F_t^{-1}f_s \, d(m_s - m_s^1)\right\|^{\alpha} .$$

By the Davis inequality ([12]) for any $\mu_t \in M_{loc}$

$$E \sup_{u \le t} \|\mu_u\|^{\alpha} \le CE\left\{\langle\mu^c\rangle_t + \sum_{u \le t} \|\Delta\mu_u\|^2\right\}^{\alpha/2} , \qquad \alpha \ge 1 .$$

For $\mu_t \in M_{loc}^2$ this inequality implies that

$$E \sup_{u \le t} \|\mu_u\|^{\alpha} \le CE(\langle\mu\rangle_t)^{\alpha/2} .$$

Let us apply now this inequality to the first term in (23).

Denote

$$\mu_u = \int_0^u I(\tau_N > s)F_t^{-1}f_s \, dm_s^1 .$$

Since on the set $(\tau_N > t)$ we have $\frac{d\langle m^1\rangle_t}{da_t} \le N$, then for $u \le t$

$$\langle\mu\rangle_u = \int_0^u I(\tau_N > s)\|F_t^{-1}f_s\|^2 \, d\langle m^1\rangle_s \le N\int_0^t \|F_t^{-1}f_s\|^2 \, da_s .$$

On the other hand, by the definition of F_t

$$\int_0^t \|F_t^{-1} f_s\|^2 \, da_s = Sp\left(\int_0^t F_t^{-1} f_s f_s' F_t^{-1} \, da_s\right)$$

$$= Sp(F_t^{-1}) \leq \frac{k}{\lambda_1(F_t)} \tag{24}$$

Combining these estimates, we obtain

$$E\left\|\int_0^t I(\tau_N > s) F_t^{-1} f_s \, dm_s^1\right\|^\alpha \leq C[\lambda_1(F_t)]^{-\alpha/2} .$$

In the case of pure discontinuous martingales Davis' inequality implies also that

$$E \sup_{u \leq t} \|\Delta\mu_u\|^\alpha \leq CE \sum_{u \leq t} \|\Delta\mu_u\|^\alpha , \qquad 1 \leq \alpha \leq 2$$

Now we apply this inequality to the martingale

$$\mu_u = \int_0^u I(\tau_N > s) F_t^{-1} f_s \, d(m_s - m_s^1) .$$

Since $\frac{d}{da_t} \int_0^t \int I(\|x\| > 1) \|x\|^\alpha \, dq \leq N$ on the set $(\tau_N > t)$, we have

$$\sum_{u \leq t} \|\Delta\mu_u\|^\alpha \leq C\int_0^t I(\tau_N > s) \|F_t^{-1} f_s\|^\alpha \int I(\|x\| > 1) \|x\|^\alpha \, dq .$$

Now by the Hölder inequality and by (24)

$$\int_0^t \|F_t^{-1} f_s\|^\alpha \, da_s \leq C\left(\frac{(a_t)^{2/\alpha - 1}}{\lambda_1(F_t)}\right)^{\alpha/2} .$$

Combining these estimates, we obtain

$$EI(\tau_N > t)\|\theta_t - \theta\|^\alpha \leq C\left(\frac{1}{\lambda_1(F_t)} + \frac{(a_t)^{2/\alpha-1}}{\lambda_1(F_t)}\right)^{\alpha/2} .$$

This estimate and the inequality (22) imply Theorem 3. ///

<u>COROLLARY 3</u>. In the model (1) let $\lim \lambda_1(F_t) = \infty$, let F_t be a deterministic matrix, and let v_t be a martingale difference such that

$$E(\|v_t\|^\alpha \mid G_{t-1}) \leq \xi \qquad \text{for some} \quad \alpha \in [1,2], \quad t \in Z^+ .$$

Then

$$\lim \frac{(t)^{2/\alpha-1}}{\lambda_1(F_t)} = 0 \Rightarrow \text{P-lim}\,\|\theta_t - \theta\| = 0 .$$

Note that the result similar to Corollary 3 for the case of the independent and one-dimensional v_t was obtained by Kaffes and Rao [18].

REFERENCES

[1] Linnik, Yu.V. *Metod naimen'shikh kvadratov i osnovy matematicheskoj statistiki* (Least Squares Methods and Fundamentals of Mathematical Statistics). 2nd edition. Moskva: Nauka, 1962.

[2] Rao, C.R. *Linear Statistical Inference and Its Applications.* 2nd edition. New York: Wiley & Son, 1973.

[3] Seber, G.A.F. *Linear Regression Analysis.* New York: Wiley & Son, 1977.

[4] Drygas, H. "Weak and Strong Consistency of the Least Squares Estimates in Regression Models." *Z. Wahrscheinlichkeitstheorie und Verw. Gebiete*, vol.34 (1976): 119-127.

[5] Anderson, T.W., and Taylor, J.B. "Strong Consistency of Least Squares Estimates in Normal Linear Regression." *Ann. Statist.*, vol.4, no.4 (1976): 788-790.

[6] Lai, T.L., and Wei, C.Z. "Least Squares Estimates in Stochastic Regression Models with Applications to Identification and Control of Dynamic Systems." *Ann. Statist.*, vol.10, no.1 (1982): 154-166.

[7] Barabanov, A.E. "An Employment of the Least Squares Method for Designing Adaptive Optimal Control Systems for a Linear Dynamic Process. " *Avtomatika i telemekhanika*, no.12 (1983): 57-65.

[8] Arato, M. Linear Stochastic Systems with Constant Coefficients. A Statistical Approach. *Lecture Notes in Control and Information Sci.*, vol.45. Berlin Heidelberg New York: Springer-Verlag Inc., 1983.

[9] Novikov, A.A. "Ob otsenkakh parametrov diffuzionnykh protsessov (On the Estimation of Parameters of Diffusion Processes). *Studia Scientiarum Mathematicarum Hungarica*, vol.7 (1972): 201-209.

[10] Le Breton, A., and Musiela, M. "Estimation des paramètres pour les diffusions gaussiennes homogènes hypoelliptiques." *C.R. Acad. Sci. Paris, Comptes rendus, Serie 1, Math.*, vol. 294, no.10 (1982): 341-344.

[11] Bellach, B. "Parameter Estimation in Linear Stochastic Differential Equations and Their Asymptotic Properties." *Math. Operationsforsch. Statist., ser. Statistics*, vol.14, no.1 (1983): 141-191.

[12] Jacod, J. Calcul stochastique et problèmes des martingales. *Lecture Notes in Math.*vol.714. Berlin Heidelberg New York: Springer-Verlag Inc., 1979.

[13] Kabanov, Yu.M., Lipčer, R.S., and Širjaev, A.N. "Absolute Continuity and Singularity of Locally Absolutely Continuous Probability Distributions." *Math. USSR, Sbornik*, vol.35, no.5 (1979): 631-680.

[14] Elliot, R.J. *Stochastic Calculus and Applications*. Berlin Heidelberg New York: Springer-Verlag Inc., 1982.

[15] Gikhman, I.I., and Skorokhod, A.V. *Stokhasticheskie differentsial'nye uravneniya i ikh primeneniya*. Kiev: Naukova Dumka, 1982.

[16] Lipčer, R.Sh. "A Strong Law of Large Numbers for Local Martingales." *Stochastics*, vol.3, no.3 (1980): 217-228.

[17] Lepingle, D. "Sur le comportement asymptotique des martingales locales." *Lecture Notes in Math.*, pp. 148-161, vol.649. Berlin Heidelberg New York: Springer-Verlag Inc., 1976/77.

[18] Kaffes, D., and Rao, M.B. "Weak Consistency of Least Squares Estimates in Linear Models." *J. Mult. Anal.*, vol.12, no.2 (1982): 186-198.

B.L. ROZOVSKIJ

NONNEGATIVE L_1-SOLUTIONS OF SECOND ORDER STOCHASTIC PARABOLIC EQUATIONS WITH RANDOM COEFFICIENTS

1. INTRODUCTION

In this paper we consider the Cauchy problem for the following linear second-order stochastic partial differential equation:

$$du(t,x) = L^*(t,x)\, u(t,x)\, dt + \sum_{\ell=1}^{d_1} M^{\ell *}(t,x)\, u(t,x)\, dw^\ell(t),$$

$$(t,x) \in \,]0,T] \times R^d, \qquad (1)$$

$$u(0,x) = \phi(x), \qquad x \in R^d, \qquad (2)$$

where

$$L(t,x)\cdot := \sum_{j,i=1}^{d} a^{ij}(t,x) \frac{\partial^2}{\partial x^i \partial x^j}\cdot + \sum_{i=1}^{d} b^i(t,x) \frac{\partial}{\partial x^i}\cdot + c(t,x)\cdot \,,$$

$$M^\ell(t,x)\cdot = \sum_{i=1}^{d} \sigma^{i\ell}(t,x) \frac{\partial}{\partial x^i}\cdot + h^\ell(t,x)\cdot \,,$$

L^* and $M^{\ell *}$ are operators formally conjugated to L and M^ℓ, correspondingly and $w := (w^1,\dots,w^{d_1})$ is a d-dimensional Wiener process.

Coefficients of the operators L^*, $M^{\ell *}$ are supposed to be random, predictable. Equations of this type arise in non-linear filtering for diffusion processes and in many other problems (see, for example, [1], [2]). In particular, if the matrix (a^{ij}) is symmetric, the initial function ϕ is a density of some probabilistic measure with respect to Lebesgue measure and $M^{\ell} = 0$, then the problem (1), (2) is the Kolmogorov forward equation. It is well known that the solution of this equation is also a density of some probabilistic measure. To be more specific, under appropriate smoothness conditions imposed on the coefficients of L^*, the solution of the Kolmogorov forward equation

$$u(t,x) = \frac{P(Y(t) \in dx)}{ds} ,$$

where $Y(t)$ is a solution of the Itô equation

$$Y^i(t) = \xi^i + \int_{[0,t]} b^i(s, Y(s))\, ds + \sum_{\ell=0}^{d_1} \int_{[0,t]} \hat{\sigma}^{i\ell}(s, Y(s))\, d\hat{w}^{\ell}(s) .$$

Here $(\hat{\sigma}^{ij})$ is the $d \times d$-dimensional square root of the matrix $2(a^{ij})$, ξ is such a random variable that $\sigma = \frac{P(\xi \in dx)}{dx}$ and $\hat{w}$ is a Wiener process independent of ξ.

In the present work, we aim at obtaining a statement which is an analogy of this result mentioned above for the equation (1) in the general case $(M^{\ell} \neq 0)$.

It was shown in [2] - [4] that the behavior of the equation (1) depends on the matrix

$$(A^{ij}) := \left(2a^{ij} - \sum_{i=1}^{d_1} \sigma^{i\ell}\sigma^{j\ell}\right)$$

In [2], examples are given to show that the condition

- A. $\sum_{i,j=1}^{d} A^{ij}\xi^i\xi^j \geq 0, \quad \forall\ \xi \in R^d,$

is necessary in some sense for the solubility of the problem (1), (2) in the space of square integrable functions.

Say that equation (1) is degenerate parabolic if condition A is fulfilled and non-degenerate parabolic if:

- A'. there exists such a $\delta > 0$ that

$$\sum_{i,j=1}^{d} A^{ij}\xi^i\xi^j \geq \delta \sum_{i=1}^{d} |\xi^i|^2, \quad \forall\ \xi \in R^d.$$

In this paper we are mainly concerned with the degenerate case. An analytical treatment of the problem (1), (2) has been presented in [3], [4].

2. DEFINITIONS AND MAIN RESULTS

Let R^d be a d-dimensional Euclidean space, T a fixed positive number, (Ω, F, P) a complete probability space, $\{F_t\}$, $t \in [0,T]$, an increasing family of right-continuous complete σ-algebras contained in F, $\{w(t), F_t\}$ a d-dimensional Wiener process, F_t^w a σ-algebra generated by $w(s)$, $s \leq t$, and completed with respect to P, and $\mathcal{P}$ a σ-algebra of predictable sets relative to $\{F_t\}$.

If $(S, \mathcal{S})$ is a measurable space, μ is some measure on $(S, \mathcal{S})$, Z is a Banach space and $P \in [1, \infty[$ as we denote the space of all measurable functions $u: S \to Z$ with the finite norm

$$\left(\int_S \|u(s)\|_Z^P \, d\mu(s)\right)^{1/P}$$

by $L_P(S,Z)$.

We denote the Sobolev space $W_2^m(R^d)$ by W_2^m, $W_2^0 = L_2(R^d;R^1)$ by L_2, the scalar product and the norm in W_2^m by $(\cdot,\cdot)_m$ and $\|\cdot\|_m$, respectively.

We denote the space of continuous functions $u: S \to Z$ by $C(S,Z)$. As usual, we denote by C^m the space of continuous real-valued functions on R^d having derivatives up to order m, by C_0^∞ the subspace of C^∞ that consists of the function with compact support and by L^∞ the space of bounded real-valued functions on R^d.

WARNING. In the sequel, the superscripts i, j and ℓ denote the differentiation with respect to the coordinates with the corresponding numbers. For example, $f_{ij}(x): = \partial^2 f(x)/\partial x^i \partial x^j$.

Repeated indices i, j and ℓ in monomials are summed over. For example,

$$\int_{[0,t]} h^\ell(\tau)\, dw^\ell(\tau): = \int_{[0,t]} \sum_{\ell=1}^{d_1} h^\ell(\tau)\, dw^\ell(\tau) \quad . \quad ///$$

Throughout the remainder of the paper we shall make the following assumptions:

• 1. functions $a^{ij}(t,x)$, $b^i(t,x)$, $c(t,x)$ $\sigma^{i\ell}(t,x)$, $h^\ell(t,x)$ are real valued functions defined on $[0,T] \times R^d \times \Omega$ */);

*/ The argument ω $(\Omega = \cup\, \omega)$ has been omitted, as a rule. Where there is no danger of confusion other arguments may be also omitted. Sometimes in writing a product of several functions that depend on the same argument, this argument is written only after the last function of the product.

•2. these functions are bounded, measurable in (t,x,ω) and P-measurable for every $x \in R^d$.

Unless otherwise stipulated we assume also that:

•3. condition A holds true for every t, x, ω;

•4. a^{ij} and $\sigma^{i\ell} \in C_b^3$, b^i and $h^\ell \in C_b^2$, $c \in C_b^1$ for every (t,ω), $\phi \in W_2^1$. */

It is known (see [4]) that under these assumptions the problem (1), (2) has a unique generalized solution

$$u \in L_2([0,t] \times \Omega;\ W_2^1) \cap L_2(\Omega;\ C([0,T];\ L_2)) .$$

(We say that u is a generalized solution of the problem (1), (2) if u is P-measurable mapping $[0,T] \times \Omega \to L_2$ which is continuous in t, belongs to W_2^1 for almost all (t,ω) and for all $\eta \in C_0^\infty$ satisfies the equation

$$(u(t),\eta)_0 = (\phi,\eta)_0 - \int_{[0,t]} [(a^{ij}u_j(s) + a_j^{ij}u(s) + b^i u(s),\ \eta_i)_0 - (cu(s),\ \eta)_0]\, ds$$

$$- \int_{[0,t]} [(\sigma^{i\ell}u(s),\ \eta_i)_0 - (h^\ell u(s),\ \eta)_0] dw^\ell(s) \quad).$$

Under some additional smoothness condition on the coefficients and ϕ, this generalized solution becomes also a classical one (for details, see [4]).

Let us denote by $(\hat{\sigma}^{ij})$ the d×d-square root of the matrix (A^{ij}). We can (see [5]) and shall assert that the elements $\hat{\sigma}^{ij}$ are Lipschitz continuous in x and uniformly bounded.

*/ C_b^m is the subspace of C^m that consists of bounded functions with bounded derivatives.

We take Σ for the matrix $(\Sigma^{i\ell})$, $i = 1/d$, $\ell = 1/d+d_1$, where

$$\Sigma^{i\ell}: = \begin{cases} \sigma^i & \ell = 1 \div d \quad , \\ \sigma^{i(\ell-d)} , & \ell = d+1 \div d+d_1 \quad , \end{cases}$$

and B for the vector (B^i), where $B^i: = b^i - \sigma^{i\ell}h^\ell$, $i = 1 \div d$. We introduce also a new d-dimensional Wiener process $\{w(t), F_t\}$ and denote $\nu(t): = (\hat{w}(t), w(t))$.

Let us consider the system of Itô's equations:

$$X^i(t) = \xi^i + \int_{[0,t]} B^i(s, X(s))\, ds + \int_{[0,t]} \Sigma^{i\ell}(s, X(s))\, d\nu^\ell(s) ,$$
$$s \in [0,t], \quad i = 1 \div d , \quad (3)$$

where ξ is an F_0-measurable random variable in R^d such that

$$P(\xi \in \Gamma) = \int_\Gamma \phi(x)\, dx , \qquad \Gamma \in \mathcal{B}(R^d). \text{ */} \qquad (4)$$

Apparently, under our conditions this system has a trajectorily unique strong solution.

We put

$$\zeta(t): = \exp\left\{ \int_{[0,t]} c(s, X(s))\, ds + \int_{[0,t]} h^\ell(s, X(s))\, dw^\ell(s) - \tfrac{1}{2} \int_{[0,t]} h^\ell h^\ell(s, X(s))\, ds \right\} .$$

The basic result of the present work consists in the following:

THEOREM. If condition (4) holds and u is the generalized

*/ Here and below $\mathcal{B}$ is a Borel σ-algebra.

solution of the problem (1), (2), then for every $\psi \in L_\infty$, $t \in [0,T]$

$$\int_{R^d} \psi(x)\, u(t,x)\, dx = E[\psi(X(t))\rho(t) \mid F_t^W] \quad P\text{-a.s.} \quad /// \tag{5}$$

From (5) we easily derive the following.

COROLLARY. Under the assumptions of the Theorem for every t the generalized solution $u(t)$ of the problem (1), (2) belongs to the cone of nonnegative functions from $L_1(R^d,R^1)$ (a.s. P).///

Some additional corollaries of the Theorem can be found in [6].

REMARK 1. Having in mind applications to the filtering problem, it is useful to generalize the results presented above in the following way.

Let G_0 be some sub-σ-algebra of F_0. Denote by G_t the minimal σ-algebra generated by G_0 and F_t^W, by $\mathcal{P}(G)$ the σ-algebra of predictable sets relative to $\{G_t\}$ and put $F(G) := (\Omega, F, \{G_t\}, P)$. Note that $\{v(t), G_t\}$ is also a Wiener process.

Suppose now that the coefficients of equation (1) are $\mathcal{P}(G)$-measurable for every $x \in R^d$, ϕ is a G_0-measurable random variable taking values in W_2^1 and belonging to $L_2(\Omega; W_2^1)$. Also denote by $P_{G_0}(\xi \in \cdot)$ the regular conditional distribution of ξ with respect to G_0.

The statements of the Theorem and the Corollary remain true if condition (4) is replaced by

$$P_{G_0}(\xi \in \Gamma) = \int_\Gamma \phi(x)\, dx\,, \quad \forall\ \Gamma \in \mathcal{B}(R^d).\ /// \tag{4'}$$

REMARK 2. If instead of condition A condition A' is fulfilled and we require only that a^{ij} and $\sigma^{i\ell} \in C_b^1$, b^i, h^ℓ and $c \in L_\infty$ and ϕ takes values in L_2 and belongs to $L_2(\Omega; L_2)$, the statements of the Theorem, the Corollary and Remark 1 are still valid.

The proof of the Theorem is presented below in Sections 3-5. The proof of generalization of the results mentioned in Remark 1 is very much the same, so we omit it.

Looking somewhat ahead, we mention here that the smoothness conditions of the coefficients and ϕ are needed basically to prove existence and uniqueness of a solution of the problem (1), (2) and, furthermore, that these conditions are different in the degenerate and the nondegenerate cases (see [3], [4]). The justification is obvious from Remark 2.

3. AUXILIARY RESULTS

First of all let us cite the following well-known result (see for example, [1], [7]).

LEMMA 1. Let g be a predictable random process in R^{d+d_1} and

$$\int_{[0,T]} E\|g(s)\|^2_{R^{d+d_1}} \, ds < \infty .$$

Then

- a. for every $t \in [0,T]$

$$E\left[\int_{[0,t]} g^\ell(s)\, d\hat{w}^\ell(s) \mid F_t^W\right] = 0 \qquad \text{a.s. } P ;$$

- b. there exists such a predictable process

$$E_w g(t) := (E_w^1 g(t), E_w^2 g(t), \ldots, E_w^{d_1} g(t))$$

that $E_w^\ell g(t) = E[g^{d+\ell} \mid F_t^W]$ for almost all t, ω and for every $t \in [0,T]$

$$E\left[\int_{[0,t]} g^{d+\ell}(s)\, dw^\ell(s) \mid F_t^W\right] = \int_{[0,t]} E_w^\ell g(s)\, dw^\ell(s) \quad \text{a.s. P.}$$

///

Let $\psi(t,x)$ be a real-valued continuous function on $[0,T] \times R^d$ having bounded continuous derivatives in x up to order 2 and in t up to the first order.

PROPOSITION 1. There exist predictable versions $\Phi_t[\psi]$, $\Phi_t[L\psi + \frac{\partial}{\partial t}\psi]$ and $\Phi_t[M^\ell\psi]$ of conditional expectations

$$E[\psi(t,X(t))\rho(t) \mid F_t^W],$$

$$E[(L\psi(t,X(t)) + \frac{\partial}{\partial t}\psi(t, X(t)))\rho(t) \mid F_t^W],$$

$$E[M^\ell\psi(t,X(t))\rho(t) \mid F_t^W],$$

respectively, such that

$$\Phi_t[\psi] = E\psi(0,\xi) + \int_{[0,t]} \Phi_s[L\psi + \frac{\partial}{\partial s}\psi]\, ds + \int_{[0,t]} \Phi_s[M^\ell\psi]\, dw^\ell(s),$$

$$\forall (t,\omega) \in [0,T] \times \Omega', \quad P(\Omega') = 1. \;/// \tag{6}$$

Different modifications of this statement are well known (see, for example, [7]), so we confine ourselves to the following sketch of the proof.

Let us apply the Itô formula to the product $\psi(t,X(t))\rho(t)$, then take conditional expectation of the both parts of the equality obtained at the first step, use Lemma 1 and we are done.

It is clear that for any $f_1, f_2 \in C_b^2$ and $\kappa, \delta \in R^1$

$$\Phi_t[\delta f_1 + \kappa f_2] = \delta\Phi_t[f_1] + \kappa\Phi_t[f_2] \quad \text{P-a.s.} \quad . \tag{7}$$

However, the ω-set for which the equality (7) holds depends on f_1, f_2, δ and κ. Later on we shall need a version of $\Phi_t[\psi]$ possessing the stochastic differential (6) and being a linear functional on C_b.*/ The latter means that the equality (7) holds on a ω-set of probability one, not depending on f_1, f_2, δ and κ. Using the same arguments as those used in Appendix of [8], we can easily obtain the following lemma.

LEMMA 2. For every $(t,\omega) \in [0,T] \times \Omega$ there exists a linear continuous functional $\tilde{\Phi}_t[\cdot](\omega)$ on C_b possessing the following properties:

• a. for every bounded Borel measurable function $\psi: [0,T] \times R^d \times \Omega \to R^1$, which is continuous in x for every $(t,\omega) \in [0,T] \times \Omega$, and P-measurable for every $x \in R^d$, $\tilde{\Phi}_t[\psi]$ is a P-measurable version of $E[\psi(t, X(t))\rho(t) \mid F_t^W]$ and also

$$|\tilde{\Phi}_t[\psi](\omega)| \leq \sup_{x\in R^d}|\psi(t,x,\omega)|\tilde{\Phi}_t[1] < \infty \quad , \tag{8}$$

$$\forall\ (t,\omega) \in [0,T] \times \Omega;$$

• b. if ψ does not depend on ω and has bounded continuous derivatives up to order two in x and up to the first order in t, then $\tilde{\Phi}_t[\psi]$ possesses stochastic differential (6).

---

*/ $C_b := C_b^0$.

4. THE PROOF OF THE THEOREM (THE NONDEGENERATE CASE, SMOOTH COEFFICIENTS)

In this Section we prove the Theorem in the case when coefficients are smooth and assumption A' holds true.

To be more specific let us denote by $]d[$ the smallest integer greater than $d/2$ and assume throughout the remainder of this Section that in addition to the assumption of the Theorem:

♦3'. for all $(t,\omega) \in [0,T]\times\Omega$, $b^i(t,\cdot,\omega)$, $c(t,\cdot,\omega)$, $\Sigma^{ik}(t,\cdot,\omega)$, $h^\ell(t,\cdot,\omega)$ have bounded derivatives in x up to order $]d[+ 2$;

♦4'. assumption A' holds true (for all t, x, ω), ϕ and ψ belong to C_b^∞.

The following statement is the starting point of our argument.

<u>PROPOSITION 2</u>. There exists such a $\mathcal{P}$-measurable function v: $[0,T]\times\Omega \to W_2^{]d[}$ that $\sup_t E\|v(t)\|^2_{]d[} < \infty$ and the function $\tilde{\Phi}_t[\psi]$ introduced in Lemma 2 possesses the following representation:

$$\tilde{\Phi}_t[\psi] = (\psi, v(t))_{]d[}, \qquad \forall (t,\omega). \quad /// \tag{9}$$

The proof of this Proposition is similar to the one of Theorem 2.2 of [8] and we leave it to the reader.

From Proposition 1, Lemma 2 and (9) it follows that for all $(t,\omega) \in [0,T]\times\Omega'$, where $\Omega' \subset \Omega$ and $P(\Omega') = 1$, the function $v(t)$ satisfies the equation

$$(v(t),\ \psi)_{]d[} = (v(0),\ \psi)_{]d[} + \int_{[0,t]} (v(s),\ L(s)\psi)_{]d[}\, ds$$

$$+ \int_{[0,t]} (v(s),\ M^{\ell}(s)\psi)_{]d[}\, dw^{\ell}(s) \quad . \qquad (10)$$

Let Δ be the Laplace operator on R^d and $\Lambda := (1-\Delta)^{\frac{1}{2}}$. Denote $\tilde{v}(t) := \Lambda^{-2} v(t)$. It is well known that $\Lambda^k : W_2^m \to W_2^{m-k}$ and

$$(\eta,\ \Lambda^k \zeta)_m = (\Lambda^k \eta,\ \zeta)_m = (\Lambda^{2m+k}\eta,\ \zeta)_0 \ {}^{*/} \qquad (11)$$

for any integers k, m and sufficiently smooth η, ζ.

From this it goes straight that $\tilde{v}(t)$ is a P-measurable function in $W_2^{]d[+2}$ and $\sup_t E\|\tilde{v}(t)\|^2_{]d[+2} < \infty$.

Applying (11) we obtain from (10) that (a.s. P)

$$(\tilde{v}(t), \psi)_{]d[+1} = (\tilde{v}(0),\ \psi)_{]d[+1} + \int_{[0,t]} (\tilde{v}(s),\ L(s)\psi)_{]d[+1}\, ds$$

$$+ \int_{[0,t]} (\tilde{v}(s),\ M^{\ell}(s)\psi)_{]d[+1}\, dw^{\ell}(s) \quad . \qquad (12)$$

Note that for every $\eta \in W_2^{]d[+2}$ for all (t,ω)

$$(\eta,\ L(t,\omega)\psi)_{]d[+1}$$

$$= -(\eta_j,\ a^{ij}(t,\omega)\psi_i)_{]d[+1}$$

$$+ (\eta,\ (b^i(t,\omega) - a^{ij}_j(t,\omega))\psi_i + c(t,\omega)\psi)_{]d[+1} :$$

$$= (\hat{L}(t,\omega)\eta,\ \psi)_{]d[+1} \quad .$$

*/ To be more precise, we can and shall select such a norm in every Sobolev space W_2^{λ} and that (11) holds true (for details see [6]).

Since $\partial/\partial x^i$ is a bounded linear operator from $W_2^{]d[}$ to $W_s^{]d[+1}$, applying the Schwarz inequality we easily find that there exists such a constant N that for all $(s,\omega) \in [0,T] \times \Omega$ and $\eta, \zeta \in W_2^{]d[+2}$

$$|(\hat{L}(s,\omega)\eta,\ \zeta)_{]d[+1}| \leq N\|\eta\|_{]d[+2} \cdot \|\eta\|_{]d[+2} . \tag{13}$$

Therefore, for every $(s,\omega) \in [0,T] \times \Omega$ there exists such a linear operator $A(s,\omega)$: $W_2^{]d[+2} \to W_2^{]d[}$ that

$$\langle A(s,\omega)\eta, \zeta\rangle = (\hat{L}(s,\omega)\eta,\ \zeta)_{]d[+1} \tag{14}$$

$$\forall\ \eta, \psi \in W_2^{]d[+2} .$$

Moreover, it follows from (13) that

$$\|A(s,\omega)\|_{]d[} \leq N\|\eta\|_{]d[+2}, \qquad \forall\ \eta \in W_2^{]d[+2}, \tag{15}$$

$$(s,\omega) \in [0,T] \times \Omega .$$

It is easy to prove that for every $\eta,\ \psi \in W_2^{]d[+2}$

$$(\eta,\ M^{\ell}(s,\omega)\psi)_{]d[+1} = -(\eta_i,\ \sigma^{i\ell}(s,\omega)\psi)_{]d[+1}$$

$$+ (\eta,\ (h^{\ell} - \sigma_i^{i\ell})(s,\omega)\psi)_{]d[+1} ,$$

$$\forall\ (s,\omega) \in [0,T] \times \Omega .$$

We can derive from this equality that for every $(s,\omega) \in [0,T] \times \Omega$ there exists such a linear continuous operator

$$B(s,\omega) := (B^1(s,\omega),\ \ldots,\ B^{d_1}(s,\omega))$$

---

*/ Here and below $\langle\cdot,\cdot\rangle$ is a duality between $W_2^{]d[+2}$ and $W_2^{]d[}$ (for details see [6]).

that $B^{\ell}(s,\omega): W^{]d[+2} \to W^{]d[+1}$, $\ell = 1 \div d_1$, where B^{ℓ} is determined by the equality

$$(B^{\ell}(s,\omega)\eta,\ \psi)_{]d[+1}: \ = \ (\eta_i,\ \sigma^{i\ell}(s,\omega)\psi)_{]d[+1} + (\eta,\ (h^{\ell} - \sigma_i^{i\ell})(s,\omega)\psi)_{]d[+1}\ ,$$

$$\forall\ \eta,\ \psi \in W_2^{]d[+2}\ .$$

As usual, making use of the Pettis theorem (see, for example, [9]), it is easy to prove that $A(s,\omega)$ and $B(s,\omega)$ are predictable processes. Thus, applying terminology of [2], we can say that $\tilde{v}(t)$ is a solution of stochastic evolution equation

$$u(t) = \phi + \int_{[0,t]} Au(s)\, ds + \int_{[0,t]} Bu(s)\, dw(s) \qquad (16)$$

in the triplet (V,H,V^*), where $V: = W_2^{]d[+2}$, $H: = W_2^{]d[+1}$, $V^*: = W_2^{]d[}$.

It follows from Corollary II.2.1 of [2] that a solution of (16) has a strongly continuous in t (P-a.s.) version taking values in $W_2^{]d[+1}$. So we shall assume below that $\tilde{v}(t)$ possesses this property.

Let us prove now that $\tilde{v}(t)$ is the unique solution of (16).

For this purpose we can use Theorems II.2.1 - II.2.3 of [2]. To apply these theorems we have to prove that operators A and B possess properties $A_1 - A_4$ from Section II.2 (for $p = 2$, $f = 0$) of [2].

The validity of A_1 and A_4 follows from (15). Since A and B are linear operators, A_3 implies A_2.

By the same reasoning as in the proof of Theorem 2.2 of [3] we can check that A_3 holds true.

Thus the results of [2] imply that the solution $\tilde{v}(t)$ of (16), or its equivalent (12), is unique.

On the other hand, from Proposition 2 and (11) it follows that

$$(\phi,\psi)_0 = (v(0),\ \psi)_{]d[} = (\tilde{v}(0),\ \psi)_{]d[+1}$$

$$= (\Lambda^{-2(]d[+1)}\phi,\ \psi)_{]d[+1} \ .$$

From this equality and arbitrariness of ψ we obtain that

$$\tilde{v}(0) = \Lambda^{-2(]d[+1)} \ . \tag{17}$$

As was mentioned above, the problem (1), (2) has a unique solution $u \in L_2([0,T]\times\Omega;\ W_2^1) \cap L^2(\Omega;\ C([0,T];\ L_2))$. Consequently, from (11) and (17) it follows that

$$\tilde{u}: = \Lambda^{-2(]d[+1)}\, u$$

$$\in L_2([0,T]\times\Omega;\ W_2^{2]d[+3}) \cap L_2(\Omega;\ C([0,T];\ W_2^{2]d[+2})$$

and satisfies the equality (12), which implies that $\tilde{u} = \tilde{v}$ and

$$\Lambda^{2]d[+2}\tilde{v} = u \ . \tag{18}$$

From Proposition 2, (11) and (18) we obtain that for every $t \in [0,T]$, $\psi \in C_b^\infty$

$$E[\psi(X(t))\rho(t) \mid F_t^W] = (\psi,\ \tilde{v}(t))_{]d[+1} = (\psi,\ u(t))_0$$

P-a.s.

///

5. THE PROOF OF THE THEOREM: THE GENERAL CASE

Apparently, it is sufficient to prove the Theorem when ψ is smooth. Therefore, we shall assume that $\psi \in C_b^\infty$.

Let $\{\tilde{w}(t), F_t\}$ be a d-dimensional Wiener process independent of ν, and $\tilde{\nu}(t) := (\tilde{w}(t), \nu(t))$.

We choose a function $\zeta \in C_0^\infty$ such that $(\zeta, 1)_0 = 1$. For $\varepsilon > 0$ define T_ε by

$$T_\varepsilon f(x) := \varepsilon^{-d} \int_{R^d} f(y)\, \zeta\left(\frac{x-y}{\varepsilon}\right) dy , \qquad \forall\, x \in R^d.$$

Denote by $({}_\varepsilon\Sigma^{i\ell})$ the $d \times (2d+d_1)$-matrix, where ${}_\varepsilon\Sigma^{i\ell}$ is equal to $\varepsilon\delta^{i\ell}$ and $\delta^{i\ell}$ is the Kronecker symbol if $\ell = 1 \div d$, to $T_\varepsilon \hat{\sigma}^{i\ell}$ if $\ell = (d+1) \div 2d$ and to $T_\varepsilon \sigma^{i\ell}$ if $\ell = (2d+1) \div (2d+d_1)$. Consider the diffusion process ${}_\varepsilon X(t)$, which is the solution of the following system of the Itô equations:

$${}_\varepsilon X(t) = \xi^i + \int_{[0,t]} (T_\varepsilon B)^i(s, {}_\varepsilon X(s))\, ds + \int_{[0,t]} {}_\varepsilon\Sigma^{i\ell}(s, {}_\varepsilon X(s))\, d\tilde{\nu}^\ell(s),$$

$$i = 1 \div d , \qquad t \in [0,T].$$

By virtue of the well-known properties of averaging operator T_ε (see, for example, [10]) we get that functions $(T_\varepsilon B)^i(s,x,\omega)$ and ${}_\varepsilon\Sigma^{i\ell}(s,x,\omega)$ belong to C_b^∞ uniformly in ε, s, ω.

Denote $a^{ij} = \frac{1}{2}\, {}_\varepsilon\Sigma^{i\ell}\, {}_\varepsilon\Sigma^{j\ell}$ and consider the problem

$$du^{\varepsilon}(t,x) = [({}_{\varepsilon}a^{ij}u^{\varepsilon}(t,x))_{ij} - ((T_{\varepsilon}b)^{i}u^{\varepsilon}(t,x))_{i} + (T_{\varepsilon}c)u^{\varepsilon}(t,x)]dt$$
$$+ [-((T_{\varepsilon}\sigma^{i\ell})u^{\varepsilon}(t,x))_{i} + (T_{\varepsilon}h^{\ell})u^{\varepsilon}(t,x)]\, dw^{\ell}(t) ,$$
$$(t,x) \in]0,T] \times R^{d} , \qquad (19)$$

$$u^{\varepsilon}(0,x) = T_{\varepsilon}\phi(x) , \qquad x \in R^{d} , \qquad (20)$$

The coefficients of equation (19) and $T_{\varepsilon}\phi(x)$, as well as the $(T_{\varepsilon}B)^{i}$ and ${}_{\varepsilon}\Sigma^{i\ell}$, belong to C_{b}^{∞} uniformly with respect to ε, s, ω. Furthermore, making some elementary calculations and applying Jensen's inequality, we see that equation (19) is non-degenerate parabolic. We have thus proved that the problem (19), (20) satisfies the assumptions made in the preceding section, and therefore the equality

$$E[\psi({}_{\varepsilon}X(t))\,{}_{\varepsilon}\rho(t) \mid F_{t}^{w}] = (\psi, u^{\varepsilon}(t))_{0}$$
$$t \in [0,T], \quad \text{a.s. } P,$$

where

$${}_{\varepsilon}\rho(t) := \exp\left\{\int_{[0,t]} (T_{\varepsilon}c - \tfrac{1}{2}(T_{\varepsilon}h^{\ell})(T_{\varepsilon}h^{\ell})(s, {}_{\varepsilon}X(s))ds + \int_{[0,t]} T_{\varepsilon}h^{\ell}(s, {}_{\varepsilon}X(s))\, dw^{\ell}(s)\right\} ,$$

is fulfilled.

Then, taking the limit as $n \to \infty$ and applying Theorem 3.2 of [4], we find that formula (5) holds true. That completes the proof.

REFERENCES

[1] Liptser, R.Sh., and Shiryayev, A.N. *Statistics of Random Processes*. New York Berlin Heidelberg: Springer-Verlag Inc., 1978.

[2] Krylov, N.V., and Rozovskii, B.L. "Stochastic Evolution Equations." *Journal of Soviet Math.*, vol.16 (1981): 1233-1276.

[3] Krylov, N.V., and Rozovskii, B.L. "On the Cauchy Problem for Stochastic Linear Partial Differential Equations." *Math. USSR Izvestiya*, vol.11 (1977): 1276-1284.

[4] Krylov, N.V., and Rozovskij, B.L. "On Characteristics of the Degenerate Parabolic Itô Equations of the Second Order." *Petrovskij seminar*, vol.8 (1982): 153-168.

[5] Freidlin, M.J. "On the Factorization of Nonnegative Definite Matrices." *Theory Probab. Applications*, vol.13 (1968): 354-356.

[6] Rozovskij, B.L. *Evolyutsionnye stokhasticheskie sistemy* (Stochastic Evolution Systems). Moskva: Nauka, 1983.

[7] Kunita, H. "Stochastic Partial Differential Equations Connected With Nonlinear Filtering." *Proceedings of C.J.M.E. Session of Stochastic Control and Filtering*, Cortona, 1981.

[8] Krylov, N.V., and Rozovskii, B.L. "On Conditional Distributions of Diffusion Processes." *Math. USSR Izvestiya*, vol.12 (1978): 336-356.

[9] Ioshida, K. *Functional Analysis*. Berlin Heidelberg New York: Springer-Verlag Inc., 1965.

[10] Bers, L., John, F., and Schechter, M. *Partial Differential Equations*. New York: Interscience Publishers, 1964.

M.G. SHUR

STRONG LIMIT THEOREMS FOR SELF-ADJOINT TRANSITION OPERATORS

INTRODUCTION

1

As is seen from the title of this article, our task consists in obtaining strong ratio limit theorems for self-adjoint transition operators. For the case of discrete state spaces, the well-known result of that kind has been found by S. Orey [5]. We shall consider general state spaces. However, we shall not restrict ourselves to conservative transition operators (see Theorems 0.1 and 0.2). Our methods (especially, the methods used in Sections 2 and 3) and results are close to those of [9], but it should be emphasized that in contrast to [9] the Orey-Molchanov condition (see (iii) in [9]) has not been imposed. Various examples of strong ratio limit theorems are given, for instance, in [2 - 5,9].

In Subsection 1 of the Introduction, the reader can get acquainted with Theorems 0.1 and 0.2 containing the main results of the paper. Theorem 1.1 also seems to be of interest (cf. [2]); it is given in Section 1 in which several results concerning positive self-adjoint operators are presented. Theorems 0.1 and 0.2 are proved in Section 3.

2

We shall use the following notation: X is a fixed locally compact separable Hausdorff space x, $\mathcal{X}$ is the σ-algebra of the Borel subset X, $\mathcal{M}$ is the space of Radon measures on X, $\mathcal{M}^+ = \{\nu \in \mathcal{M}: \nu \not\equiv 0\}$, and C_0^+ is the space of continuous functions $f: X \to [0,\infty)$ with compact supports $(f \not\equiv 0)$. Let $p(x,E)$, $x \in X$, $E \in \mathcal{X}$ be a sub-Markov transition probability on $(X,\mathcal{X})$ [5,7] and let $P_n(x,E)$ be n-step transition probabilities generated by $P(x,E)$ $(n \geq 1;\ P_1(\cdot,\cdot) \equiv P(\cdot,\cdot))$. The kernel $P(x,E)$ induces in the usual manner a transition operator $P: f \to Pf$ acting in the space of nonnegative Borel functions f on X and an operator $P: \nu \to \nu P$ acting on elements of $\mathcal{M}$ (it is convenient to use the same notation for the operators; the first operator will sometimes be extended to another function space). For $S > 0$ and $x \in X$ we set

$$G_S(x,dy) = \sum_n S^n P_n(x,dy)$$

and

$$G_S f(x) = \int f(y)\, G_S(x,dy)$$

if $f \in C_0^+$.

We need the following assumptions (cf. [9]):

•(i). Pf is continuous if $f \in C_0^+$ and $P1 > 0$;

•(ii). for every compact $K \subset X$ there exists such $b > 0$ and natural α and β that

$$P_\alpha(x,\cdot) \geq b\, P_\beta(x,\cdot) \qquad (0.1)$$

if $x,y \in K$.

The last condition (see Remark 0.1) implies the irreducibility of P in the same sense that

•(iii). there exists $\nu \in M^+$ such that $G_1(\cdot, A) > 0$ if $\nu(A) > 0$ and $G_1(\cdot, A) = 0$ a.e. */ if $\nu(A) = 0$.

Supplementing (iii) by the requirement of positiveness of ν on nonempty open sets in X, we obtain the useful condition (iii'). In accordance with (i) and (iii) the operator P is either R-conservative for some $R \geq 1$ (i.e., $G_S f < \infty$ and $G_R f(x) = \infty$ for $S < R$, $f \in C_0^+$ and ν, almost all x) or R-dissipative (i.e., $G_R f < \infty$ a.e. and $G_S f = \infty$ for $S > R$, $f \in C_0^+$); see [6,8,11]. A one-conservative (respectively, one-dissipative) operator is also called conservative (dissipative).

As a rule, we shall regard P to be a self-adjoint operator with respect to measure $\xi \in M^+$; this means that $(f,Pg) = (Pf,g)$ for Borel functions $f,g \geq 0$, where, by definition, $(f,g) = \int fg \, d\xi$. For such an operator the above conditions imply the equivalence of ν and ξ (see Remark 0.3), and for this reason our abbreviation "a.e." also means "ξ almost everywhere."

The first basic results of the paper consist in the following.

<u>THEOREM 0.1</u>. Let P be conservative and self-adjoint with respect to $\xi \in M^+$. If P satisfies conditions (i), (ii) and (iii'), then for $f,g \in C_0^+$, $x,y \in X$ and $\ell = 0,1,\ldots$

$$\lim_{n\to\infty} \frac{P^n f(x)}{P^{n+\ell} g(y)} = \frac{h(x)}{h(y)} \times \frac{\int f \, d\mu}{\int g \, d\mu} \quad , \qquad (0.2)$$

where the Borel function h $(0 < h < \infty)$ and the measure $\mu \in M^+$ are invariant for P:

$$Ph = h \qquad \text{and} \qquad \mu P = P \, . \qquad (0.3)$$

---

*/ This abbreviation means "almost everywhere relative to ν."

Now we shall attempt to omit the conservativeness condition of P. It should be noted that under the conditions of Theorem 0.1:

•(iv). up to scalar factors, theoperator P^2 has one invariant function h $(0 < h < \infty)$ and one invariant Borel measure at most (see our Lemma 2.2).

In contrast to the "Liouville's conditions" (cf. [3,4,8,9]), the requirement (iv) is formulated in terms of P^2 instead of P. It shall be considered together with the following assumption:

•(v). X is noncompact; for every $f \in C_0^+$ there exists $\phi \in C_0^+$ and natural S and m for which

$$\lim_{x\to\infty} \left[\frac{P^S f(x)}{\sum_{1\le i\le m} P^i \phi(x)} \right] = 0 \tag{0.4}$$

(here we set $\frac{0}{0} = 0$; the symbol ∞ denotes the points at infinity of the standard one-point compactification of X).

Condition (v) is a little stronger than (vi) from [9]; it is satisfied, e.g. if the functions $P^n f$ $(n \ge 1,\ f \in C_0^+)$ have compact supports.

<u>THEOREM 0.2</u>. The assertion of Theorem 0.1 still holds if instead of conservativeness of P we require that the equality $R=1$ and conditions (iv) and (v) be satisfied.

<u>REMARK</u>. Suppose P is not only self-adjoint with respect to ξ but also positive in the sense that $(f,Pf) \ge 0$ for any continuous function f with the compact support (this occurs each time when P equals an even power of the self-adjoint transition operator). Then the assertion of Theorem 0.1 holds also if $R = 1$ and if the conservativeness assumption of P is replaced by condition

(v) together with the analog of (iv), with P instead of P^2.

3

A few remarks concerning the conditions imposed are in order.

REMARK 0.1. Let (ii) be given and let $P1 > 0$. Then

- (a). the triple (b,α,β) in (0.1) can be replaced by any triple $(b,\alpha+k,\beta+k)$ with natural k;
- (b). α and β can be assumed even;
- (c). condition (iii) holds true.

The Assertion (a) is obvious. By (a), from (0.1) we have $P_{k+2\gamma}(x,\cdot) \geq bP_{k+\gamma}(x,\cdot) \geq b^2P_k(y,\cdot)$ for $x,y \in K$, where $\gamma = \alpha-\beta$ and K is sufficiently large. Thus, to prove assertion (b) one must consider the triple $(b^2,k+2\gamma,k)$ with sufficiently large even k. In order to demonstrate assertion (c), we choose compacts k_n, $n \geq 1$, jointly exhausing X and having a fixed common point y. Let (a_n,α_n,β_n) be an analog of (a,α,β) in the case of k_n and let the probability measure $P_n(\cdot)$ be equivalent to $P_{\beta_n}(y,\cdot)$. Then, $G_1(\cdot,A) > 0$ if

$$\tilde{\nu}(A) \equiv \sum_n 2^{-n} P_n(A) > 0 \quad , \qquad A \in \mathcal{X} \quad ,$$

and we need only to make reference to [11, p. 842].

REMARK 0.2. Conditions (i), (ii) and (iii') imply the following:

- (vi). if $f \in C_0^+$ and the compact $K \subset X$ are fixed, then $P^n f > 0$ on K for every sufficiently large n.

Extending K, we can always find and fix $y \in K$ for which $P(y,K) > 0$. By (iii') there exist infinitely many m such that $P^m f(y) > 0$; and, choosing (b,α,β) in accordinace with (0.1), we fix one of these $m \geq \alpha$. If $\gamma \equiv \alpha - \beta > 0$ and $n = m + k$, $k \geq 1$, then, by virtue of (ii), $P^n f > 0$ on K. Consequently, $P^n f(y) > 0$ if $n = m + \gamma k + 1$ and again (ii) shows that $P^n f > 0$ on K is $n = m + (k+1)\gamma + 1$. Repeating this argument, we convince ourselves of the validity of the last inequality for all sufficiently large members of any progression $\{m+k\gamma+\ell;\ k \geq 1\}$, where $\ell = 1,\ldots,\gamma$. We have thus proved (vi).

The case $\gamma = 0$ is as simple as the previous case. In the less important case of $\gamma < 0$, a similar reasoning leads to the required inequality for $n \in [n_i+\gamma,\ n]$ and for suitably given $n_1 < n_2 < \cdots$, and it remains again to use (ii).

REMARK 0.3. Under the same conditions as those in Remark 0.2, our measures ν and ξ are equivalent.

Here the relation $\nu \ll \xi$ arises from (iii) and from the equalities $(G_1(\cdot,A),\ 1) = (1_A,\ G_i 1) = 0$, where we assume $\nu(A) = 0$ and denote the indicator of A by 1_A. On the contrary, if $A \subset X$ is a compact and $\nu(A) = 0$, then $P_n(\cdot,A) = 0$, $n \geq 1$, ν almost everywhere (see (iii)) and hence $P_n(\cdot,A) \equiv 0$ on any fixed compact K with $\nu(K) > 0$ if n is sufficiently large (see assertion (a) in Remark 0.1). Thus

$$(1_A,\ P^n f) \;=\; (f,\ P_n(\ ,A)) \;=\; 0$$

for $f \in C_0^+$ and large n. Since, by Remark 0.2, we can assume that $P^n f > 0$ on A, these relations ensure the equality $\xi(A) = 0$. Thus $\xi \gg \nu$.

1. SOME RESULTS ON POSITIVE OPERATORS

In this Section we suggest some preliminary results that we shall need later on. These results concern mainly positive self-adjoint operators, and the basic assertions of this Section are stated in Theorem 1.1 and Lemma 1.4.

First we dwell on a simple Lemma. We shall include a Borel function f in the family $K(P)$ iff $f_1 \le f \le P^S f_2$ for some $f_1, f_2 \in C_0^+$ and $S = 0,1,\ldots$.

LEMMA 1.1. Let P satisfy (i) and (vi). Then $K(P) \supset C_0^+$. Moreover, for fixed $f, w \in K(P)$ and sufficiently large natural r there exists $c = c(r) > 0$ such that

$$P^r f \ge cw \quad . \tag{1.1}$$

Proof. Indeed, the functions $P^n f$ $(f \in C_0^+,\ n \ge 1)$ are lower semi-continuous and the desired inequality for $f, w \in C_0^+$ follows from (vi). Thus $K(P) \supset C_0^+$. Let $f \ge f_1$, $P^t v \ge w$ and $P^S f_1 \ge cv$, where $f, w \in K(P)$, $f_1, v \in C_0^+$, $t, S \in \{0,1,\ldots\}$ and $c > 0$. Then the relations $P^{S+t} f_1 \ge cP^t v \ge cw$ lead to (1.1) with $r = S + t$.

Now we shall occupy ourselves with the preparation for the use of the Hilbert space theory. If $\xi \in M^+$ is subinvariant for P (i.e., $\xi P \le \xi$), the Schwarz inequality implies the following estimate:

$$\int (Pf)^2 d\xi \le \int (P1)^2 P(f^2) d\xi \le \int P(f^2) d\xi \le \int f^2 d\xi$$

for Borel functions f that allow us to extend P on the complex Hilbert space $H = L^2(\xi)$. The norm and the inner product in H will be denoted by $\|\cdot\|$ and $(\cdot,\cdot)$, respectively, the second notation being compatible with the former meaning of $(\cdot,\cdot)$. The extended operator has the norm ≤ 1 and it will be denoted either by $\tilde{P}$ or (with special reserve) by P. It is clear that P is self-adjoint with respect to ξ iff $\tilde{P}$ is self-adjoint in H, and if P is self-adjoint, then its positiveness (see Introduction) is equivalent to the positiveness of $\tilde{P}$ in H (the latter implies that $(\tilde{P}f, f) \geq 0$ for $f \in H$).

LEMMA 1.2. Let condition (ii) be satisfied and let $R = 1$. If $\xi \in M^+$ is subinvariant, then $\|\tilde{P}\| = 1$.

Proof. It is easy to prove the inequality $R \geq \|\tilde{P}\|^{-1}$, which involves our Lemma for $R = 1$ and $\|\tilde{P}\| \leq 1$.

In the remainder of this Section the following condition is accepted without special mention:

•(vii). P is positive and self-adjoint relative to $\xi \in M^+$; it satisfies (iii') and (vi); finally, $R = 1$.

LEMMA 1.3. Let $f \in K(P)$ (see the definition of $K(P)$ before Lemma 1.1). Then

$$\lim_{n\to\infty} \frac{(P^n f, f)}{(P^{n+1} f, f)} = 1 . \tag{1.2}$$

Proof. Here $\tilde{P}$ will stand for P. In view of the previous Lemma the spectral expansion of P^n, $n \geq 1$, has the form

$$P^n = \int_0^{1+0} \lambda^n \, dE_\lambda ,$$

where $\{E_\lambda ; \lambda \in (-\infty, \infty)\}$ is the spectral family of projectors corresponding to P (for the sake of definiteness, this family is assumed left continuous; see the necessary information on self-adjoint operators, e.g., in [1]). The inequality $\lambda^n \geq \lambda^{n+1}$ $(0 \leq \lambda \leq 1)$ implies the positiveness of $P^n - P^{n+1}$ so that

$$(P^n f, f) \geq (P^{n+1} f, f) , \qquad f \in H \tag{1.3}$$

Similar simple arguments allow us to write

$$(S^{n+1} f, f) \geq \gamma(S^n f, f) , \qquad f \in H \tag{1.4}$$

where $T = \int_0^{\gamma+0} \lambda \, dE_\lambda$, $0 < \gamma < 1$ and $S = P - T$.

Now let $f \in K(P)$. By Theorem 2 in [10] $\lim_{n\to\infty} (P^n w)^{1/n} = 1$, a.e. for some function $w \in C_0^+$. Therefore

$$\liminf_{n\to\infty} \nu(A_n) > 0 ,$$

where

$$A_n = \bigcap_{m\geq n} \{x: P^m w(x) \geq \alpha^{-m}\}$$

and $\alpha \in (0,1)$ is fixed, and for every $x \in X$ there exists $r = r(x)$ and $K = K(x)$ such that $p_r(x, A_k) > 0$. Hence $\liminf_{n\to\infty} (P^n w)^{1/n} \geq \alpha$ everywhere, and here we can write 1 instead of α in view of arbitrariness of $\alpha \in (0,1)$. This fact together with (1.1) implies that

$$\liminf_{n\to\infty} (P^n f, f)^{1/n} \geq 1 . \tag{1.5}$$

Combined with the obvious relation $||T||^n \leq \gamma^n$, the obtained inequality ensures the tendency of $(S^n f, f)/(P^n f, f)$ to 1. Recalling (1.4), where γ, $0 < \gamma < 1$, is arbitrary, we deduce from here that the upper limit of the ratio $(P^n f, f)/(P^{n+1} f, f)$ does not exceed 1. This proves (1.2) in view of (1.3).

THEOREM 1.1. For $f, g \in K(P)$

$$\lim_{n\to\infty} \left[\frac{(P^n f, g)}{(P^{n+1} f, g)}\right] = 1 . \tag{1.6}$$

Proof. First we shall prove the inequalities:

$$0 < \liminf_{n\to\infty} \left|\frac{(P^n f,g)}{(P^n f,f)}\right| \le \limsup_{n\to\infty} \left|\frac{(P^n f,g)}{(P^n f,f)}\right| < \infty \quad . \tag{1.7}$$

For suitable S and $a > 0$ the estimate $P^S g \ge af$ of the kind (1.1) is valid so that

$$(P^n f,g) = (P^{n-S}f,P^S g) \ge a(P^{n-S}f,f) , \qquad n > S .$$

As a result, we deduce from (1.2) the first inequality in (1.7). But exactly in the same way, $P^r f \ge cg$ for corresponding r and $c > 0$, and the last inequality in (1.7) follows from (1.2) as well.

Next we fix the r and c and set $v = cg$ and $P^r f \not\equiv v$ since otherwise (1.6) is equivalent to (1.2). Now each of the functions $F = P^r f$, $f_1 = F + v$ and $f_2 = F - v$ belongs to $K(P)$. By (1.2), $(P^{K+1}f_1,f_1) \le \lambda(P^K f_1,f_1)$ and $(P^{K+1}f_2,f_2) \ge \mu(P^k f_2,f_2)$, provided that $\lambda > 1$ and μ, $0 < \mu < 1$, are fixed and $n > r$ is sufficiently large, where $K = n - r$. Therefore, for the same n and $\varepsilon = \lambda - \mu$ we find

$$\begin{aligned} 4(P^{n+1}f,v) &= (P^{K+1}f_1,f_1) - (P^{K+1}f_2,f_2) \\ &\le \lambda[(P^K f_1,f_1) - (P^K f_2,f_2)] + \varepsilon(P^K f_2,f_2) \\ &\le 4\lambda(P^n f,v) + \varepsilon(P^n f,f) \quad ; \end{aligned} \tag{1.8}$$

here the second inequality follows from a relation similar to the first equality in (1.8) and from the definition of f_2. Dividing (1.8) by $(P^n f,v)$ and taking into account (1.7) and (1.2), we conclude from (1.8) that the upper limit of the ratio on the left-hand side of (1.6) does not exceed $\lambda + c\varepsilon$, provided that c is a suitable constant and n is large.

Similar arguments give the estimate

$$4(P^{n+1}f,v) \geq 4\mu(P^n f,v) - \varepsilon(P^n f,f)$$

for sufficiently large n and thus the lower limit of this ratio is not less than $\mu - c\varepsilon$. Since λ and μ can be chosen arbitrarily close to 1, our Theorem is proved.

LEMMA 4.1 (cf. [9; Lemma 2]). Let (besides (vii)) condition (ii) hold. We have the following "Harnack's inequality": if a compact $K \subset X$, $f \in C_0^+$ and entire $\ell \geq 0$ are fixed, then

$$\sup_{x,y \in K} \left[\frac{P^{n\pm\ell} f(x)}{P^n f(y)}\right] < c \tag{1.9}$$

for some constant c and each sufficiently large n.

Proof. Symbols c, c_1 and so on will stand for some positive constants. We fix $g \in C_0^+$ and suppose that the support K_1 of g contains K. Let b, α and β be such that (0,1) is satisfied with K_1 instead of K. Relation (1.1) with $w = f$ implies $P^{n+r} f \geq cP^n f$, $n > 1$, if natural r exceeds some $r_0 > 0$, where $c = c(r) > 0$, and we can assume $r_0 > \max(\alpha, \beta)$. Hence for $x \in K_1$ and large n

$$\max_{K_1} P^n f < b^{-1} P^{n+\gamma} f(x) < c_1 P^{n+m} f(x) \leq c_2 \max_{K_1} P^{n+t} f(x) , \tag{1.10}$$

where $\gamma = \alpha - \beta$, m passes through the family of natural numbers lying between $\gamma + r_0$ and $t - r_0$ and natural $t > 2r_0 + \gamma$ is chosen such that this family contains at least two numbers $\tilde{m}$ and $\tilde{m}+1$, which will be fixed later. It is convenient to require that $t = Sk$ with natural S and $k = r + \gamma$, $r > r_0$.

Our choice of c, α, β, k permits us to write

$$P^{n+k}f(x) \geq cP^{n+\gamma}f(x) = cP^{\alpha}P^{n-\beta}f(x) \geq c_3P^{\beta}P^{n-\beta}f(y)$$
$$= c_3P^n f(y)$$

for $x,y \in K_1$ and $n > \beta$. Consequently,

$$(P^n f,g) \leq c_4 \max_{K_1} P^n f \leq c_5 \min_{K_1} P^{n+k}f \leq c_6(P^{n+k}f,g)$$
$$\leq c_7 \max_{K_1} P^{n+k}f \leq c_8(P^{n+2k}f,g) \tag{1.11}$$

(here the last inequality follows from the previous ones). By Theorem 1.1., the ratio of any two inner products in (1.11) tends to a finite limit as $n \to \infty$. Thus the ratio of the maxima in (1.11) together with that in (1.10) vary between two positive constants if n is sufficiently large. By virtue of (1.10) the same can be said about the ratio $P^{n+\tilde{m}}f(x) / P^{n+\tilde{m}+1}f(y)$, $x,y \in K_1$.

The Lemma is proved for $\ell = 0$ or ± 1 and therefore for other ℓ as well.

2. WEAK LIMIT POINTS FOR NORMALIZED FUNCTIONS $P^n f$

Now our objective is to get some information about functions which can be limited in some sense to the function $P^n f/a_n$, provided that $f \in C_0^+$, a_n are appropriate constants and n tends to infinity through a sequence (see Theorem 2.1). But to avoid interrupting our reasoning thereupon, we start with the following.

<u>LEMMA 2.1</u>. Let (ii) and (iii) hold. If the Borel function h satisfies a.e. the condition

$$Ph = h, \qquad 0 < h < \infty, \tag{2.1}$$

then there exists its regularization (i.e., a functional equal to

h a.e. and satisfying (2.1) everywhere).

Proof. Let

$$A = \{x: h(x) = P^n h(x),\ n = 1,2,\ldots\} .$$

Since $\nu(A^c) = 0$, where $A^c = X\backslash A$, we have $P_n(y,A^c) = 0$ for some $y \in X$ and all $n \geq 1$. By applying (ii) to a two-point compact $K = \{x,y\}$, $x \in X$, we find that $P_n(x,A^c) = 0$ for all n exceeding some number $n_0(x)$. For these n and all $k \geq 1$

$$P^n h(x) = P^n[1_A h](x) = P^n[1_A P^k h](x) = P^{n+k} h(x) \tag{2.2}$$

so that the function $\hat{h}(x) = \lim_{n\to\infty} P^n h(x)$ coincides with any member of (2.2). Besides $h = \hat{h}$ on A.

The Fatou lemma shows that $P\hat{h} \leq \lim_{n\to\infty} P^{n+1} h = \hat{h}$, whence $P^n\hat{h} \leq P\hat{h} \leq \hat{h}$. But

$$P^n\hat{h}(x) = P^n[1_A h](x) = P^n h(x) = \hat{h}(x)$$

if $n > n_0(x)$. Thus $\hat{h} = P\hat{h}$.

The inequalities $0 < \hat{h} < \infty$ are easy to prove due to (ii).

REMARK. In the above proof condition (ii) was used only for the case of two-point compacts.

Throughout the remainder of this Section let us admit that the following hypothesis holds:

- (viii). conditions (i), (ii), (iii') are satisfied and $R = 1$.

The next lemma will allow us to apply the main results of Section 1 to P^2.

LEMMA 2.2. The operator $\mathcal{P} = P^2$ satisfies the condition (iii') with the measure ν corresponding to P, as well as conditions (ii) and (vii). If P is conservative, then $\mathcal{P}$ is also conservative.

Proof. Condition (ii) for $\mathcal{P}$ follows from assertion (b) of Remark 0.1. Thus condition (iii) for $\mathcal{P}$ holds with a measure $\tilde{\nu} \in M^+$ (see assertion (c) of Remark 0.1). By Lemma 1.1 we, in fact, have (iii') so that $\tilde{\nu}$ is equivalent to ξ and therefore to ν (see Remark 0.3). As a result, we have the right to assume $\tilde{\nu} = \nu$. Note also that Lemma 1.1 implies condition (vi) for $\mathcal{P}$.

Furthermore, for given $S > 0$, $x \in X$ and $f \in C_0^+$ the series $\sum_n S^n P^n f(x)$ and $\sum_n S^{2n} P^{2n} f(x)$ either converge or diverge simultaneously. Indeed, the convergence of the former series implies the convergence of the latter. If the series $\sum_n S^{2n} P^{2n} f(x)$ converges, then by Lemma 1.1 the series $\sum_n S^{2n+1} P^{2n+1} f(x)$ converges too, but this ensures the convergence of the series $\sum_n S^n P^n f(x)$. Therefore the conservativeness of P guarantees the conservativeness of $\mathcal{P}$.

Using Lemma 2.2, Theorem 1.1 and Lemma 1.4, we easily come to the following assertions.

COROLLARY 2.1. For $f,g \in C_0^+$

$$\lim_{n\to\infty} \left| \frac{(P^n f,g)}{(P^{n+2} f,g)} \right| = 1 \quad . \tag{2.3}$$

It is necessary to verify whether $b_{2n}(f,g)$ and $b_{2n+1}(f,g) = b_{2n}(f,Pg)$ tend to 1 as $n \to \infty$, where

$$b_n(f,g) = \left| \frac{(P^n f,g)}{(P^{n+2} f,g)} \right| \quad .$$

But this is obvious by virtue of Theorem 1.1.

COROLLARY 2.2. In the present context the statement of Lemma 1.4 remains valid.

By applying Lemma 1.4 to the operator $\mathcal{P}$ we shall obtain part of the required statement for even n and ℓ only. But any other case can be reduced to this situation. For instance, for sufficiently large odd n the quotient on the left-hand side of (1.9) does not exceed $cP^{n+r}f(x) / P^{n-r}f(y)$ if $c > 0$ and odd r are properly chosen (see Lemma 1.1); as a result, we find ourselves in the situation studied above.

Now it is necessary to recall two useful definitions. Let a measure $\kappa \in M$ and functions $g_n \in L(\kappa)$, $n \geq 1$, be given. The sequence $\{g_n\}$ is called κ-equi-integrable if for every $\varepsilon > 0$ we are able to indicate $\delta > 0$ and a compact $K \subset X$ such that $\int_A g_n \, d\kappa < \varepsilon$ when $\kappa(A) < \delta$ and $\int_{X \setminus K} g_n \, d\kappa < \varepsilon$ $(n = 1,2,\ldots)$. On the other hand, the sequence $\{g_n\}$ is called weakly convergent in $L(\kappa)$ to a function $\phi \in L(\kappa)$ if $\int g_n w \, d\kappa \to \int gw \, d\kappa$ for any bounded Borel function w as $n \to \infty$.

<u>LEMMA 2.3</u>. Let $f \in C_0^+$ and $Q_n = P^n f/a_n$ if $a_n \equiv (P^n f, f) > 0$ and let $Q_n = 0$ otherwise. Any increasing sequence N of natural numbers contains a subsequence N_1 such that for all even $\ell \geq 0$ (odd $\ell \geq 1$) and all compacts $K \subset X$ the functions $1_K P^\ell Q_n$ weakly tend in $L(\xi)$ to the function $1_K h_0$ (respectively, $1_K h_1$) as $n \in N_1$ tends to infinity, where the functions $h_i \geq 0$ satisfy the inequality $h_i \geq Ph_i$ a.e. $(i = 0,1)$, in addition to $(h_i, f) = 1$.

<u>Proof</u>. Let $K_1 \subset K_2 \subset \cdots$ be a sequence of compacts and let $\bigcup_n K_n = X$. Corollary 2.2 implies the ξ-equi-integrability of $\{1_{K_m} P^\ell Q_n;\ n \geq 1\}$ for fixed $m \geq 1$ and $\ell \geq 0$. Thus, by Dunford-Pettis's criterion [1], there exists a sequence $N_1 \subset N$ such that the functions $1_{K_m} P^\ell Q_n$ weakly tend in $L(\xi)$ to a function

$h_{\ell m} \geq 0$ as $n \in N_1$ tends to infinity. The sequence N_1 can be chosen to be independent of $m \geq 1$. Then from the definition of the equi-integrability it is easy to derive that $h_{\ell m} = h_{\ell,m+1}$ a.e. on K_m and therefore each function $h_{\ell m}$, $m \geq 1$, coincides with the function h_ℓ a.e. on K_m, the last function being independent of m.

We assume without loss of generality, that all members of N_1 are of the same evenness. By the definition of h_ℓ, the limit of $(P^\ell Q_n, g)$, $g \in C_0^+$, is equal to (h_ℓ, g), where $n \in N_1$ tends to ∞, but at the same time it coincides with (h_0, g), if ℓ is even, or with (h_1, g) if ℓ is odd (see Corollary 2.1). This reasoning gives us the opportunity to take either $h_\ell \equiv h_0$ or $h_\ell \equiv h_1$, depending on the evenness or the oddness of ℓ.

Let ℓ be even and let the functions $w_r \in C_0^+$, $r \geq 1$ be such that $w_r \uparrow 1$ as $r \to \infty$ and $w_r \equiv 1$ on any given compact for sufficiently large r. If $\ell \geq 0$, $\varepsilon > 0$, $g \in C_0^+$ and r are fixed and if $(h_0, (1-w_r)\mathcal{P}g) < \varepsilon$, we obtain[*/]

$$(\mathcal{P}h_0, g) < \varepsilon + (h_0, w_r \mathcal{P}g) = \varepsilon + \lim_{N_1} (Q_n, w_r \mathcal{P}g)$$

$$\leq \varepsilon + \lim_{N_1} (Q_n, \mathcal{P}g) = \varepsilon + \lim_{N_1} a_n^{-1}(P^{n+2} f, g) = \varepsilon + (h_0, g),$$

where Corollary 2.1 is used once more. We then have the inequality $h_0 \geq \mathcal{P}h_0$ a.e. since $\varepsilon > 0$ and $g \in C_0^+$ are arbitrary. The similar relation for h_1 is proved in the same way.

The last assertion of the Lemma is obvious.

[*/] The symbol $\lim\limits_{N_1}$ denotes a limit when n tends to infinity through N_1.

COROLLARY 2.3. The functions h_0 and h_1 from the preceding Lemma satisfy a.e. the inequalities

$$h_0 \geq Ph_1 \geq P^2h_0 \geq P^3h_1 \geq \cdots , \qquad (2.4)$$

$$h_1 \geq Ph_0 \geq P^2h_1 \geq P^3h_0 \geq \cdots . \qquad (2.5)$$

In fact, if $g,w \in C_0^+$ and $w \leq Pg$, then

$$(PQ_nf,g) = (Q_nf,Pg) \geq (Q_nf,w) , \qquad (2.6)$$

whence, by Lemma 2.3, $(h_1f) \geq (h_0,w)$. In (2.6) w can be replaced by Pg since Pg admits an approximation from below by means of the functions $w \in C_0^+$. Thus $(h_1,g) \geq (Ph_0,g)$ for all $g \in C_0^+$ so that $h_1 \geq Ph_0$ a.e. In the same way relations (2.6) with PQ_nf instead of Q_nf give us the first inequality in (2.4). Recalling (iii) and taking into account the equivalence of ξ and ν, we easily get (2.4) and (2.5).

THEOREM 2.1. Let the conditions of either Theorem 0.1 or Theorem 0.2 hold and let h_0 and h_1 be chosen according to Lemma 2.3. Then

$$h_0 = h_1 = h \qquad \text{a.e.} , \qquad (2.7)$$

where h is an invariant function for P $(0 < h < \infty)$.

Proof. •1. Let P be conservative. Lemma 2.2 ensures the conservativeness of $\mathcal{P}$ and according to [11; Proposition 3.3] the last assertion of Lemma 2.3 gives the inequalities

$$h_i = \mathcal{P}h_i , \qquad i = 0,1 , \qquad (2.8)$$

(a.e.). The equivalence of ξ and η and the existence of regularizations of h_i, $i = 0,1$, enable us to regard the obtained equalities as identities. But then all the functions from (2.4) are equal a.e. to each other and the same is true for the functions from (2.5). Moreover, $h_i = c_i h$, where $c_i > 0$, h is the same as in (0.2) and $i = 0,1$. These remarks show that $h_0 = c_0 h = c_1 Ph$ and $h_1 = c_1 h = c_0 Ph$ a.e. From here we have $c_0 = c_1$ and $h = Ph$. So the case where the operator P is conservative, is finished.

•2. In this part of the proof the conditions of Theorem 0.2 are imposed and we reduce our study to the case when (2.8) is valid a.e. However, first we must verify the ξ-equi-integrability of the families $\{Q_n P^S g\}$ and $\{PQ_n P^S g\}$, where S is the same as in condition (v), $n \in N_1$ and $g \in C_0^+$. Let $\varepsilon > 0$. By condition (v), $P^S g(x) \leq \varepsilon \sum_{1 \leq i \leq m} P^i \phi$ outside some compact K, provided that $S, m \geq 1$ and $\phi \in C_0^+$ are properly chosen. Let the functions w_r, $r \geq 1$, be such that $w_r \uparrow 1$ and $w_r \equiv 1$ on K for all large r. Then

$$(Q_n, (1-w_r)P^S g) \leq \varepsilon \sum_{1 \leq i \leq m} (Q_n, P^i \phi) \leq \varepsilon \sum_{1 \leq i \leq m} (P^i Q_n, \phi)$$

and for sufficiently large $n \in N_1$ the last sum does not exceed $m(h_0 + h_1, \phi)$. As a result, the ξ-equi-integrability of the first of the above families is established. A similar argument proves this property of the second family.

Now we pick a function $g \in C_0^+$, whose support includes a prescribed open set U, with the closure of U being compact, and choose S in the same way as in the previous subsection. Let, for example, S be odd. Then the passage to the limit in the equalities $(P^S Q_n, g) = (Q_n, P^S g)$ and $(P^{S+1} Q_n, g) = (PQ_n, P^S g)$ gives us $(h_1, g) = (h_0, P^S g)$ and $(h_0, g) = (h_1, P^S g)$, or

$(h_1 - P^S h_0, g) = (h_0 - P^S h_1, g) = 0$, where Lemma 2.3 and the material of the preceding subsection are used. By virtue of (2.4) and (2.5) this means that $h_1 = Ph_0$ and $h_0 = Ph_1$, a.e. on U and hence the same is true a.e. in X. This immediately implies (2.8). A similar but simpler reasoning also leads to (2.8) if S is even.

So (2.8) is in force. To complete the proof it remains to repeat the arguments from the first part of the proof.

The following Lemma is a simple corollary of the preceding results.

LEMMA 2.4. Let the conditions of either Theorem 0.1 or Theorem 0.2 be satisfied. Let $k \geq 1$ and $w \in C_0^+$ be fixed and let the measure η be defined as follows:

$$\eta(A) = \int_A P^k w \, d\xi = \int P_k(x,A) \, w(dx) \, \xi(dx) , \quad A \in \mathcal{X} . \qquad (2.9)$$

Then the family $\{Q_n ; n \in N_1\}$ is η-equi-integrable, where N_1 is the same as in Lemma 2.3.

Proof. Theorem 2.1 and Lemma 2.3 give us, first, η-integrability of h since

$$\int h \, d\eta = \int w P^k h \, d\xi = \int wh \, d\xi < \infty$$

and, second, η-equi-integrability of the family $\{1_K Q_n ; n \geq 1\}$, provided that $K \subset X$ is compact. Thus it remains only to find a function $v \in C_0^+$, $0 \leq v \leq 1$, such that $\int (1-v) Q_n \, d\eta < \varepsilon$ for sufficiently large n if $\varepsilon > 0$ is given (here we consider $n \in N_1$ only). It turns out that the function $v \in C_0^+$, $0 \leq v \leq 1$, for which $\int (1-v) h \, d\eta < \varepsilon$ is the required one. Indeed,

$$\int (1-v) Q_n \, d\eta = \int w P^k Q_n \, d\xi - \int v Q_n P^k w \, d\xi \equiv I - J ,$$

where $I \to \int wh\, d\xi = \int h\, d\eta$ and $J \to \int vhP^k w\, d\xi = \int vh\, d\eta$ as $n \to \infty$. Now, to complete the proof, we need only recall the choice of v.

3. THE PROOF OF THEOREMS 0.1 and 0.2

Theorems 0.1 and 0.2 will be proved simultaneously. Let the conditions of any of them be satisfied. As has been noted in the Introduction, here we do without conditions like the Orey-Molcanov ones (see, e.g., (IV) and (IV') in [8] or (iii) in [9]). A similar condition was used in [8,9] with the only objective to prove the formula

$$\lim_{n\to\infty} \left[\frac{P^n f(x)}{P^{n+1} f(x)}\right] = 1 \tag{3.1}$$

for points of some invariant set with the function $f \in C_0^+$ regarded as fixed. Therefore, if we assume the validity of (3.1) to be proved for all $x \in X$, we could apply the results of [8] to the present situation.*/ In particular, we could affirm that $P^n f / P^{n+\ell} g$ converges to $\int f\, d\mu / \int g\, d\mu$ everywhere in X, and this fact would permit us to restrict ourselves to the proof of the relation

$$\lim_{n\to\infty} \left[\frac{P^n f(x)}{P^n f(y)}\right] = \frac{h(x)}{h(y)} . \tag{3.2}$$

As a result, Theorem 0.1 and 0.2 will be proved as soon as (3.1) and (3.2) are obtained. Consider, first of all, an increasing sequence N of natural numbers and extract from N a subsequence N_1 with the properties described in Lemma 2.3 and 2.4. Fix points $x,y,z \in X$ contained in a compact K with $\xi(K) > 0$

*/ For the detailed version of [8] see one of the forthcoming issues of "Teoriya veroyatn. i ee prim." (Probab. Theory and Applications).

and a function $w \in C_0^+$ nonvanishing on K. If b, α, β are chosen according to (ii), then $P_\alpha(\tilde{x},\cdot) \geq bP_\beta(z,\cdot)$ for all $\tilde{x} \in K$. Supposing η is defined by means of (2.9) with $k=\alpha$, we find

$$\eta(A) = \int P_\alpha(\tilde{x},A)\, w(\tilde{x})\, \xi(d\tilde{x}) \geq aP_\beta(A) ,$$

where $a \equiv b\int_K w\, d\xi > 0$ and $P_\beta(A) \equiv P_\beta(z,A)$. Thus Lemma 2.4 yields the P_β-equi-integrability of the family $\{Q_n;\ n \in N_1\}$ and, by Lemma 2.3 and Theorem 2.1,

$$h(z) = P^\beta h(z) = \lim_{N_1} a_n^{-1} P^\beta P^n f(z) = \lim_{N_1} a_n^{-1} P^{n+\beta} f(z) \qquad (3.3)$$

if the function h is normalized so that $h_1 = h_2 = h$ (as in Section 2, we put $a_n = (P^n f, f)$). Of course, (3.3) is also valid with z replaced by x or y so that (3.3) leads to the relation

$$\lim_{N_1} \left[\frac{P^{n+\beta} f(x)}{P^{n+\beta} f(y)}\right] = \frac{h(x)}{h(y)} . \qquad (3.4)$$

Note that in this case the triplet (b,α,β) can be replaced by $(b,\alpha+1,\beta+1)$ (see (Remark 0.1) and consequently we can use (3.3) and (3.4) with β replaced by $\beta+1$.

Now the verification of (3.2) and (3.1) becomes a simple task. Indeed, if (3.2) is not satisfied, e.g., for x and y chosen above, we can find an increasing sequence N of natural numbers such that $P^{n-\beta} f(x) / P^{n-\beta} f(y)$ tends to a limit different from $h(x)/h(y)$ when $n \in N$ tends to infinity. But then (3.4) does not hold for every subsequence and a contradiction arises.

Proceed to prove (3.1). Since

$$a_{n+1} = (P^{n+1} f, f) = a_n (PQ_n f, f) ,$$

Lemma 2.3 and Theorem 2.1 imply the existence of the following limit:

$$\lim_{n\to\infty} \frac{a_{n+1}}{a_n} = \lim_{n\to\infty} (PQ_n f, f) = (h,f) \quad . \tag{3.5}$$

On the other hand, we can write (3.3) with β replaced by $\beta+1$, as is noted above. Thus every increasing sequence N of natural numbers contains a subsequence N_1 such that

$$\lim_{N_1} \frac{a_n^{-1} P^{n+\beta} f(x)}{a_{n+1}^{-1} P^{n+\beta+1} f(x)} = \frac{P^{\beta} h(x)}{P^{\beta+1} h(x)} = 1$$

for given x. This fact combined with (3.5) establishes the validity of the equality

$$\lim_{N_1} \frac{P^{n+\beta} f(x)}{P^{n+\beta+1} f(x)} = (h,f)^{-1} \quad ,$$

which ensures (3.1) with $(h,f)^{-1}$ instead of 1 on the righthand side (cf. the preceding section). We obtain the desired result by noting that, in view of the equality $R=1$, the lefthand side of (3.1) can have the only value 1 (we could, however, refer to the last conclusion of Lemma 2.3, too).

REFERENCES

[1] Dunford, N., and Schwartz, J.T. *Linear Operators*. Part I. General Theory. New York: Interscience Publishers, 1963.

[2] Lin, M. "Strong Ratio Limit Theorems for Mixing Markov Operators." *Ann. Inst. H. Poincaré*. Sec.B, vol.12, no.2 (1976): 181-191.

[2] Narimanian, S.M. "A Ratio Limit Theorem for Random Walks on Groups." *Vestnik Moskovskogo Gosudarstvennogo Universiteta*. Ser. matem. mekh., no.6 (1975): 17-24.

[4] Molchanov, S.A. "A Limit Theorem for Quotients of Transition Probabilities of Markov Chains." *Uspekhi matem. nauk*, vol.22, no.2 (1967): 124-125.

[5] Orey, S. *Lecture Notes on Limit Theorems for Markov Transition Probabilities.* London: Van Nostrand, 1971.

[6] Pollard, D.B., and Tweedie, R.L. "R-Theory for Markov Chains on a Topological Space. II." *Z. Wahrscheinlichkeitstheorie und Verw. Gebiete*, vol.34, no.2 (1976): 269-278.

[7] Revuz, D. *Markov Chains*. Amsterdam Oxford: North-Holland, 1975.

[8] Shur, M.G. "On the Asymptotic Behavior of Powers of Positive Operators." *Funktsional'nyj analiz i prilozheniya*, vol.16, no.2 (1982): 91-93.

[9] Sur, M.G. "Strong Ratio Limit Theorems." *Lecture Notes in Math.*, pp. 647-654. Vol.1021. Berlin Heidelberg New York: Springer-Verlag, 1983.

[10] Shur, M. "Asymptotic Behavior of Multistep Transition Probabilities." *Lithuanian Math. Journal*, vol.20, no.4 (1980): 368-372.

[11] Tweedie, R.L. "R-Theory for Markov Chains on a General State Space. Parts I, II." *Ann. Probab.*, vol.2, no.5 (1974): 840-878.

A.R. STEFANYUK

A METHOD FOR ESTIMATING PROBABILITY DENSITY

1. INTRODUCTION

Let

$$t_1,\ t_2,\ \ldots,\ t_n \tag{1}$$

be the sample of observations resulting from random and independent trials of a one-dimensional random variable ζ with density function $p_0(t)$. As noted in [1], the problem of estimating the function $p_0(t)$ on the basis of empirical data (1) should be viewed as an ill-posed problem. The method of solving it using the structural risk-minimization principle has been suggested. This method consists in the following.

Suppose that the random variable ζ lies in the interval [0,1]. Divide the interval into the $\ell+1$ parts (where ℓ is some positive integer) by the points

$$\tau_i = \frac{i}{\ell+1}, \qquad i = 1,2,\ldots,\ell . \tag{2}$$

Measure at these points the value of the sample distribution function $F_n(t)$:

$$y_i = F_n(\tau_i) .$$

We shall seek the density-function estimate in the form of the following expansion:

$$p(t,\lambda) = \sum_{k=1}^{N} \lambda_k \phi_k(t) , \qquad t \in [0,1], \qquad (3)$$

where $\{\phi_k(t)\}$ is some complete (in $L_2[0,1]$) orthonormal set of functions. We find the appropriate number of N and the coefficients λ_k, using the structural risk-minimization method, i.e., by minimizing (on $N = 1,2, \ldots$ and λ) the functional

$$W(\lambda) = \frac{\frac{1}{\ell}(y - F(\lambda))^T R_y^{-1}(y - F(\lambda))}{1 - \sqrt{\frac{(N+1)\left(1 + \ln \frac{\ell}{N+1}\right) - \ln \eta}{\ell}}} , \qquad (4)$$

where

$$y = \{y_i\} ; \qquad F(\lambda) = \{F_i^\lambda\} ;$$

$$F_i^\lambda = \int_0^{\tau_i} p(t,\lambda)\, dt \quad ;$$

R_y is the covariance matrix of the random vector y; $(1-\eta)$ is the reliability.

The matrix R_y^{-1} is tri-diagonal symmetrical (see [1, 2]). The principal diagonals are

$$\rho_i = n \frac{F_0(\tau_{i+1}) - F_0(\tau_{i-1})}{(F_0(\tau_{i+1}) - F_0(\tau_i))(F_0(\tau_i) - F(\tau_{i-1}))} ,$$

$$i = 1,\ldots,\ell; \qquad (5)$$

the side diagonals are

$$r_i = -\frac{n}{F_0(\tau_{i+1}) - F_0(\tau_i)}, \qquad i = 1,\ldots,\ell-1. \quad (6)$$

Since they depend on the unknown distribution function F_0, we suggest to estimate the density function $p_0(t)$ in two stages:

- 1. the preliminary rough estimation of the distribution function $F_0(t)$ and (by (5) and (6)) of the matrix R_y^{-1};
- 2. using the estimate found above, more accurate estimation of the density $p_0(t)$ by minimizing on N and λ the functional (4).

The computational practice shows high efficiency of the method suggested and its superiority over the traditional ones (see [2, 3]). But the question of convergence of the estimates obtained needs a special study.

It will be shown in this paper that for each N the density estimate, as n increases, converges with probability one to the best (in the sense of the functional (11) minimum) approximation of the function $p_0(t)$ by the first N basis functions $\phi_1, \ldots, \phi_N$.

2. DENSITY ESTIMATES CONVERGENCE

We assume the basis functions $\phi_k(t)$ to be continuous on $[0,1]$ and to have bounded derivatives. For example, these may be the functions

$$1;\ \sqrt{2}\cos\pi t;\ \sqrt{2}\cos 2\pi t;\ \ldots, \qquad t \in [0,1]. \quad (7)$$

They also may be regarded as a canonical example throughout the

Let

$$m_k = \sup_t |\phi_k(t)| , \qquad M_k = \sup_t |\phi_k'(t)| ,$$

$$\psi_k(t) = \int_0^t \phi_k(\tau)\, d\tau , \qquad t \in [0,1]. \tag{8}$$

In estimating the matrix R_y^{-1} (the first stage), the following distribution-function estimate us used:

$$\Phi_n(t) = \begin{cases} \frac{1}{2n}\left(\frac{t}{t_1}\right) & \text{for} \quad 0 \le t < t_1; \\ \frac{k-\frac{1}{2}}{n} + \frac{1}{n}\left(\frac{t - t_k}{t_{k+1} - t_k}\right) & \text{for} \quad t_k \le t \le t_{k+1}, \\ & \qquad k = 1,\ldots,n-1; \\ \frac{n-\frac{1}{2}}{n} + \frac{1}{2n}\left(\frac{t - t_n}{1 - t_n}\right) & \text{for} \quad t_n < t \le 1 , \end{cases} \tag{9}$$

where $t_1,\ldots,t_n$ are the sample points (1). It is easy to see that for arbitrary $\tau_i, \tau_j \in [0,1]$

$$|(\Phi_n(\tau_i) - \Phi_n(\tau_j)) - (F_n(\tau_i) - F_n(\tau_j))| \le \frac{1}{n} . \tag{10}$$

Define

$$\Lambda_N^1 = \left\{\lambda: \int_0^1 \left(\sum_{k=1}^N \lambda_k \phi_k(t)\, dt\right) = 1\right\} .$$

Denote by $\tilde{\lambda}_{(N)} = \{\tilde{\lambda}_1,\ldots,\tilde{\lambda}_N\} \in \Lambda_N^1$ the parameter vector, minimizing for fixed N the numerator of (4) with the matrix R_t^{-1} estimate found by the formulas (5) and (6) with the help of the function (9). Denote also by $\lambda^*_{(N)} = \{\lambda^*_1, \ldots, \lambda^*_N\}$ the parameter vector, minimizing in Λ_N^1 the functional

$$I^*(\lambda) = \int_0^1 \frac{\left\{\sum_{k=1}^{N} \lambda_k \phi_k(t) - p_0(t)\right\}^2}{p_0(t)} dt \quad . \tag{11}$$

<u>THEOREM</u>. Suppose the unknown density function $p_0(t)$ is continuous on $[0,1]$, has a bounded derivative, and does not vanish. Also, as the sample size n tends to infinity, let us increase the number ℓ (see (2)) in such a manner that

$$\ell = \ell(n) \xrightarrow[n\to\infty]{} \infty, \qquad \ell(n) \frac{\ln n}{n} \xrightarrow[n\to\infty]{} 0 \quad . \tag{12}$$

Then for the arbitrary fixed N with probability one

$$\lim_{n,\ell\to\infty} \|p(t,\tilde{\lambda}_{(N)}) - p(t,\lambda^*_{(N)})\|_{L_2} = 0 \quad . \tag{13}$$

<u>REMARK</u>. By the property of completeness of $\{\phi_k\}$ we have

$$\|p(t,\lambda^*_{(N)}) - p_0(t)\|_{L_2} \xrightarrow[N\to\infty]{} 0 \quad .$$

The assertion of the Theorem implies that for any $\varepsilon > 0$ there exists the number N and the values of n and ℓ such that

$$\|p(t,\tilde{\lambda}_{(N)}) - p_0(t)\|_{L_2} < \varepsilon$$

with arbitrary probability $P < 1$.

Before we prove the Theorem, it is convenient to make additional transformations. Let $\tau_0 = 0$, $\tau_{\ell+1} = 1$ (see (2)). For the sake of simplicity denote

$$\Delta^i F_0 = F_0(\tau_{i+1}) - F_0(\tau_i) \;,$$

$$\Delta^i F_n = F_n(\tau_{i+1}) - F_n(\tau_i) \;, \tag{14}$$

$$\Delta^i \Phi_n = \Phi_n(\tau_{i+1}) - \Phi_n(\tau_i) \;, \qquad i = 0,1,\dots,\ell.$$

Using the expressions (5) and (6) for the elements of the matrix R_y^{-1} and taking into account the fact that function Φ_n is used for estimating the matrix, it is easy to see that the numerator on the right-hand side of (4) can be represented as

$$I(\lambda) = \frac{n}{\ell} \sum_{i=0}^{\ell} \frac{\left[\Delta^i F_n - \int_{\tau_i}^{\tau_{i+1}} \left\{\sum_{k=1}^{N} \lambda_k \phi_k(t)\right\} dt\right]^2}{\Delta^i \Phi_n} \;. \tag{15}$$

Note that the equality

$$\int_0^1 \left\{\sum_{k=1}^{N} \lambda_k \phi_k(t)\right\} dt = 1$$

implies that

$$\lambda_1 = \frac{1 - \sum_{k=2}^{N} \lambda_k \psi_k(1)}{\psi_1(1)}$$

(if $\psi_1(1) \neq 0$). Hence for all $\lambda \in \Lambda_N^1$

$$\sum_{k=1}^{N} \lambda_k \phi_k(t) = \hat{\phi}_1(t) + \sum_{k=2}^{N} \lambda_k \hat{\phi}_k(t) \;, \qquad t \in [0,1],$$

where

$$\hat{\phi}_1(t) = \frac{\phi_1(t)}{\psi_1(1)} \;, \qquad \hat{\phi}_k(t) = \phi_k(t) - \frac{\psi_k(1)}{\psi_1(1)}\, \phi_1(t) \tag{16}$$

and $\lambda_2, \ldots, \lambda_N$ could be chosen arbitrarily.

Now according to the mean-value theorem for integrals with any $\lambda \in \Lambda_N^1$

$$\int_{\tau_i}^{\tau_{i+1}} \left(\sum_{k=1}^{N} \lambda_k \phi_k(t) \right) dt = \int_{\tau_i}^{\tau_{i+1}} \hat{\phi}_1(t)\, dt + \sum_{k=2}^{N} \lambda_k \int_{\tau_i}^{\tau_{i+1}} \hat{\phi}_k(t)\, dt$$

$$= \frac{1}{\ell+1} \left(\hat{\phi}_1(\xi_i^1) + \sum_{k=2}^{N} \lambda_k \hat{\phi}_k(\xi_i^k) \right) ,$$

where ξ_i^k is some point of the interval $[\tau_i, \tau_{i+1}]$.

Consequently the numerator of the expression (4) has the form

$$I(\lambda) = \frac{n}{\ell} \sum_{i=1}^{\ell} \frac{\left| \Delta^i F_n - \frac{1}{\ell+1} \left\{ \hat{\phi}_1(\xi_i^1) + \sum_{k=2}^{N} \lambda_k \hat{\phi}_k(\xi_i^k) \right\} \right|^2}{\Delta^i \Phi_n} . \tag{17}$$

The minimum condition for the functional (17) on $\lambda \in \Lambda_N^1$ yields a system of linear equations. In matrix form the solution of the system is

$$\tilde{\lambda}_{(N)} = \tilde{A}^{-1} \tilde{b} , \tag{18}$$

where

$$\tilde{A} = \{\tilde{a}_{jk}\} ; \qquad \tilde{b} = \{\tilde{b}_j\} ; \qquad j,k = 2,3,\ldots,N;$$

$$\tilde{a}_{jk} = \frac{1}{\ell+1} \sum_{i=0}^{\ell} \frac{\hat{\phi}_j(\xi_i^j)\hat{\phi}_k(\xi_i^k)}{(\ell+1)\Delta^i \Phi_n} ; \tag{19}$$

$$\tilde{b}_j = \frac{1}{\ell+1} \sum_{i=0}^{\ell} \hat{\phi}_j(\xi_i^j) \frac{\Delta^i F_n - \frac{\hat{\phi}_1(\xi_i^1)}{\ell+1}}{\Delta^i \Phi_n} .$$

It is easy to see that the parameter vector

$\lambda^*_{(N)} = \{\lambda^*_1, \dots, \lambda^*_N\}$ minimizing in Λ^1_N the functional (11) is given by

$$\lambda^*_{(N)} = A^{-1} b \quad , \tag{20}$$

where

$$A = \{a_{jk}\} \; ; \qquad b = \{b_j\} \; ; \qquad j,k = 2,3,\dots,N \; ;$$

$$a_{jk} = \int_0^1 \frac{\hat{\phi}_j(t)\hat{\phi}_k(t)}{p_0(t)} \, dt \quad ; \tag{21}$$

$$b_j = \int_0^1 \hat{\phi}_j(t) \frac{p_0(t) - \hat{\phi}_1(t)}{p_0(t)} \, dt \quad .$$

For each N the matrix A is obviously nonsingular (for the linear independence of $\hat{\phi}_k$ and for the condition $p_0(t) > 0$).

Denote

$$D_\ell = \tilde{A} - A = \{d_{jk}\} \; ;$$

$$\delta_\ell = \tilde{b} - b \quad . \tag{22}$$

To prove the Theorem the following lemmas will be required.

<u>LEMMA 1</u>. If the conditions of the Theorem are satisfied, then for any fixed N with probability one

$$\lim_{n,\ell\to\infty} \|D_\ell\| = 0 \quad ,$$

$$\lim_{n,\ell\to\infty} \|\delta_\ell\| = 0 \quad , \tag{23}$$

$$\lim_{n,\ell\to\infty} \|(E + A^{-1}D_\ell)^{-1}\| \le 3$$

(E is the $(N-1)\times(N-1)$ identity matrix).

LEMMA 2. There exists a constant C_λ such that the value of $\|\lambda^*_{(N)}\|$ does not exceed it for any N.

PROOF OF LEMMA 1. By (22), (19) and (21)

$$d_{jk} = \frac{1}{\ell+1}\sum_{i=0}^{\ell}\left[\frac{\hat{\phi}_j(\xi_i^j)\hat{\phi}_k(\xi_i^k)}{(\ell+1)\Delta^i\Phi_n} - \frac{\hat{\phi}_j(\theta_i)\hat{\phi}_k(\theta_i)}{p_0(\theta_i)}\right], \tag{24}$$

where θ_i is some point of the interval $[\tau_i,\tau_{i+1}]$ (the mean-value theorem for integrals has been used). Then note that by (8) and (16)

$$\sup_t |\hat{\phi}_k(t)| \le m_k + \frac{\psi_k(1)}{\psi_1(1)} m_1 = \hat{m}_k ,$$

$$\sup_t |\hat{\phi}'_k(t)| \le M_k + \frac{\psi_k(1)}{\psi_1(1)} M_1 = \hat{M}_k ,$$

whence

$$|\hat{\phi}_j(\xi_i^j)\hat{\phi}_k(\xi_i^k) - \hat{\phi}_j(\theta_i)\hat{\phi}_k(\theta_i)| \le \frac{\hat{m}_j\hat{M}_k + \hat{m}_k\hat{M}_j}{\ell + 1} \tag{25}$$

and also

$$|(\ell+1)\Delta^i\Phi_n - p_0(\theta_i)| \le (\ell+1)|\Delta^i\Phi_n - \Delta^i F_0| + |(\ell+1)\Delta^i F_0 - p_0(\theta_i)| . \tag{26}$$

Taking into account that in the second term $\Delta^i F_0 = \frac{p_0(\mu_i)}{\ell + 1}$, $\mu_i \in [\tau_i,\tau_{i+1}]$, we obtain

$$|(\ell+1)\Delta^i F_0 - p_0(\theta_i)| \le \frac{K}{\ell+1} ; \qquad K = \sup_t |p'_0(t)| .$$

Besides,

$$|\Delta^i\Phi_n - \Delta^i F_0| \le |\Delta^i F_n - \Delta^i\Phi_n| + |\Delta^i F_n - \Delta^i F_0| \le \frac{1}{n} + |\Delta^i F_n - \Delta^i F_0| ,$$

whence

$$|(\ell+1)\Delta^i\Phi_n - p_0(\theta_i)| \le \frac{K}{\ell+1} + (\ell+1)\left\{\frac{1}{n} + |\Delta^i F_n - \Delta^i F_0|\right\} . \qquad (27)$$

Combining (24), (25) and (26), we find that

$$\begin{aligned}|d_{jk}| \le \frac{1}{\ell+1} \sum_{i=1}^{\ell} \Bigg[& V\left\{\frac{\hat{m}_j\hat{M}_k + \hat{m}_k\hat{M}_j}{\ell+1}\right\} \\ & + \hat{m}_j\hat{m}_k\left(\frac{K}{\ell+1} + (\ell+1)\left\{\frac{1}{n} + |\Delta^i F_n - \Delta^i F_0|\right\}\right)\Bigg] \\ & \times \left|v^2 - V\left(\frac{K}{\ell+1} + (\ell+1)\left\{\frac{1}{n} + |\Delta^i F_n - \Delta^i F_0|\right\}\right)\right|^{-1} ,\end{aligned} \qquad (28)$$

where

$$V = \max_t |p_0(t)|, \qquad v = \min_t |p_0(t)| .$$

Suppose that

$$\ell + 1 > \frac{3KV}{v^2} , \qquad (29)$$

$$V(\ell+1)\left\{\frac{1}{n} + \sup_i |\Delta^i F_n - \Delta^i F_0|\right\} \le \frac{v^2}{3} . \qquad (30)$$

Then the denominator in (28) is not less than $v^2/3$ and therefore */

$$\begin{aligned}\|D_\ell\| &\le \sqrt{\sum_{j=2}^{N}\sum_{k=2}^{N} |d_{jk}|^2} \qquad (31) \\ &\le N\left(\frac{6\,V\,m\,M}{v^2(\ell+1)} + \frac{3m^2}{v^2}\left\{\frac{K}{\ell+1} + \frac{\ell+1}{n} + (\ell+1)\sup_i |\Delta^i F_n - \Delta^i F_0|\right\}\right) .\end{aligned}$$

Since $\ell \to \infty$, the condition (29) will be satisfied sooner or

*/ It is designated: $m = \max\{\hat{m}_1,\dots,\hat{m}_N\}$; $M = \max\{\hat{M}_1,\dots,\hat{M}_N\}$.

later. As for the condition (30), it will be convenient to designate it by A and to write

$$P\{\|D_\ell\|>\varepsilon\} = P\{\|D_\ell\|>\varepsilon \mid A\}P\{A\} + P\{\|D_\ell\|>\varepsilon \mid \bar{A}\}P\{\bar{A}\} \tag{32}$$

$$\leq P\{\|D_\ell\|>\varepsilon \mid A\}P\{A\} + P\{\bar{A}\} .$$

Using the inequality

$$P\{B \mid A\}P\{A\} \leq P\{B\} ,$$

we obtain from (31):

$$P\{\|D_\ell\|>\varepsilon \mid A\}P\{A\}$$

$$\leq P\left\{\frac{6VmM}{v^2(\ell+1)} + \frac{3m^2}{v^2}\left(\frac{K}{\ell+1} + \frac{\ell+1}{n} + (\ell+1)\sup_i |\Delta^i F_n - \Delta^i F_0|\right) > \frac{\varepsilon}{N}\right\}$$

$$= P\left\{\sup_i |\Delta^i F_n - \Delta^i F_0| > \frac{1}{\ell+1}\left(\frac{\varepsilon v^2}{3m^2N} - \frac{6VM}{3m(\ell+1)} - \frac{K}{\ell+1}\right) - \frac{1}{n}\right\} .$$

By the condition of the Lemma $\ell \to \infty$ and $(n/\ell) \to \infty$; therefore sooner or later it will be

$$\ell+1 \geq \max\left\{\frac{18VmMN}{\varepsilon v^2}, \frac{9m^2KN}{\varepsilon v^2}\right\} ,$$

$$\frac{n}{\ell+1} \geq N\left(\frac{18m^2}{\varepsilon v^2}\right)$$

and next

$$P\{\|D_\ell\|>\varepsilon \mid A\}P\{A\} \leq P\left\{\sup_i |\Delta^i F_n - \Delta^i F_0| > \frac{\varepsilon v^2}{18m^2N(\ell+1)}\right\} . \tag{33}$$

Besides,

$$P\{\bar{A}\} = P\left\{\sup_i |\Delta^i F_n - \Delta^i F_0| > \frac{v^2}{3V(\ell+1)} - \frac{1}{n}\right\} .$$

Consequently, for $\frac{n}{\ell+1} > \frac{6V}{v^2}$

$$P\{\bar{A}\} \leq P\left\{\sup_i |\Delta^i F_n - \Delta^i F_0| > \frac{v^2}{6V(\ell+1)}\right\} .$$

So it is proved that for sufficiently large n and ℓ

$$P\{\|D_\ell\| > \varepsilon\} \leq P\left\{\sup_i |\Delta^i F_n - \Delta^i F_0| > \frac{c_1}{\ell+1}\right\} + P\left\{\sup_i |\Delta^i F_n - \Delta^i F_0| > \frac{c_2}{\ell+1}\right\} , \qquad (34)$$

where $c_1 = \frac{\varepsilon v^2}{18m^2 N}$, $c_2 = \frac{v^2}{6V}$ are the constants independent of n and ℓ.

Taking into consideration that $\Delta^i F_n$ is the number of the sample points falling into the interval $[\tau_i, \tau_{i+1}]$ (divided by the total number of points n), that $\Delta^i F_0$ is the probability of the event, and also using the well-known Bernstein inequality (see [4, p. 158]), it is easy to find that for any constant $c > 0$

$$\sum_{n=1}^{\infty} P\left\{\sup_i |\Delta^i F_n - \Delta^i F_0| > \frac{c}{\ell+1}\right\} < \infty \qquad (35)$$

if only (12) is satisfied.

Using (34) and the Borel-Cantelli lemma, we find

$$P\left\{\lim_{n,\ell\to\infty} \|D_\ell\| = 0\right\} = 1 .$$

The equality

$$P\left\{\lim_{n,\ell\to\infty} \|\delta_\ell\| = 0\right\} = 1$$

can be proved in a similar way.

Let us now prove the last assertion of Lemma 1. The matrix A has already been said to be nonsingular for any N and consequently to have the bounded inverse matrix: $\|A^{-1}\| = \text{const} < \infty$.

If $\|D_\ell\| \le \frac{1}{2}\|A^{-1}\|^{-1}$ then $\|A^{-1}D_\ell\| \le \frac{1}{2}$ and also (see [5, p. 230]):

$$\|(E + A^{-1}D_\ell)^{-1}\| \le \sum_{k=0}^{\infty} \|A^{-1}D_\ell\|^k$$

$$\le \sum_{k=0}^{\infty} (\tfrac{1}{2})^k = 2 .$$

This is immediately followed by

$$P\{\|(E + A^{-1}D_\ell)^{-1}\| > 3\} \le P\{\|(E + A^{-1}D_\ell)^{-1}\| > 3 \mid \|D_\ell\| \le \tfrac{1}{2}\|A^{-1}\|^{-1}\}$$
$$+ P\{\|D_\ell\| > \tfrac{1}{2}\|A^{-1}\|^{-1}\}$$
$$= P\{\|D_\ell\| > \tfrac{1}{2}\|A^{-1}\|^{-1}\} .$$

The inequality (34) and the Borel-Cantelli lemma give now

$$P\left\{\lim_{n,\ell\to\infty} \|(E + A^{-1}D_\ell)^{-1}\| \le 3\right\} = 1 .$$

This completes the proof of Lemma 1.

PROOF OF LEMMA 2. Since

$$\left\|\sum_{k=1}^{N} \lambda_k^* \phi_k(t) - p_0(t)\right\| \xrightarrow[N\to\infty]{} 0$$

(see Remark above), then

$$\left\|\sum_{k=1}^{N} \lambda_k^* \phi_k(t)\right\| \xrightarrow[N\to\infty]{} \|p_0(t)\| .$$

This implies that for every N greater than N_1

$$\left\| \sum_{k=1}^{N} \lambda_k^* \phi_k(t) \right\| \le 2\|p_0(t)\| \quad .$$

By the orthonormality property of the functions $\{\phi_k\}$

$$\left\| \sum_{k=1}^{N} \lambda_k^* \phi_k(t) \right\|^2 = \sum_{k=1}^{N} (\lambda_k^*)^2 \|\phi_k\|^2 = \|\lambda_{(N)}^*\| \quad .$$

Thus for every $N > N_1$

$$\|\lambda_{(N)}^*\| \le 2\|p_0\| \quad .$$

But for every finite N the value of $\|\lambda_{(N)}^*\|$ is surely bounded.///

<u>PROOF OF THE THEOREM</u>. By (18) and (22)

$$\tilde{\lambda}_{(N)} = (A + D_\ell)^{-1}(b + \delta_\ell) = (E + A^{-1}D_\ell)^{-1}(\lambda_{(N)}^* + A^{-1}\delta_\ell) \quad ,$$

then

$$(E + A^{-1}D_\ell)\tilde{\lambda}_{(N)} = \lambda_{(N)}^* + A^{-1}\delta_\ell$$

or (taking into account the previous equality)

$$\begin{aligned} \lambda_{(N)}^* - \tilde{\lambda}_{(N)} &= A^{-1}D_\ell\tilde{\lambda}_{(N)} - A^{-1}\delta_\ell \\ &= A^{-1}D_\ell(E + A^{-1}D_\ell)^{-1}(\lambda_{(N)}^* + A^{-1}\delta_\ell) - A^{-1}\delta_\ell \quad . \end{aligned}$$

By Lemmas 1 and 2

$$\begin{aligned} \|\lambda_{(N)}^* - \tilde{\lambda}_{(N)}\| &\le \|A^{-1}\| \, \|D_\ell\| \, \|(E + A^{-1}D_\ell)^{-1}\| \, (\|\lambda_{(N)}^*\| + \|A^{-1}\| \, \|\delta_\ell\|) \\ &\quad + \|A^{-1}\| \times \|\delta_\ell\| \\ &\xrightarrow[n,\ell\to\infty]{P=1} 0 \quad . \end{aligned}$$

Then the assertion of the Theorem follows directly since N is fixed and $\{\phi_k\}$ is the orthonormal system of functions.

REFERENCES

[1] Vapnik, V.N., and Stefanyuk, A.R. "Nonparametric Methods for Reconstructing Probability Density." *Automat. Remote Control*, vol.39, no.8 (1978): 1127-1140.

[2] Vapnic, V.N. *Estimation of Dependences Based on Empirical Data*.Berlin Heidelberg New York: Springer-Verlag Inc., 1982.

[3] Vapnik, V.N. Ed. *Algoritmy i programmy dlya otsenivaniya zavisimosti* (Algorithms and Programs for Estimation of Dependence). Moskva: Nauka, 1984.

[4] Bernstein, S.N. *Teoriya veroyatnostei* (Probability Theory). 4th ed. Moskva-Leningrad, 1946.

[5] Kolmogorov, A.N., and Fomin, S.V. *Elements of the Theory of Functions and Functional Analysis*. Rochester, NY: Graylock Press, 1957. Revised edition: *Introductory Real Analysis*. Englewood Cliffs, NJ: Prentice-Hall, 1970.

A.I. YASHIN

DYNAMICS OF SURVIVAL ANALYSIS: CONDITIONAL GAUSSIAN PROPERTY VERSUS THE CAMERON-MARTIN FORMULA

1. INTRODUCTION

The well-known Cameron and Martin formula [1, 2, 3] gives a way of calculating the mathematical expectation of the exponent which is the functional of a Wiener process. More precisely, let (Ω, H, P) be the basic probability space, let $H = (H_u)_{u\geq 0}$ be the nondecreasing right-continuous family of σ-algebras, and let H_0 be completed by the events of P-probability zero from $H = H_\infty$. Denote by W_u an n-dimensional H-adapted Wiener process and by $Q(u)$ a symmetric nonnegative definite matrix whose elements $q_{i,j}(u)$, $i,j = 1,2,\dots,n$ satisfy for some t the condition

$$\int_0^t \sum_{i,j=1}^n |q_{i,j}(u)|\, du < \infty . \tag{1}$$

The following result is known as the Cameron-Martin formula.

<u>THEOREM 1</u>. Let (1) hold. Then

$$E \exp\left[-\int_0^t (W_u, Q(u)W_u)\, du\right] = \exp\left[\tfrac{1}{2}\int_0^t \mathrm{Sp}\Gamma(u)\, du\right] , \tag{2}$$

where $(W_u, Q(u)W_u)$ is the scalar product, and $\Gamma(u)$ is a symmetric nonpositive definite matrix being the unique solution of the Riccati matrix equation

$$\frac{d\Gamma(u)}{du} = 2Q(u) - \Gamma^2(u) \quad ; \tag{3}$$

$\Gamma(t) = 0$ the zero matrix.

The proof of this formula given in [3] is based on the property of likelihood ratio for diffusion processes. The idea to use this property is due to A.A. Novikov [18]. In [4], L.E. Myers developed this approach and found the formula for averaging the exponent when, instead of a Wiener process, there is a process satisfying a linear stochastic differential equation driven by a Wiener process. (Novikov [18] obtained earlier this result in the particular case when trend and diffusion coefficients are constant.) His result may be formulated as follows.

THEOREM 2. Let $Y(t)$ be an m-dimensional diffusion process of the form

$$dY(t) = a(t)\, Y(t)\, dt + b(t)\, dW_t$$

with a deterministic initial condition $Y(0)$. Assume that the matrix $Q(u)$ has the properties described above. Then the following formula holds:

$$\begin{aligned} &E \exp \left[-\int_0^t Y^*(u)\, Q(u)\, Y(u)\, du \right] \\ &\quad = \exp \left[Y^*(0)\, \Gamma(0)\, Y(0) + \operatorname{Sp}\int_0^t b(u)\, b^*(u)\, \Gamma(u)\, du \right], \end{aligned} \tag{4}$$

where $\Gamma(u)$ is the solution of the Riccati matrix equation

$$\frac{d\partial(u)}{du} = Q(u) - (\Gamma(u) + \Gamma^*(u))a(u)$$
$$- \tfrac{1}{2}(\Gamma(u) + \Gamma^*(u))b(u)\, b^*(u)(\Gamma(u) + \Gamma^*(u)) \tag{5}$$

with the terminal condition $\Gamma(t) = 0$.

These results have a direct application to survival analysis: any exponent on the left-hand sides of (2) and (4) can be regarded as a conditional survival function in some lifecycle problem [4, 5, 6].

Such interpretation was used in some biomedical models. The quadratic dependence of risk on some risk factors was confirmed by results of the numerous studies in physiology and medicine [5]. The results are also applicable to reliability analysis.

The way of proving the Cameron-Martin formula and its generalizations given in [1, 2, 3, 18] does not invoke this kind of interpretation and, unfortunately, does not provide any physical or demographical sense to the variables (u) that appear on the right-hand side of the formulas (2) and (4). Moreover, the form of the boundary value conditions for the equations (3) and (5) on the right-hand side complicate the computing of the Cameron-Martin formula when one needs to calculate it on-line for many time moments t. These difficulties pile up when there are some additional on-line observations correlated with influential factors.

Fortunately, there is the straightforward method that allows us to avoid these complications. The approach uses the innovative transformations random intensities or compensators of a point process. Usage of this "martingale" technique allows a more general formula for averaging exponents which might be a more complex functional of a random process from a wider class.

If the functional is of a quadratic form one can get another constructive way of averaging the exponent using the conditional Gaussian property. The goal of this paper is to illustrate this approach.

2. THE FORMULATION OF RESULTS

We shall start from the following general statement.

THEOREM 3. Let $Y(u)$ be an arbitrary H-adapted random process and let $\lambda(Y,u)$ be some nonnegative H^y-adapted function such that for some $t \geq 0$

$$E\int_0^t \lambda(Y,u)\, du < \infty . \tag{6}$$

Then

$$E \exp\left[-\int_0^t \lambda(Y,u)\, du\right] = \exp\left[-\int_0^t E[\lambda(Y,u) \mid T > u]\, du\right] , \tag{7}$$

where T is the stopping time associated with the process $Y(u)$ as follows:

$$P(T > t \mid H_t^y) = \exp\left[-\int_0^t \lambda(Y,u)\, du\right] \tag{8}$$

and $H_t^y = \bigcap_{u>t} \sigma\{Y(v), v \leq u\}$ is the σ-algebra generated by the history of the process $Y(u)$ up to time t, $H^y = (H_t^y)_{t \geq 0}$.

The proof of this statement is based on the idea of "innovation," widely used in martingale approach to filtering and stochastic control problems [3, 7, 8]; it is given in the Appendix.

Another form of this idea appeared and was explored in the demographical studies of population heterogeneity dynamics [6, 9, 10]. Differences among the individuals or units in these studies were described in terms of a random heterogeneity factor called "frailty." This factor is responsible for the individual's susceptibility to death and can change over time in accordance with the changes of some external variables, influencing the individual's chances to die (or to have failure for some unit if one deals with the reliability studies).

When the influences of the external factors on the failure rate may be represented in terms of a function which is a quadratic form of the diffusion Gaussian process, the result of Theorem 3 may be developed as follows.

THEOREM 4. Let the m-dimensional H-adapted process $Y(u)$ satisfy the linear stochastic differential equation

$$dY(t) = [a_0(t) + a_1(t)Y(u)]dt + b(t)dW_t , \quad Y(0) = Y_0 ,$$

where Y_0 is the Gaussian random variable with mean m_0 and variance γ_0. Denote by $Q(u)$ a symmetric nonnegative definite matrix whose elements satisfy the condition (1). Then the following formula holds:

$$E \exp\left[-\int_0^t (Y^*(u)Q(u)Y(u))du\right] = \exp\left[-\int_0^t (m_u^*Q(u)m_u + Sp(Q(u)\gamma_u))du\right] . \tag{9}$$

The processes m_u and γ_u are solutions of the ordinary differential equations:

$$\frac{dm_t}{dt} = a_0(t) + a_1(t)m_t - 2\gamma_t Q(t)m_t \ , \tag{10}$$

$$\frac{d\gamma_t}{dt} = a_1(t)\gamma_t + \gamma_t a_1^*(t) + b(t)b^*(t) - 2\gamma_t Q(t)\gamma_t \tag{11}$$

with the initial conditions m_0 and γ_0, respectively.

The proof of this Theorem is based on the Gaussian property of the conditional distribution function $P(Y(t) \le x \mid T > t)$. This situation invokes the well-known generalization of the Kalman filtering scheme [3, 11, 12, 13] (see Appendix).

Note that a similar approach to the averaging of the survival function was studied in [5] under the assumption that the conditional Gaussian property holds. In that paper, the mortality rate was assumed to be influenced by the values of some randomly evolving physiological factors, such as blood pressure or serum cholesterol level. We will illustrate the ideas and the results by several examples.

3. EXAMPLES

3.1. Failure Rate as a Function of the Random Variable

Let (Ω,H,P) be the basic probability space and let $Y(\omega)$ and $T(\omega)$ be two random variables such that $T(\omega) > 0$ with probability one and has a continuous distribution function. $Y(\omega)$ and $T(\omega)$ will be interpreted as external environmental factor and termination (death) time, respectively.

Assume that an external factor influences a failure rate through the random variable $Z = Y^2$. Let $\sigma(Z)$ be the σ-algebra in Ω generated by the random variable Z. Denote by $F(t,Z) = P(T \le t \mid \sigma(Z))$ the $\sigma(Z)$-conditional distribution function of termination time T. Assume that $F(t,Z)$ has the form

$$F(t,Z) = 1 - \exp\left(-Z\int_0^t \lambda(u)\, du\right), \qquad (12)$$

where $\lambda(t)$, $t \ge 0$, is the deterministic function of t that may be interpreted as the age-specific mortality rate for an average (standard) individual [10].

Let $\bar{F}(t)$ denote the unconditional distribution function for $T(\omega)$

$$\bar{F}(t) = p(T < t)$$

and let $\bar{\lambda}(t)$ be determined by the equality

$$\bar{\lambda}(t) = \frac{d\bar{F}(t)}{dt}[1 - \bar{F}(t)]^{-1} .$$

This function is "called" mortality rate in [10] since it represents mortality approximated by empirical death rates which are evaluated without taking population heterogeneity into account. It can be easily shown [10] that

$$\bar{\lambda}(t) = \bar{z}(t)\lambda(t) ,$$

where

$$\bar{z}(t) = E(z \mid T > t)$$

is the conditional mathematical expectation of Z given the event $\{T > t\}$.

The form of the $\bar{\lambda}(t)$ as a function of time is determined by the conditional distribution of frailty Z and $\lambda(t)$. It turns out that if the frailty Z is generated by a Gaussian random variable Y, the analytical form for $\bar{Z}(t)$ might be easily found. Moreover, this conditional distribution of Y is Gaussian, as shown in the following.

PROPOSITION 1. Let $Z = Y^2$, where Y is a Gaussian random variable with mean a and variance σ^2. Then the conditional distribution of Y given the event $\{T > t\}$ is also Gaussian, with mean m_t and variance γ_t that satisfy the equations

$$\frac{dm_t}{dt} = -2\lambda(t) m_t \gamma_t \ , \qquad m_0 = a \ , \tag{13}$$

$$\frac{d\gamma_t}{dt} = -2\lambda(t)\gamma_t^2 \ , \qquad \gamma_0 = \sigma^2 \ . \tag{14}$$

The result of this statement follows from Theorem 4. It can also be proved independently using the Bayes rule. According to this rule the conditional density of the random variable Y may be represented in the form

$$g(x,t) = h(x)\, P(T > t \mid Y = x)[P(T > t)]^{-1} \ , \tag{15}$$

where (from the definitions of Z and T)

$$h(x) = \frac{1}{(2\pi\gamma_0)^{\frac{1}{2}}} \exp\left\{-\frac{(x-m_0)^2}{2\gamma_0}\right\} \ ,$$

$$P(T > t \mid x) = \exp\left\{-x^2 \int_0^t \lambda(u)\, du\right\}$$

and

$$g(x,t) = \frac{d}{dx} P(Y \le x \mid T > t) \ .$$

Substitution of the formulas for $h(x)$ and $P(T > t)$ into the equation for $P(x \mid T > t)$ leads to

$$g(x,t) = f(t) \exp \left\{ \frac{-[x(2\sigma^2\Lambda(t)+1) - a]^2}{2\sigma^2(2\sigma^2\Lambda(t)+1)} \right\} ,$$

where

$$\Lambda(t) = \int_0^t \lambda(u)\, du$$

and $f(t)$ is some function that does not depend on x and acts as a normalizing factor. It is evident that this form of the conditional density $g(x,t)$ corresponds to a Gaussian distribution with $a[2\sigma^2\Lambda(t)+1]^{-1}$ and $\sigma^2[2\sigma^2\Lambda(t)+1]^{-1}$ as mean and variance, respectively. Substituting these values for m_t and γ_t, it is not difficult to check that they satisfy the equations given in the Theorem.

REMARK. Note that results of this Theorem may be represented by the following averaging formula:

$$E \exp \left[-Y^2 \int_0^t \lambda(u)\, du \right] = \exp \left[-\int_0^t (m_u^2 + \gamma_u)\lambda(u)\, du \right] , \quad (16)$$

which is similar to Cameron-Martin's result.

3.2. Mortality in a Structurized Population

Assume that some population may be represented as a collection of several groups of individuals (men and women, ethnic groups, etc.). Introduce a random variable Z taking a finite number of possible values $(1,2,\ldots,K)$ with a priori probabilities $p_1, p_2, \ldots, p_K$. Let the age-specific mortality rate of the average individual depend on the value of the random variable Z -- this will be associated with a particular social group. Assume that the survival probability of a person from group j with a history H_t^y of environmental or physiological characteristics up to time t may be written as follows:

$$P(T > t \mid H_t^y, \{z = j\}) = \exp\left[-\int_0^t Y^2 \lambda(j,u)\, du\right] ,$$

where $Y(t)$ is the process described in the formulation of Theorem 4.

If the observer takes into account the differences between the persons belonging to different social groups, he should produce K different patterns of age-specific mortality rates $\bar{\lambda}(i,t)$, $i = \overline{1,K}$.

<u>PROPOSITION 2</u>. The mortality rates corresponding to the conditional survival probabilities

$$P(T > t \mid Z = i) = \exp\left(-\int_0^t \bar{\lambda}(i,u)\, du\right) , \quad i = \overline{1,K} ,$$

are given by the formulas

$$\bar{\lambda}(i,t) = \lambda(i,t)(m_t^2(i) + \gamma_t(i)) , \qquad i = \overline{1,K} ,$$

where K different estimates $m_t(i)$, $\gamma_t(i)$ are the solutions of the following equations:

$$\frac{dm_t(i)}{dt} = a_0(t) + a_1(t)m_t(i) - 2m_t(i)\gamma_t(i)\lambda_t(i,t) ,$$

$$m_0(i) , \quad i = \overline{1,K} ,$$

$$\frac{d\gamma_t(i)}{dt} = 2a_1(i,t)\gamma_t(i) + b^2(i,t) - 2\lambda(i,t)\gamma_t^2(i) ,$$

$$\gamma_0(i) , \quad i = \overline{1,K} .$$

If the observer does not differentiate between persons from different groups, the observed age-specific mortality rate $\bar{\lambda}(t)$ will depend on the ratio $\pi_i(t)$, $i = \overline{1,K}$, of individuals in the different groups. These ratios coincide with the conditional probabilities of the events $\{Z=i\}$, $i = \overline{1,K}$, given $\{T>t\}$, and can be shown to satisfy the following equations:

$$\pi_j(t) = \pi_j(0) + \int_0^t \pi_j(u)\left(\bar{\lambda}(j,u) - \sum_{i=1}^{i=K} \bar{\lambda}(i,u)\pi_i(u)\right) du ,$$

where $\pi_j(0) = p_j$. In this case $\bar{\lambda}(t)$ may be represented as follows:

$$\bar{\lambda}(t) = \sum_{i=1}^{i=K} \bar{\lambda}(i,u)\,\pi_i(t) . \tag{17}$$

3.3. Evaluation of a Mortality Rate in the Multistate Demography

Assume that Z_t is a finite state, continuous time, Markov process with vector initial probabilities $p_1, \ldots, p_K$ and intensity matrix

$$R(t) = ||r_{i,j}(t)|| , \quad i,j = \overline{1,K} , \quad t \geq 0$$

with bounded elements for any $t \geq 0$. The process Z_t can be interpreted as a formal description of the individual's transition from one state to another in the multistate population model. Denote $H_t^Z = \sigma\{Z_u, u \leq t\}$. The following statement is a direct corollary of Theorem 3.

PROPOSITION 3. Let the process Z_t be associated with the death time T as follows:

$$P(T > t \mid H_t^Z) = \exp\left[-\int_0^t \lambda(z_u, u)\,du\right] .$$

Then the following formula is true:

$$E\exp\left[-\int_0^t \lambda(z_u,u)\,du\right] = \exp\left[-\int_0^t \sum_{i=1}^{i=K} \lambda(i,u)\pi_i(u)\,du\right] ,$$

where the $\pi_i(t)$ is the solution of the system of ordinary differential equations:

$$\frac{d\pi_j(t)}{dt} = \sum_{i=1}^{i=K} \pi_i(t) r_{i,j}(t) + \pi_j(t)\left[\bar{\lambda}(j,t) - \sum_{i=1}^{i=K} \bar{\lambda}(i,t)\pi_i(t)\right]$$

with $\pi_j(0) = p_j$.

The variables $\pi_j(t)$, $j = \overline{1,K}$ can be interpreted as the proportions of the individuals in different groups at time t.

4. APPENDIX

4.1. The Proof of Theorem 3

Let $H = (H_t)_{t \geq 0}$ be a nondecreasing right-continuous family of σ-algebras in Ω and let H_0 be completed by sets of P-zero measure from $H = H_\infty$.

Denote by $Y(t)$, $t \geq 0$, the continuous time, H-adapted process defined on (Ω, H, P) that describes the evolution of these factors. Denote by H^y the family of σ-algebras in Ω generated by the values of the random process $Y(u)$:

$$H^y = (H_t^y)_{t \geq 0}, \qquad H_t^y = \bigcap_{u>t} \sigma\{Y(v), v \leq u\} .$$

Assume that the H_t^y-conditional distribution function of death time T may be represented by the formula

$$P(T \leq t \mid H_t^y) = 1 - \exp\left\{-\int_0^t \lambda(y,u)\, du\right\}, \qquad (A1)$$

where $\lambda(Y,u)$ was introduced before.

Using the terminology of martingale theory [3, 14] and the recent compensator representation results [15], one can say that the process

$$A(t) = \int_0^{t \Delta T} \lambda(Y,u)\, du$$

is an H^{xy}-predictable compensator of the lifecycle process

$$X_t = I(T \leq t), \qquad t \geq 0,$$

where $H^{xy} = (H_t^{xy})_{t \geq 0}$, $H_t^{xy} = H_t^x \nabla H_t^y$, $H_t^x = \sigma\{X_u, u \leq t\}$. This means that the process

$$M_t = I(T \leq t) - A(t), \qquad t \geq 0$$

is an H_{xy}-adapted martingale. If the termination time T is viewed as the time of death, the process $\lambda(Y,u)$, $0 \leq u \leq t$, may be regarded as the age-specific mortality rate for an individual with history $Y_0^t = \{Y(u), 0 \leq u \leq t\}$.

Let $H^x = (H_t^x)_{t \geq 0}$. Denote by $\bar{A}(t)$ the H^x-predictable compensator of the lifecycle process X_t. Using the definition of the compensator and the compensator representation results

[3, 16], one can write

$$\bar{A}(t) = \int_0^{t\Delta T} \frac{dP(T \geq u)}{P(T \geq u)} = \int_0^{t\Delta T} \bar{\lambda}(u)\, du \quad .$$

The formula for $\bar{\lambda}(u)$ is the result of the following lemma.

LEMMA 1. Let $Y(t)$ and T be related as is described by the formula (17). Then

$$\bar{\lambda}(t) = E[\lambda(Y,t) \mid T \geq t] \quad .$$

Proof. Note that the process

$$\bar{M}_t = E(M_t \mid H_t^x) , \qquad t \geq 0$$

is an H^x-adapted martingale that can be represented in the form

$$\bar{M}_t = I(T \leq t) - \int_0^{t\Delta T} E[\lambda(Y,u) \mid H_u^x]\, du + N_t \quad ,$$

where

$$N_t = E\left[\int_0^{t\Delta T} \lambda(Y,u) du \;\middle|\; H_t^x\right] - \int_0^{t\Delta T} E[\lambda(Y,u) \mid H_u^x]\, du \quad .$$

The process N_t seems to be an H^x-predictable martingale. To prove that, it is enough to check the martingale property

$$E(N_t \mid H_v^x) = N_v$$

that easily follows from the equality

$$E\left[\int_{v\Delta T}^{t\Delta T} \lambda(Y,u) du \;\middle|\; H_v^x\right] - E\left[\int_{v\Delta T}^{t\Delta T} E[\lambda(Y,u) \mid H_u^x] du \;\middle|\; H_v^x\right]$$

and that the process

$$I(T \le t) - \int_0^{t \Delta T} E[\lambda(Y,u) \mid H_u^x]\, du$$

is an H^x-adapted martingale. Note further that the σ-algebra H_u^x has the form $\{T > u\}$ [17] and consequently

$$\int_0^{t \Delta T} E[\lambda(Y,u) \mid H_u^x] du = \int_0^{t \Delta T} E[\lambda(Y,u) \mid T > u]\, du \quad .$$

The nondecreasing process on the right-hand side of this equality is H^x-adapted and continuous and, consequently, H^x-predictable. The uniqueness of the H^x-predictable compensator implies the formula

$$\bar{A}(t) = \int_0^{t \Delta T} E[\lambda(Y,u) \mid T > t]$$

and consequently

$$\bar{\lambda}(t) = E[\lambda(Y,t) \mid t > t] \quad .$$

In particular cases when $\lambda(Y,u) = Y^*(u)Q(u)Y(u)$, where $Q(u)$ is the matrix with property (1), for $\bar{\lambda}(t)$ we have the formula

$$\bar{\lambda}(t) = m_t^* Q(t)\, m_t + Sp(Q(t)\gamma_t) \quad ,$$

where $m_t = E[Y(t) \mid t > t]$ and $\gamma_t = E[(Y(t)-m_t)(Y(t)-m_t)^* \mid T > t]$.

4.2. The Proof of Theorem 4

Introduce the conditional characteristic function $f_t(\alpha)$ defined as follows:

$$f_t(\alpha) = E(\exp \{i\alpha^* Y(t) \mid T > t\} \quad , \qquad t \ge 0 \; .$$

According to the Bayes rule this can be approximated by

$$f_t(\alpha) = E'(\exp\{i\alpha^*Y(t)\}\phi(t))$$

where

$$\phi(t) = \exp\left\{-\int_0^t (Y^*(u)Q(u)Y(u) - \overline{Y^*(u)Q(u)Y(u)})\,du\right\} ,$$

and E' denotes the mathematical expectation with respect to marginal probability measure corresponding to the trajectories of the Wiener process W_u, $0 \le u \le t$, and

$$\overline{Y^*(u)Q(u)Y(u)} = E(Y^*(u)Q(u)Y(u) \mid T > u) .$$

Using Itô's differential rule, one can rewrite the product $\exp\{i\alpha^*Y(t)\}\phi(t)$ as follows:

$$\begin{aligned}
&\exp\{i\alpha^*Y(t)\}\phi(t) \\
&\quad = \exp\{i\alpha^*Y(0)\} \\
&\qquad + \int_0^t i\alpha^* \exp\{i\alpha^*Y(u)\}\,\phi(u)\,(a_0(u)+a_1(u)Y(u))\,du \\
&\qquad + \int_0^t i\alpha^* \exp\{i\alpha^*Y(u)\}\,\phi(u)\,b(u)\,dW_u \\
&\qquad - \tfrac{1}{2}\int_0^t \exp\{i\alpha^*Y(u)\}\,\phi(u)\,\alpha^*b(u)\,b^*(u)\,\alpha\,du \\
&\qquad + \int_0^t \exp\{i\alpha^*Y(u)\}\,\phi(u)\,[Y^*(u)Q(u)Y(u) - Y^*(u)Q(u)Y(u)]\,du.
\end{aligned}$$

Taking the mathematical expectation E' of both sides of this equality leads to

$$f_t(\alpha) = f_0(\alpha) + i\alpha^* \int_0^t a_0(u) f_u(\alpha)\, du$$

$$+ i\alpha \int_0^t a_1(u)\, E'[\exp\{i\alpha^* Y(u)\}\, \phi(u)\, Y(u)]\, du$$

$$+ \tfrac{1}{2} \int_0^t \alpha^* b(u)\, b^*(u) \alpha\, f_u(\alpha)\, du$$

$$- \int_0^t E'[\exp\{i\alpha^* Y(u)\}\, \phi(u)\, Y^*(u) Q(u) Y(u)]\, du$$

$$+ \int_0^t f_u(\alpha)\, \overline{Y^*(u)Q(u)Y(u)}\, du \quad .$$

Notice that $f_0(\alpha)$ has the form

$$f_0(\alpha) = \exp\{i\alpha^* m_0 - \tfrac{1}{2}\alpha^* \gamma_0 \alpha\} \quad .$$

This particular form and the equation for $f_t(\alpha)$ generate the idea that one should look for an $f_t(\alpha)$ in the same form:

$$f_t(\alpha) = \exp\{i\alpha^* m_t - \tfrac{1}{2}\alpha^* \gamma_t \alpha\} \quad , \tag{A2}$$

where m_t and γ_t satisfy some ordinary differential equations

$$\frac{dm_t}{dt} = g(t) \ , \qquad m_0 \ , \tag{A3}$$

$$\frac{d\gamma_t}{dt} = G(t) \ , \qquad \gamma_0 \ . \tag{A4}$$

(We assume that the equations for m_t and γ_t have unique solutions.) The vector function $g(t)$ and matrix $G(t)$ can be found from the equation for $f_t(\alpha)$. In order to do this, note that the following equalities hold:

$$f'_{\alpha t} = E'(i \exp\{i\alpha^* Y(t)\} \phi(t) Y(t)) ,$$

$$f''_{\alpha\alpha t} = -E'(\exp\{i\alpha^* Y(t)\} \phi(t) Y(t) Y(t)^*(t)) ,$$

where $f'_{\alpha t}$ and $f''_{\alpha\alpha t}$ denote respectively the vectors of the first-order derivatives of the function $f_t(\alpha)$ with respect to α.

Applying these formulas to the equation for $f_t(\alpha)$, we obtain (omitting the dependence of $f_t(\alpha)$ on α for simplicity):

$$f_t = f_0 + i\alpha^* \int_0^t a_0(u) f_u \, du + \alpha^* \int_0^t f'_{\alpha u} a_1(u) \, du$$

$$- \tfrac{1}{2} \int_0^t f_u \alpha^* b(u) b^*(u)\alpha \, du + \int_0^t Sp(Q(u) f''_{\alpha\alpha u}) \, du$$

$$+ \int_0^t [m_u^* Q(u) m_u + Sp(Q(u)\gamma_u)] f_u \, du .$$

The derivatives $f'_{\alpha t}$ and $f''_{\alpha\alpha t}$ can be calculated from the equation (A2):

$$f'_{\alpha t} = f_t(im_t - \tfrac{1}{2}\alpha^*\gamma_t - \tfrac{1}{2}\gamma_t\alpha) ,$$

$$f''_{\alpha\alpha t} = f_t(im_t - \tfrac{1}{2}\alpha^*\gamma_t - \tfrac{1}{2}\gamma_t\alpha)(im_t - \tfrac{1}{2}\alpha^*\gamma_t - \tfrac{1}{2}\gamma_t\alpha)^* - f_t\gamma_t .$$

Substituting these derivatives into the equation for $f_t(\alpha)$, differentiating with respect to t and using the equations (A3) and (A4) for m_t and γ_t, we obtain:

$$\begin{aligned} f_t[i\alpha^* g(t) - \tfrac{1}{2}\alpha^* G(t)\alpha] &= i\alpha^* a_0(t) f_t + \alpha^* f_t(im_t - \tfrac{1}{2}\alpha^*\gamma_t - \tfrac{1}{2}\gamma_t\alpha) a_1(t) \\ &- \tfrac{1}{2} f_t \alpha^* b(t) b^*(t) \\ &+ f_t Sp\{Q(t)[(im_t - \tfrac{1}{2}\alpha^*\gamma_t - \tfrac{1}{2}\gamma_t\alpha)(im_t - \tfrac{1}{2}\alpha^*\gamma_t - \tfrac{1}{2}\gamma_t\alpha)^* - \gamma_t]\} \\ &+ f_t[m_t^* Q(t) m_t + Sp(Q(t)\gamma_t] . \end{aligned}$$

Taking the real and imaginary parts of this equality yields

$$g(t) = a_0(t) + a_1(t)m_t - 2\gamma_t Q(t)m_t \quad , \tag{A5}$$

$$G(t) = a_1(t)\gamma_t + \gamma_t a_1^*(t) + b^2(t) - 2\gamma_t Q(t)\gamma_t \quad , \tag{A6}$$

which leads to the equations for m_t and γ_t described in the Theorem.

Notice that the form of the $f_t(\alpha)$ noted above corresponds to the Gaussian law for conditional distribution of the $Y(t)$ given the event $\{T > t\}$.

It remains to show that the equation (A4) with $G(t)$ given by (A6) has a unique solution. One can easily do this following the approach developed in [3, Chapter 12].

REFERENCES

[1] Cameron, R., and Martin, W. "The Wiener Measure of Hilbert Neighborhoods in the Space of Real Continuous Functions." *J. Math. Physics*, 23 (1944): 195-209.

[2] Cameron, R., and Martin, W. "Transformation of Wiener Integrals by Nonlinear Transformations." *Trans. Amer. Math. Soc.*, vol.58 (1945): 184-219.

[3] Liptser, R.Sh., and Shiryayev, A.N. *Statistics of Random Processes*. I, II. Berlin Heidelberg New York: Springer-Verlag Inc., 1978.

[4] Myers, L.E. "Survival Functions Induced by Stochastic Covariate Processes." *J. Appl. Probab.*, vol.18 (1981): 523-529.

[5] Woodbury, M.A., and Manton, K.G. "A Random Walk Model of Human Mortality and Aging." *Theoretical Population Biology*, vol.11 (1977): 37-48.

[6] Yashin, A.I. *Chances of Survival in a Chaotic Environment*. WP-83-100. International Institute for Applied System Analysis. Laxemburg (Austria), 1983.

[7] Yashin, A.I. "Filtering of Jumping Processes." *Avtomatika i Telemekhanika*, 5 (1970): 52-58.

[8] Brémaud, P. *Point Processes and Queues*. Berlin Heidelberg New York: Springer-Verlag Inc., 1981.

[9] Vaupel, J.V., Manton, K.G., and Stallard, E. "The Impact of Heterogeneity in Individual Frailty on the Dynamic of Mortality." *Demography*, 16 (1979): 439-454.

[10] Vaupel, J.V., and Yashin, A.I. *The Deviant Dynamics of Death in Heterogeneous Populations*. RR-83-1. International Institute for Applied System Analysis. Laxemburg (Austria), 1982.

[11] Liptser, R.Sh. "Gaussian Martingales and a Generalization of the Kalman-Bucy Filter. *Theory Probab. Applications*, vol.20, no.2 (1975): 285-301.

[12] Yashin, A.I. "Conditional Gaussian Estimation of Characteristics of the Dynamic Stochastic Systems." *Avtomatika i Telemekhanika*, 5 (1980): 57-67.

[13] Yashin, A.I. "A New Proof and New Results in Conditional Gaussian Estimation Procedures." *Proceedings of the 6th European Conf. on Cybernetics and System Research*, April 1982, pp. 205-207. Amsterdam New York: North-Holland Pub. Co., 1982.

[14] Jacod, J. Calcul stochastique et problemès des martingales. *Lecture Notes in Math.*, vol.714. Berlin Heidelberg New York: Springer-Verlag Inc., 1979.

[15] Yashin, A.I. *Hazard Rates and Probability Distributions: Representation of Random Intensities*. WP-84-21. International Institute for Applied System Analysis. Laxemburg (Austria), 1984.

[16] Jacod, J. "Multivariate Point Processes: Predictable Projection, Radon-Nikodym Derivatives, Representation of Martingales." *Z. Wahrscheinlichkeitstheorie und Verw. Gebiete*, vol.31 (1975): 235-253.

[17] Dellacherie, C. *Capacités et processus stochastiques*. Berlin Heidelberg New York: Springer-Verlag Inc., 1972.

[18] Novikov, A.A. "On Estimation of Parameters of Diffusion Processes." *Studia Scientarium Mathematicarum Hungarica*, 7 (1972): 201-209. (In Russian.)

O.K. ZAKUSILO

MARKOV DRIFT PROCESSES

0

The appearance of this report was stimulated by the work of Wobst [6], where "jump processes with drift" had been considered. Wobst investigated the right-continuous time-homogeneous Markov processes x_t having the property:

> there exists a family $g_y(t)$ of functions such that P_x - almost surely, for any $s \geq 0$ one can find a $\delta(\omega) > 0$ satisfying the equality
> $x_t = g_y(t-s)$, $(s \leq t < s+\delta)$ where $y = x_s$.

These processes, having reached some state x, move along the curve $g_x(\cdot)$ for a random time, then jump randomly, etc.

The present report suggests another generalization of the notion of the classical jump process, which includes the processes of Wobst. As far as possible we follow the line and notations of Dynkin, [1,2], Gikhman and Skorokhod [3] and Wobst [6].

1. PRELIMINARY DEFINITIONS AND NOTATION

Let F be a σ-field and let P be a measure on F. We say that sub-σ-fields $F_1 \subset F$ and $F_2 \subset F$ coincide P-a.s. iff

- a. $\forall A_1 \in F_1 \quad \exists A_2 \in F_2$: $P(A_1 \Delta A_2) = 0$ and
- b. $\forall A_2 \in F_2 \quad \exists A_1 \in F_1$: $P(A_1 \Delta A_2) = 0$.

We use the following notation:

$F_1 = F_2 \quad (A_1 = A_2, \ \xi_1 = \xi_2) \quad (\text{mod } P)$ the coincidence of σ-fields (sets, random variables) P-a.s.,

X the Polish space,

B σ-field of Borel subsets of X,

Γ the Borel subset of X,

(X_t, F_t, P_X) the nonterminating homogeneous Markov process with the state space (X,B) defined on the measurable space $(\Omega, F_{\geq 0})$,

$\sigma\{\cdot\}$ σ-field generated by whatever appears between the braces,

$F_t = \sigma\{x_s, \ s \leq t\}$,

$F_{\geq 0} = \sigma\{x_t, \ t \geq 0\}$,

$F_{t+0} = \bigcap_{\varepsilon > 0} F_{t+\varepsilon}$,

τ a stopping time with respect to the flow F_t,

$F_{\tau-0} = \sigma\{\{x_t \in \Gamma, \ t < \tau\}, \ t \geq 0, \ \Gamma \in B\}$,

θ_t, θ_t^{-1} shift operators

$T = [0,\infty)$,

$T^* = [0,\infty]$,

$\mathcal{T}(\mathcal{T}^*)$ σ-field of Borel subsets of $T(T^*)$,

$I(A)$ the indicator of A.

2. THE MAIN DEFINITION

A Markov drift process is a sample right-continuous process (x_t, F_t, P_t) having a stopping time τ such that for each $x \in X$ the following conditions hold:

- a. $P_X(\{\tau > 0\}) = 1$,
- b. $F_{\tau-0} = \sigma\{\tau\}$ (mod P_X). (1)

3. EXAMPLES

3.1

Jump process. One can take τ = the moment of the first jump.

3.2

Consider a particle moving uniformly to the right along T and assume that in the case where its initial position is the origin, the particle is delayed for a random time. If the delay time has an exponential distribution then the position x_t of the particle at the instant t is a Markov process. (This process was considered by Dynkin [1], who proved the strong Markov property of x_t.

Dynkin [1] and Itô and McKean [4] show that x_t does not possess strong Markov property with respect to the flow F_{t+0}).

Fix $a > 0$ and put

$$\tau = \begin{cases} \min \{t: x_t = a\}, & \text{if } x_0 < a , \\ \infty , & \text{if } x_0 \geq a . \end{cases}$$

The stopping time τ obviously satisfies conditions (1) and x_t turns out to be Markov drift process.

This example shows that many different variants in constructing stopping time τ satisfying conditions (1), are possible. Besides the absence of strong Markov property with respect to F_{t+0} apparently makes it difficult to check the strong Markov property with respect to F_t in the general case.

3.3

The generalized Poisson process with drift and the process of virtual waiting time in the queueing system $M|G|1$. In both cases one can choose τ = the moment of first jump.

3.4

The process $x_t \in R^1$, $x_t = x_0 + t$ with $\tau \equiv 1$. Contrary to the previous examples, τ is not a terminal time (the equality $\theta_h \tau = \tau - h$ does not hold on the event $\{\tau > h\}$).

4. DRIFT FUNCTIONS

The following lemma explains our terminology.

LEMMA 1. If x_t is a Markov drift process with a suitable stopping time τ then for any $s \in T$, $x \in X$ the equality

$$x_s = g_x(s,\tau) \qquad (\text{mod } P_x) \tag{2}$$

holds on the set $\{\tau > s\}$.

The function $g_x(s,t)$ is defined for $x \in X$, $0 \le s < t$, t lying in the support of P_x-distribution of τ. It can be chosen $(B \times \mathcal{T} \times \mathcal{T})$-measurable and right continuous in s.

Proof. Take any injection $f\colon X \to [0,1]$ such that

- a. f is one-to-one and
- b. both f and f^{-1} are measurable functions.

Then for every $x \in X$

$$I(\{\tau > s)\}f(x_s) = E_x\left(I_{(\{\tau>s\})}f(x_s) \,\middle|\, F_{\tau-0}\right)$$

$$= E_x\left(I(\{\tau > s\})f(x_s) \,\middle|\, \sigma\{\tau\}\right) = I(\{\tau > s\})E_x\left(f(x_s) \,\middle|\, \sigma\{\tau\}\right)$$

$$= I(\{\tau > s\})q_x(s,t) \qquad (\text{mod } P_x) \quad ,$$

where $q_x(s,\tau) = E_x\left(f(x_s) \,\middle|\, \sigma\{\tau\}\right)$.

If $s > \tau$ then $f(x_s) = q_x(x,\tau)$ $(\text{mod } P_x)$ and

$$x_s = f^{-1}(q_x(s,\tau)) = g_x(s,\tau) \qquad (\text{mod } P_x).$$

Using the martingale convergence theorem (see [3]) we can choose a suitable variant of $q_x(s,t)$ in the form of

$$q_x(s,t) = \lim_{h \downarrow 0} \frac{E_x I(\{t < \tau \le t+h\})f(x_s)}{P_x(\{t < \tau \le t+h\ )} \quad . \tag{3}$$

(If the denominator in the right-hand side of (3) vanishes then the ratio is assumed to vanish too. It is known [3] that the limit in (3) exists for almost all values of t with respect to P_x-distribution of τ.) Both the numerator and the denominator in the right-hand side of (3) are measurable in x (see [5]) and right-continuous in t. Thus, both their ratio and limit in (3) are $B \times \mathcal{T}$-measurable functions in (x,t). Equality (2) permits the assumption of right-continuity in s of the function $g_x(s,t)$ and, as a consequence, its $B \times \mathcal{T} \times \mathcal{T}$-measurability. The assertion is proved.

5. SHIFTED PROCESSES

In what follows we assume that the sure event Ω satisfies the condition:

$$\text{for every} \quad h \geq 0, \qquad \theta_h \Omega = \Omega . \tag{4}$$

Each of the following suggestions is sufficient for the validity of (4):

- a. the set of sample paths of x_t coincides with the set of all right-continuous functions;
- b. the set of sample paths is closed under delays, i.e., for every $\omega \in \Omega$ there exists $\omega' \in \Omega$ such that $x_t(\omega') = x_{(t-h)\vee 0}(\omega)$.

In this Section we consider the shifted process (x'_t, F'_t, P'_x), where

$$x'_t = \theta_h x_t = x_{t+h} ,$$

$$F'_t = \theta_h^{-1} A , \qquad A \in F_t ,$$

P'_x is a measure on $F'_{\geq 0}$ defined by the equality
$P'_x(A') = P_x(A)$, if $A' = \theta_h^{-1}A$, $A \in F_{\geq 0}$.

We should check the correctness of this definition.

LEMMA 2. If $A' = \theta_h^{-1}A = \theta_h^{-1}B$, $A \in F_{\geq 0}$, $B \in F_{\geq 0}$, and the condition (4) holds then $A = B$.

Proof. If the condition (4) holds, then for every $\omega \in A \cup B$ the set $\theta_h^{-1}\omega \in A'$ is non-empty.

For any $\omega' \in A'$ consider the system of subsets of Ω which contain either all values of $\theta_h\omega'$ or none of them. This system S is a σ-field and $S \supset F_{\geq 0}$. To prove the last inclusion, put $\omega^1 = \theta_h^{(1)}\omega'$, $\omega^2 = \theta_h^{(2)}\omega'$, where $\omega' \in A'$. Then

$x_t(\omega^1) = x_{t+h}(\omega') = x_t(\omega^2)$, and $\{x_t \in \Gamma\} \in S$ for every $\Gamma \in B$.
Therefore, $S \supset F_{\geq 0}$. These arguments lead to the equality
$A = B = \theta_h A'$, if $A \in F_{\geq 0}$, $B \in F_{\geq 0}$.

REMARK. The measure $P'_x(A')$ is a regular variant of the conditional distribution $P_a(A/x_h)$ as $x_h = x$.

LEMMA 3. The process (x'_t, F'_t, P'_x) is a Markov drift process connected with $\tau' = \theta_h\tau$. Its drift function $g'_x(s,t)$ coincides with $g_x(s,t)$.

Proof. The Markov property, time-homogeneity and right-continuity of x_t are obvious. Further, τ' is a stopping time with respect to the flow F'_t, since

$$\{\tau' < s\} = \{\omega\colon \tau(\theta_h\omega) \leq s\} = \theta_h^{-1}\tau^{-1}([0,s]) \in F'_s .$$

Besides, $P'_x(\{\tau' > 0\}) = P_x(\{\tau > 0\}) = 1$. To check the condition (1), we are left to show that $\sigma\{\tau'\} = F'_{\tau'-0}$ (mod P'_x).

If $A = \{x_t \in \Gamma,\ \tau > t\}$, then one can take $C = \{\tau \in K\}$ such

$P_x(A\Delta C) = 0$. Putting $A' = \theta_h^{-1}A = \{x'_t \in \Gamma,\ \tau' > t\}$, $C' = \theta_h^{-1}C = \{\tau' > K\}$, we obtain the equality

$$P'_x(A'\Delta C') = P_x(A\Delta C) = 0 .$$

Since the σ-field $F'_{\tau'-0}$ is generated by the events of the form $\{x'_t \in \Gamma,\ \tau' > t\}$, and since $\sigma\{\tau'\} \subset F'_{\tau'-0}$, the last equality implies $\sigma\{\tau'\} = F'_{\tau'-0}$ (mod P'_x).

The coincidence of the drift functions follows from the definitions of $g_x(s,\tau)$, x_s, τ' and P_x.

COROLLARY. For every $x \in X$ the equality $g_x(h+t,\tau) = g_{g(h,\tau)}(t,\tau')$ (mod P_x) holds on the event $\{\tau > t+h,\ \tau' > t\}$.

Proof. By using Lemma 3, we have

$$\begin{aligned} P_x(\{\tau > h+t,\ \tau' > t\}) &= P_x(\{\tau > h+t,\ \tau' > t,\ x'_t = x_{t+h}\}) \\ &= P_x(\{\tau > h+t,\ \tau' > t,\ g_{x'_0}(t,\tau') = g_x(t+h,\tau)\}) \\ &= P_x(\{\tau > h+t,\ \tau' > t,\ g_{x_h}(t,\tau') = g_x(t+h,\tau)\}) \\ &= P_x(\{\tau > h+t,\ \tau' > t,\ g_{g_x(h,\tau)}(t,\tau') = g_x(t+h,\tau)\}) \ . \end{aligned}$$

6. THE DISTRIBUTIONS OF τ, x_τ AND (τ, x_τ)

Since $\tau \in T^*$, we should extend the phase space to define x_τ correctly for any $\omega \in \Omega$. Take any points $* \bar{\in} x$, $\omega^* \bar{\in} \Omega$ and put

$$x^* = x \cup \{*\} \ ,$$

$$B^* = \sigma\{B, \{*\}\} \ ,$$

$$\Omega^* = \Omega \cup \{\omega^*\} ,$$

$$F^*_0 = \sigma\{F_0, \{\omega^*\}\} ,$$

$$F^*_t = \sigma\{F_t, \{\omega^*\}\} ,$$

$$x_t(\omega^*) = * ,$$

$$x_\infty(\omega) = * ,$$

$$P_x(\{\omega^*\}) = 0 ,$$

$$P_*(\{\omega^*\}) = 1 .$$

One can easily turn X^*, T^* and $T^* \times X^*$ into Polish spaces. Since x_t is right-continuous, it is measurable, and x_τ is a random variable. We denote by Q^1_x, Q^2_x and Q_x the P_x-distributions of τ, x_τ and (τ, x_τ) respectively. On the basis of Theorem 3 from [3, p. 53], there exists a regular variant $\Pi(x,\tau,\Gamma)$ of conditional distribution $P_x(\{x_\tau \in \Gamma\}/\tau)$. In just the same way as in the proof of Lemma 1, one can show that Q_x - almost everywhere

$$\Pi(x,t,\Gamma) = \lim_{h \downarrow 0} \frac{Q_x((t,t+h] \times \Gamma)}{Q^1_x((t,t+h])} \tag{5}$$

and that $\Pi(x,t,\Gamma)$ may be assumed to be a $B^* \times \mathcal{T}^*$ - measurable function in (x,t) and a probabilistic measure on B^*.

In the sequel we omit the superscripts $*$ by setting that the original spaces X and Ω were replaced by X^* and Ω^* from the very beginning.

7. ITERATIONS OF τ

Assume that the set of sample paths of x_t is closed under stoppings, i.e., that for every $\omega \in \Omega$, $h \geq 0$ there exists $\omega' \in \Omega$ with the property $x_t(\omega') = x_{t \wedge h}(\omega)$. (This assumption can be made without loss of generality).

For any countable ordinal number α put

$$\tau_0 = 0 , \qquad \tau_{\alpha+1} = \tau_\alpha + \theta_{\tau_\alpha} \tau , \qquad \tau_{\alpha+1} = \infty ,$$

if $\tau_\infty = \infty$,

$$\tau_\alpha = \sup_{\beta<\alpha} \tau_\beta ,$$

if α is a limit ordinal number.

<u>LEMMA 4</u>. The random variables τ_α, $\alpha \in \omega_1$, are stopping times.

<u>Proof</u>. The assertion of the Lemma follows from Galmarino's criterion (see [4]) by using transfinite induction. (This criterion needs closeness of Ω under stoppings).

In the rest of the report we restrict ourselves by condition:

R. all τ_α, $\alpha < \omega_1$ are regeneration moments.

<u>LEMMA 5</u>. If the condition R holds then for every $x \in X$ there exists a countable ordinal number $\alpha = \alpha(x)$ such that $\tau_\alpha = \infty$ (mod P_x).

<u>Proof</u>. The condition R implies $P_x(\{\tau_{\alpha+1} > \tau_\alpha\} / \{\tau_\alpha < \infty\}) = 1$. This is followed by $\tau_\alpha > \tau_\beta$ (mod P_x), if $\alpha > \beta$ and $\tau_\beta < \infty$. Now the assertion follows from the arguments given by Dynkin [2, p. 44].

<u>LEMMA 6</u>. For any countable ordinal numbers α and β,

$$\theta_{\tau_\alpha}\theta_\beta = \tau_{\alpha+\beta} - \tau_\alpha$$

$$\theta_{\tau_\alpha} x_{\tau_\beta} = x_{\tau_{\alpha+\beta}} .$$

Proof. Write the first equality in the equivalent form:

$$\tau_\alpha = \tau_\beta + \theta_{\tau_\beta}\tau_{\alpha-\beta} \quad \text{for all} \quad \beta < \alpha \tag{6}$$

We prove (6) by transfinite induction. If $\alpha = 2$ then (6) coincides with the definition of τ_2. Assume that (6) holds for $\alpha < \tau$. If $\gamma = \delta + 1$ then by the definition of τ_γ,

$$\tau_\gamma = \tau_\delta + \theta_{\tau_\delta}\tau . \tag{7}$$

By the assumption of induction, $\tau_\delta = \tau_\beta + \theta_{\tau_\beta}\tau_{\delta-\beta}$ for all $\beta < \delta$ and

$$\begin{aligned}
\tau_\gamma &= \tau_\beta + \theta_{\tau_\beta}\tau_{\delta-\beta} + \theta_{\tau_\beta+\theta_{\tau_\beta}\tau_{\delta-\beta}}\tau \\
&= \tau_\beta + \theta_{\tau_\beta}\tau_{\delta-\beta} + \theta_{\tau_\beta}\theta_{\tau_{\delta-\beta}}\tau \\
&= \tau_\beta + \theta_{\tau_\beta}\left(\tau_{\delta-\beta} + \theta_{\tau_{\delta-\beta}}\tau\right) \\
&= \tau_\beta + \theta_{\tau_\beta}\tau_{\delta+1-\beta} \\
&= \tau_\beta + \theta_{\tau_\beta}\tau_{\gamma-\beta}
\end{aligned} \tag{8}$$

Equalities (7) and (8) imply (6) for $\alpha = \gamma$.

If γ is a limit ordinal number and if $\beta < \gamma$ then

$$\tau_\gamma = \sup_{\beta \le \alpha < \gamma} \tau_\alpha = \sup_{\beta \le \alpha < \gamma} \left(\tau_\beta + \theta_{\tau_\beta} \tau_{\alpha-\beta}\right)$$

$$= \tau_\beta + \theta_{\tau_\beta} \sup_{\beta \le \alpha < \gamma} \tau_{\alpha-\beta} = \tau_\beta + \theta_{\tau_\beta} \tau_{\gamma-\beta}$$

Now one can easily verify the second equality

$$x_{\tau_{\alpha+\beta}} = x_{\tau_\alpha + \theta_{\tau_\alpha} \tau_\beta} = \theta_{\tau_\alpha} x_{\tau_\beta} \quad .$$

<u>LEMMA 7</u>. For any countable limit number $\alpha < \alpha(x)$,

$$F_{\tau_\alpha - 0} = \sigma\{F_{\tau_\beta}, \ \beta < \alpha\} \quad .$$

<u>Proof</u>. Since τ_β is a stopping time and $\{x_t \in \Gamma\} \cap \{\tau \le \tau_\beta\} \in F_t$, and since $F_{\tau_\alpha - 0}$ and $\sigma\{F_{\tau_\beta}, \ \beta < \alpha\}$ are generated by the events $\{x_t \in \Gamma\} \cap \{t < \tau_\alpha\}$ and $\{x_t \in \Gamma\} \cap \{t \le \tau_\beta\}$, $\beta < \alpha$, respectively, the statement follows from the relations:

$$\{x_t \in \Gamma\} \cap \{t < \tau_\alpha\} = \bigcup_{\beta < \alpha} \{x_t \in \Gamma\} \cap \{t \le \tau_\beta\} \in \sigma\{F_{\tau_\beta}, \ \beta < \alpha\} \quad ,$$

$$\{x_t \in \Gamma\} \cap \{t \le \tau_\beta\} = \{x_t \in \Gamma\} \cap \{t \le \tau_\beta\} \cap \{t < \tau_\alpha\} \in F_{\tau_\alpha - 0} \quad .$$

8. REGENERATION FUNCTIONS (See [6])

Let $\alpha < \alpha(x)$ be a limit ordinal number. The sequence $P_x(\{x_{\tau_\alpha} \in \Gamma\} | F_{\tau_\beta})$ is a martingale as $\beta \uparrow \alpha$. Using Lemma 7 and a martingale convergence theorem

$$\lim_{\beta \uparrow \alpha} P_x \left(\{x_{\tau_\alpha} \in \Gamma\} \middle| F_{\tau_\beta}\right) = P_x \left(\{x_{\tau_\alpha} \in \Gamma\} \middle| F_{\tau_\alpha - 0}\right) \quad (\text{mod } P_x) \ .$$

Using Lemma 6 and the property R we obtain

$$\begin{aligned} P_x\left(\{x_{\tau_\alpha} \in \Gamma\} \mid F_{\tau_\alpha - 0}\right) &= \lim_{\beta \uparrow \alpha} P_x\left(\{x_{\tau_\alpha} \in \Gamma\} \mid x_{\tau_\beta}\right) \\ &= \lim_{\beta \uparrow \alpha} P_x\left(\{\theta_{\tau_\beta} x_{\tau_{\alpha-\beta}} \in \Gamma\} \mid x_{\tau_\beta}\right) \\ &= \lim_{\beta \uparrow \alpha} P_{x_{\tau_\beta}}\left(\{x_{\tau_{\alpha-\beta}} \in \Gamma\}\right) \quad (\text{mod } P_x) . \end{aligned} \tag{9}$$

The difference $\alpha - \beta$ stabilizes for sufficiently large values of β, i.e., $\alpha - \beta = \alpha^*$ where $\alpha^* = \alpha^*(\alpha)$ is an indecomposable limit number. Thus

$$P_x\left(\{x_{\tau_\alpha} \in \Gamma\} \mid F_{\tau_\alpha - 0}\right) = \lim_{\beta \uparrow \alpha} P_{x_{\tau_\beta}}\left(\{x_{\tau_{\alpha^*}} \in \Gamma\}\right) \quad (\text{mod } P_x) . \tag{10}$$

A slight modification of the arguments given above in (9) leads us to the equality

$$P_x\left(\{x_{\tau_\alpha} \in \Gamma\} \mid F_{\tau_\alpha - 0}\right) = P_x\left(\{x_{\tau_\alpha} \in \Gamma\} \mid x_{\tau_\beta}, \ \beta < \alpha\right) \quad (\text{mod } P_x) . \tag{11}$$

Combining (10) and (11) and assuming $P_x(\{x_{\tau_\alpha} \in \Gamma\} / F_{\tau_\alpha - 0})$ to be a regular conditional distribution, we obtain

$$P_x\left(\{x_{\tau_\alpha} \in \Gamma\} \mid F_{\tau_\alpha - 0}\right) = \Pi^*\left[\left(x_{\tau_\beta}, \ \beta < \alpha\right), \ \alpha^*, \ \Gamma\right] \quad (\text{mod } P_x) ,$$

where $\Pi^*((y_\beta, \ \beta < \alpha), \ \alpha^*, \ \Gamma)$ is a measurable function in $(y_\beta, \ \beta < \alpha)$ and measure in Γ, and $\alpha^* = \alpha^*(\alpha) = \min_{\beta < \alpha} (\alpha - \beta)$ is an indecomposable limit ordinal number. Besides, $\Pi^*((y_\beta, \beta < \alpha), \alpha^*, \Gamma)$ depends only on the tail of the sequence $(y_\beta, \beta < \alpha)$.

We will call Π^* a regeneration function (see [6]).

9. FINITE-DIMENSIONAL DISTRIBUTIONS

Wobst [6] proved that the finite-dimensional distributions of jump process with drift are uniquely determined by their drift functions, distributions of the drift time, jump distributions and regeneration functions. His proof is applicable in the more general case of Markov drift processes.

THEOREM. The finite-dimensional distributions of Markov drift process x_t are uniquely determined by its drift function $g_x(s,t)$, distribution Q_x^1 of stopping time τ, conditional distribution Π and regeneration function Π^*.

Proof. We give only a sketch of the proof. We have (see Lemma 5):

$$P(t,x,\Gamma) = P_x\left(\{x_t \in \Gamma\}\right)$$

$$= \sum_{\alpha<\alpha(x)} P_x\left(\{x_t \in \Gamma\},\ \tau_\alpha \le t < \tau_{\alpha+1}\}\right) \tag{12}$$

$$= \sum_{\alpha<\alpha(x)} P_x\left(\{g_{x_{\tau_\alpha}}\left(t-\tau_\alpha,\ \tau_{\alpha+1}-\tau_\alpha\right) \in \Gamma,\ \tau_\alpha \le t < \tau_{\alpha+1}\}\right).$$

By the use of Lemma 7 we obtain that the couple $\bar{y}_\alpha = (\tau_\alpha, x_{\tau_\alpha})$ forms "a Markov chain":

$$P^* = P_x\left(\{\bar{y}_\alpha \in K\} \mid \bar{y}_\gamma,\ \gamma \le \beta\right) = P_{x_{\tau_\beta}}\left(\{\bar{y}_{\alpha-\beta} \in K\}\right) \quad (\text{mod } P_x).$$

By transfinite induction one can show that the distribution of the chain $\bar{y}_\alpha$ is uniquely determined by the measures Q_x and Π^* (or by Q_x^1, Π and Π^*). Now the statement of the theorem follows from (12).

10. THE RELATIONSHIP BETWEEN CHARACTERISTICS

The Markov property and homogeneity of the process (x_t, F_t, P_t) are equivalent to the condition:

$$P_x\left(\{x_{t_1} \in \Gamma_1, \ldots, x_{t_n} \in \Gamma_n\}\right) = \int_{\Gamma_1} P(t_1, x, dy_1) \int_{\Gamma_2} \cdots \int_{\Gamma_n} P(t_n - t_{n-1}, y_{n-1}, dy_n) \ . \tag{13}$$

Both sides of (13) can be expressed in terms of basic characteristics. Such representation for the transition function $P(t,x,\Gamma)$ is indicated in the theorem proved above. The representation for $P_x(\{x_{t_1} \in \Gamma_1, \ldots, x_{t_n} \in \Gamma_n\})$ can be obtained in the same way:

$$P_x\left(\{x_{t_1} \in \Gamma_1, \ldots, x_{t_n} \in \Gamma_n\}\right) = \sum_{0 \le \alpha_1 \le \cdots \le \alpha_n \le \alpha(x)} P_x\left(\left\{t_1 \in [\tau_{\alpha_i}, \tau_{\alpha_i+1}), \quad g_{x_{\tau_{\alpha_i}}}\left(t_i - \tau_{\alpha_i}, \tau_{\alpha_i+1} - \tau_{\alpha_i}\right) \in \Gamma_i, \quad 1 \le i \le n\right\}\right) . \tag{14}$$

Each of summands on the right-hand side of (14) can be expressed in terms of transition probabilities P^* of the chain $(\tau_\alpha, x_{\tau_\alpha})$, i.e., finally in terms of basic characteristics $g_x(s,t)$, Q_x^1, Π and Π^*. Expressing both sides of (13) in terms of these characteristics, we establish a relation between them. In the sequel we shall call it the characterizing relation.

11. THE RELATIONSHIP BETWEEN τ AND BASIC CHARACTERISTICS

In this section we show that with probability one τ coincides with $\sup\{t: x_s = g_{x_0}(s,t)$ for $s < t\}$.

Recall Galmarino's criterion (see [4]) for τ to be a stopping time with respect to the flow F_t for the process whose set of sample paths is closed under stoppings:

Let $\omega \in \Omega$, $\omega' \in \Omega$. Random variable τ is a stopping time iff the conditions $x_s(\omega) = x_s(\omega')$ for all $s \le t$, $\tau(\omega) \le t$ imply $\tau(\omega') \le t$, (in fact, $\tau(\omega) = \tau(\omega')$).

Let A be a set of total P_x-measure such that $x_s(\omega) = g_x(s, \tau(\omega))$ for $s < \tau(\omega)$, $\omega \in A$. Take any $\omega \in A$, $\omega' \in A$. If $\tau(\omega) = \tau(\omega')$ then

$$x_s(\omega) = g_x(s,\tau(\omega)) = g_x(s,\tau(\omega')) = x_s(\omega')$$

for $s < \tau(\omega)$.

Assume that $t = \tau(\omega) < \tau(\omega') = t'$ and that $x_s(\omega) = g_x(s,t')$ for $s \in [0,t]$. Then, $x_s(\omega) = x_s(\omega')$ for $s \in [0,t]$ and Galmarino's criterion implies $t = \tau(\omega) = \tau(\omega') = t'$. This contradicts our assumption $t < t'$. Hence, there exists a point $s = s(\omega)$, $s \in [0,t]$ such that $x_s(\omega) \neq g_x(s,t)$.

If $s < t$, then $g_x(s,t) \neq g_x(s,t')$. Consequently, in this case

$$g_x(s,t) = g_x(s,t') \quad \text{for all} \quad s < t \quad \text{implies}$$

$$x_t(\omega) \neq g_x(t,t') \quad . \tag{15}$$

(Remark that (15) implies the relation:

$$\text{if} \quad g_x(s,t) = g_x(s,t') \quad \text{for all} \quad s < t, \quad \text{then}$$

$$\Pi(t, x, \{g_x(t,t')\}) = 0. \quad) \tag{16}$$

These arguments show that the value of τ is determined uniquely (mod P_x) by its characterizing property

$$x_s = g_s(s,\tau) \qquad \text{for} \quad s \in [0,\tau) \quad (\text{mod } P_x) \tag{17}$$

provided the drift function $g_x(s,t)$ is known in advance.

Indeed, for every $\omega \in A$ there exists at least one value of τ satisfying equality (17). Assuming the existence of two different values τ^1 and τ^2, we get a contradiction:

if $\tau^1 < \tau^2$ then (17) implies

$$g_x(s,\tau^1) = g_x(s,\tau^2) \qquad \text{for} \quad s < \tau^1 \tag{18}$$

and

$$x_{\tau^1} = g_x(\tau^1,\tau^2) \quad . \tag{19}$$

Equalities (18) and (19) contradict the condition (15).

Thus, (17) implies $\tau = \sup\{t\colon x_s = g_{x_0}(s,t) \ \text{ for } \ s < t\}$ (mod P_x) for any $x \in X$.

12

In this section we show that the basic characteristics determine not only the finite-dimensional distributions but also a Markov process possessing property R.

Suppose we are given functions $g_x(s,t)$ satisfying requirements of Lemma 1, measures Π and Π^*, satisfying requirements Sections 7 and 9, and measures $Q_x^1(C)$, $C \in \mathcal{T}^*$, which are measurable in x for any fixed C.

Suppose also that $g_x(s,t)$, Q_x^1, Π and Π^* satisfy the characterizing relation of Section 11, and that Π satisfies condition (16).

Think of Ω as the set of all transfinite sequences $\{\omega = (t_\alpha, x_{t_\alpha}),\ \alpha \in \omega_1\}$, where $x_{t_\alpha} \in X$; $t_0 = 0$, $0 \le t_\alpha \le \infty$, $t_\alpha > t_\beta$, if $\alpha > \beta$ and $t_\beta < \infty$; $\lim_{\gamma \uparrow \alpha} t_\gamma = t_\alpha$, if α is a limit ordinal number. Think of σ-field $F_{\ge 0}$ as σ-field generated by cylinders in Ω with $\mathcal{T} \times B$-measurable bases.

Define the process x_t by the equality

$$x_t(\omega) = g_{x_{t_\alpha}}(t-t_\alpha,\ t_{\alpha+1}-t_\alpha) \quad , \qquad \text{if} \quad t \in [t_\alpha, t_{\alpha+1}) \ .$$

Defining the measures P_x on $F_{\ge 0}$ with help of transfinite induction (see Section 10) so that $y_\alpha(\omega) = (t_\alpha, x_{t_\alpha})$ form a "Markov chain" with transition probabilities P^* and so that $P_x(\{\omega\colon x_0 = x\}) = 1$, we attain fulfillment of (12).

Now Markov property and homogeneity of x_t follow from the characterizing relation connecting $g_x(s,t)$, θ_x^1, Π and Π^*.

If we put $\tau = \sup\{t\colon x_s = g_x(s,t)$ for all $s < t\}$ then, based on condition (16), τ will be determined uniquely P_x - a.s. for every $x \in X$: $\tau = t_1 \pmod{P_x}$, and we will satisfy conditions under which x_t will be a Markov drift process with characteristics $g_x(s,t)$, Q_x^1, Π and Π^*.

The author wishes to thank V.M. Shurenkov for his benevolent interest in this problem and for the helpful discussions of the material presented in this report.

REFERENCES

[1] Dynkin, E.B. *Theory of Markov Processes.* Englewood Cliffs, NJ: Prentice-Hall, 1961.

[2] Dynkin, E.B. "Markov Jump Processes." *Teoriya veroyatnostej i ee primeneniya*, vol.3, no.1 (1958): 41-60.

[3] Gikhman, I.I., and Skorokhod, A.V. *The Theory of Stochastic Processes.* Berlin Heidelberg New York: Springer-Verlag Inc., 1974-79.

[4] Itô, K., and McKean, H.P. "Any Markov Process in a Borel Space Has a Transition Function." *Theory Probab. Applications*, vol.25, no.2 (1980): 384-388.

[6] Wobst, R. On Jump Processes With Drift. *Dissertationes mathematicae*, CCII, 1983. Warszawa (Poland): Polish Scientific Publishers.

V.F. Dem'yanov, and L.V. Vasil'ev
Nondifferentiable Optimization
1985, approx. 350 pp.
ISBN 0-911575-09-X Optimization Software, Inc.
ISBN 0-387-90951-6 Springer-Verlag New York Berlin Heidelberg Tokyo
ISBN 3-540-90951-6 Springer-Verlag Berlin Heidelberg New York Tokyo

V.P. Chistyakov, B.A. Sevast'yanov, and V.K. Zakharov
Probability Theory For Engineers
1985, approx. 200 pp.
ISBN 0-911575-13-8 Optimization Software, Inc.
ISBN 0-387-96167-4 Springer-Verlag New York Berlin Heidelberg Tokyo
ISBN 3-540-96167-4 Springer-Verlag Berlin Heidelberg New York Tokyo

B.T. Polyak
Introduction To Optimization
1985, approx. 450 pp.
ISBN 0-911575-14-6 Optimization Software, Inc.
ISBN 0-387-96169-0 Springer-Verlag New York Berlin Heidelberg Tokyo
ISBN 3-540-96169-0 Springer-Verlag Berlin Heidelberg New York Tokyo

V.A. Vasilenko
Spline Functions: Theory, Algorithms, Programs
1985, approx. 280 pp.
ISBN 0-911575-12-X Optimization Software, Inc.
ISBN 0-387-96168-2 Springer-Verlag New York Berlin Heidelberg Tokyo
ISBN 3-540-96168-2 Springer-Verlag Berlin Heidelberg New York Tokyo

V.F. Kolchin
Random Mappings
1985, approx. 250 pp.
ISBN 0-911575-16-2 Optimization Software, Inc.
ISBN 0-387-96154-2 Springer-Verlag New York Berlin Heidelberg Tokyo
ISBN 3-540-96154-2 Springer-Verlag Berlin Heidelberg New York Tokyo

A.A. Borovkov, Ed.
Advances In Probability Theory:
Limit Theorems For Sums of Random Variables
1985, approx. 400 pp.
ISBN 0-911575-17-0 Optimization Software, Inc.
ISBN 0-387-96100-3 Springer-Verlag New York Berlin Heidelberg Tokyo
ISBN 3-540-96100-3 Springer-Verlag Berlin Heidelberg New York Tokyo

Continued on page 506

TRANSLATION SERIES IN MATHEMATICS AND ENGINEERING

V.V. Ivanishchev, and A.D. Krasnoshchekov
Control of Variable Structure Networks
1985, approx. 200 pp.
ISBN 0-911575-05-7 Optimization Software, Inc.
ISBN 0-387-90947-8 Springer-Verlag New York Berlin Heidelberg Tokyo
ISBN 3-540-90947-8 Springer-Verlag Berlin Heidelberg New York Tokyo

A.N. Tikhonov, Ed.
Problems In Modern Mathematical Physics and Computational Mathematics
1985, approx. 500 pp.
ISBN 0-911575-10-3 Optimization Software, Inc.
ISBN 0-387-90952-4 Springer-Verlag New York Berlin Heidelberg Tokyo
ISBN 3-540-90952-4 Springer-Verlag Berlin Heidelberg New York Tokyo

N.I. Nisevich, G.I. Marchuk, I.I. Zubikova, and I.B. Pogozhev
Mathematical Modeling of Viral Diseases
1985, approx. 400 pp.
ISBN 0-911575-06-5 Optimization Software, Inc.
ISBN 0-387-90948-6 Springer-Verlag New York Berlin Heidelberg Tokyo
ISBN 0-387-90948-6 Springer-Verlag Berlin Heidelberg New York Tokyo

V.G. Lazarev, Ed.
Processes and Systems In Communication Networks
1985, approx. 250 pp.
ISBN 0-911575-08-1 Optimization Software, Inc.
ISBN 0-387-90950-8 Springer-Verlag New York Berlin Heidelberg Tokyo
ISBN 3-540-90950-8 Springer-Verlag Berlin Heidelberg New York Tokyo

Transliteration Table

R	E	R	E
а А	a	р Р	r
б Б	b	с С	s
в В	v	т Т	t
г Г	g	у У	u
д Д	d	ф Ф	f
е Е	e	х Х	kh
ё Ё	e	ц Ц	ts
ж Ж	zh	ч Ч	ch
з З	z	ш Ш	sh
и И	i	щ Щ	shch
й Й	j	ъ Ъ	"
к К	k	ы Ы	y
л Л	l	ь Ь	'
м М	m	э Э	eh
н Н	n	ю Ю	yu
о О	o	я Я	ya
п П	p